Energy Conversion Materials/Devices and Their Applications

Energy Conversion Materials/Devices and Their Applications

Guest Editors

Bin Liu
Yaling Wang
Lei Liu

MDPI

Basel • Beijing • Wuhan • Barcelona • Belgrade • Novi Sad • Cluj • Manchester

Guest Editors
Bin Liu
School of Energy and Power Engineering
North University of China
Taiyuan
China

Yaling Wang
School of Energy and Power Engineering
North University of China
Taiyuan
China

Lei Liu
School of Energy and Power Engineering
North University of China
Taiyuan
China

Editorial Office
MDPI AG
Grosspeteranlage 5
4052 Basel, Switzerland

This is a reprint of the Special Issue, published open access by the journal *Micromachines* (ISSN 2072-666X), freely accessible at: https://www.mdpi.com/journal/micromachines/special_issues/J12R5U3939.

For citation purposes, cite each article independently as indicated on the article page online and as indicated below:

Lastname, A.A.; Lastname, B.B. Article Title. *Journal Name* **Year**, *Volume Number*, Page Range.

ISBN 978-3-7258-6356-3 (Hbk)
ISBN 978-3-7258-6357-0 (PDF)
https://doi.org/10.3390/books978-3-7258-6357-0

Contents

About the Editors

Bin Liu

Bin Liu is a professor and doctoral supervisor at North University of China, where he is also recognized as a University Youth Talent Fellow. He received his Ph.D. in 2019 from Zhejiang University and completed an academic visit in 2018 at the University of Bayreuth in Germany, working with the group of Andreas Greiner, a leading pioneer in electrospinning. His research focuses on carbon quantum dots, nonconjugated luminescent polymers, light-emitting diodes, luminescent solar concentrators, and electrocatalytic nitrogen fixation. He has published over thirty papers as first or corresponding author in prominent journals, including *Aggregate*, *Advanced Science*, *Materials Horizons*, *Green Chemistry*, and *Advanced Optical Materials*, with five of these works recognized as ESI Highly Cited Papers. He also plays an active role in the scientific community, serving on the Youth Editorial Boards of *Chinese Chemical Letters*, *cScience*, and *ChemPhysMater*, as well as acting as Guest Editor for *Micromachines* and *Discover Materials*. In addition, he is a member of the Expert Committee on Energy Materials and Devices. In 2024, he contributed a chapter to the *Encyclopedia of Aggregation-Induced Emission*.

Yaling Wang

Yaling Wang is a professor and master's supervisor at North University of China. She received her Ph.D. in 2019 from Taiyuan University of Technology, during which she conducted a two-year joint training program at the Suzhou Institute of Nano-Tech and Nano-Bionics. Her research focuses on carbon quantum dots, light-emitting diodes, and solar cells. To date, she has published more than twenty papers as first or corresponding author in leading international journals such as *ACS Applied Materials & Interfaces*, *Advanced Optical Materials*, *Journal of Materials Chemistry C*, and *Chinese Chemical Letters*. She contributes actively to the academic community as a Youth Editorial Board Member of *Chinese Chemical Letters* and as a Guest Editor for *Micromachines*. She has also authored one academic monograph.

Lei Liu

Lei Liu is a professor and doctoral supervisor at the School of Energy and Power Engineering, North University of China, China. He is dedicated to researching electrochromic materials in areas such as spectral modulation, wide-spectrum dynamic camouflage, and intelligent sensing monitoring. His primary interests also focused on transparent conductive films, WO3-, NiOx-, and MnO2-based electrochromic materials, inorganic and polymer-ion electrolytes, interface and surface properties of metal oxide thin films, and all-solid-state electrochromic energy conversion, harvesting, and storage devices. Additionally, he has expertise in a multitude of sensor and spectroscopic techniques.

Preface

This Special Issue, titled "Energy Conversion Materials/Devices and Their Applications", focuses on the research and development of advanced materials and devices for energy conversion applications. The primary goal is to address growing global energy demands and environmental concerns by exploring innovative solutions for more efficient energy conversion and storage. The Special Issue features high-quality original research articles and review articles written by leading experts in the field. The articles spotlight the latest advancements in energy conversion materials, including novel materials synthesis methods, structural characterization, and property evaluation. Additionally, the Special Issue covers the development of innovative devices, such as high-performance solar cells, fuel cells, and batteries, as well as their integration into energy systems for sustainable energy generation and storage. This Special Issue aims to inspire further research and innovations in the field, ultimately accelerating the transition towards a sustainable energy future.

Bin Liu, Yaling Wang, and Lei Liu
Guest Editors

micromachines

MDPI

Editorial

Editorial for This Special Issue on Energy Conversion Materials and Devices and Their Applications

Bin Liu *, Yaling Wang * and Lei Liu *

State Key Laboratory of Coal and CBM Co-Mining, School of Energy and Power Engineering, North University of China, Taiyuan 030051, China
* Correspondence: liubin@nuc.edu.cn (B.L.); wangyaling@nuc.edu.cn (Y.W.); liulei91@nuc.edu.cn (L.L.)

1. Introduction

The global push toward sustainable energy, driven by soaring energy demands, escalating environmental concerns, and urgent climate challenges, has catalyzed remarkable advancements in energy conversion materials and devices [1–5]. This necessitates next-generation systems that prioritize higher efficiency, reduced toxicity, enhanced stability, and sustainable material bases [6–10]. Breakthroughs in materials science and device engineering are now paving the way toward resilient and intelligent energy solutions [11–14]. This Special Issue presents cutting-edge research addressing these critical needs, exploring innovations spanning from perovskite solar cells, solid-state lithium-metal batteries, and sodium-ion batteries to electrochromic devices, organic light-emitting diodes (OLEDs), and thermally activated light emitters.

This Special Issue includes seven original research articles and three review articles. These works not only reflect the latest scientific innovations but also emphasize application-orientated strategies with potential for commercial translation. In this editorial, we will provide an integrated overview of the thematic developments within this Special Issue and highlight the broader significance of each contribution.

Energy storage lies at the heart of the clean energy transition [15–18]. Batteries are indispensable for grid stabilization, electric vehicles, and portable electronics. However, conventional lithium-ion batteries face bottlenecks in resource availability, safety, and energy density [19–22]. In this Special Issue, Li et al. [23] report on $ZnMn_2O_4/V_2CT_x$ composites via high-temperature calcination to synergize spinel-type oxides with MXene to enhance lithium storage. The unique lamellar rod-like structure achieves an improved specific discharge capacity (163 mAh g^{-1}) and superior cycle life compared to pure $ZnMn_2O_4$. The synergistic interaction between $ZnMn_2O_4$ and V_2CT_x is key to the composite's remarkable performance, significantly boosting its lithium-ion storage capacity. This demonstrates how composite architectures can enhance electrochemical kinetics and mitigate volume expansion during charge/discharge cycles.

Advancing toward the next generation of lithium-ion batteries, Lee et al. [24] comprehensively review strategies to stabilize lithium metal anodes in solid-state electrolytes (SSEs), a critical step in enabling the production of all-solid-state lithium-ion batteries. While lithium metal offers unparalleled energy density, its high reactivity and dendritic growth pose serious challenges. The authors systematically compare sulfide-, oxide-, and polymer-based SSEs, evaluating their electrochemical performance, ion conductivity, and interfacial compatibility. Sulfide SSEs, though promising due to their high conductivity, are susceptible to dendrite formation and suffer from a limited electrochemical window. In contrast, oxide SSEs (e.g., $Li_7La_3Zr_2O_{12}$) offer superior chemical stability but suffer from

 https://doi.org/10.3390/mi16080943

high interfacial resistance due to Li_2CO_3 formation. Polymer electrolytes provide flexibility but lack sufficient ion mobility.

Another dimension involves strategies to stabilize lithium metal anodes in SSEs [25–27]. Ahangari et al. [28] introduce a surface engineering approach using ultrathin Al_2O_3 coatings (via atomic layer deposition) to resolve structural instability and parasitic reactions in Ni-rich $Li(Ni_{0.8}Co_{0.1}Mn_{0.1})O_2$ cathodes. The modified Ni-rich $Li(Ni_{0.8}Co_{0.1}Mn_{0.1})O_2$ exhibits capacity retention improvements of 5%, 9.11%, and 11.28% compared to the pristine material at 4.3 V, 4.4 V, and 4.5 V, respectively. This enhances cycling stability and capacity retention, offering insights into interfacial engineering for high-capacity cathodes.

For grid-scale storage, sodium-ion systems are emerging as cost-effective alternatives [29,30]. In this Special Issue, Ahangari et al. [31] comprehensively review layered transition metal oxide cathodes for sodium-ion batteries. Their analysis focuses on structural properties, electrochemical performance, degradation mechanisms, and the intrinsic and extrinsic factors underlying instability. Furthermore, they highlight advances in doping, surface modification, and composites aimed at mitigating capacity fading and structural instability, suggesting promising prospects for this technology.

Electrochromic materials represent another cornerstone of energy-conversion applications, especially in smart windows and adaptive optics [32–34]. Two studies in this Special Issue exemplify breakthroughs in tungsten oxide (WO_3)-based electrochromic films, offering transformative insights for future applications. Zhang et al. [35] utilize ex situ spectroscopic ellipsometry to investigate Li^+ insertion mechanics in WO_3 films, revealing a layer-by-layer permeation process critical in optical modulation. Their findings provide insights into optical modulation for zoom lenses and phase modulators. Wang et al. [36] advance this further with an MXene-regulated, ethylene glycol-induced coral-like TiO_2@WO_3 heterostructure, achieving a coloring efficiency of 137.02 cm^2C^{-1} and >90% performance retention after 4000 cycles. Together, these works demonstrate how nanoscale design can unlock new paradigms in dynamic optics and energy efficiency.

Solar energy remains vital in global decarbonization strategies, but lead toxicity and stability concerns hinder the mass adoption of perovskite solar cells [37–39]. In this Special Issue, Gao et al. [40] tackle this issue by developing compact $(CH_3NH_3)_3Bi_2I_9$ perovskite films using a two-step vapor-assisted method. Their approach yields morphologically optimized films with enhanced power conversion efficiency (~1.13%) and device stability, retaining 99% performance after 60 days in the glove box environments. This suggests that with refined crystallization control and morphological optimization, bismuth-based perovskites could play a pivotal role in non-toxic photovoltaics. This work paves the way for new classes of high-performance, lead-free, solution-processable solar materials, especially in regions with environmental regulations in place regarding heavy metals.

At the energy–optoelectronics frontier, light-emitting technologies are becoming integral to energy-efficient devices [41,42]. In this Special Issue, Zhang et al. [43] review host-free white OLEDs using thermally activated delayed fluorescence (TADF). This technology enables up to 100% internal quantum efficiency and simplifies device architecture, paving the way for next-generation flexible and efficient displays and lighting systems. By comparing TADF, TADF phosphorescence, and all-TADF materials, the authors discuss the benefits and current challenges in this area.

For telecommunications, optical amplifiers must cover the full NIR window (1000–1600 nm) [44–46]. Traditional rare-earth-doped amplifiers (e.g., erbium) are limited by narrow emission bands. In this Special Issue, Song and Zhang [47] demonstrate that Bi/Cr co-doped aluminosilicate glasses can achieve broadband emissions (650–1600 nm), surpassing traditional rare-earth-doped amplifiers and enhancing bandwidth for fiber-optic networks.

The future of energy hinges not only on performance but also on scalability, safety, accessibility, and environmental impact. The studies in this Special Issue converge on a common vision: material innovation that is informed by application-specific needs. However, challenges remain. Bridging the lab-to-fab gap, ensuring material reproducibility, and integrating these devices into complex systems require interdisciplinary efforts. We hope that this Special Issue will serve as both a snapshot of current progress and a springboard for future breakthroughs.

We extend our deepest gratitude to all the authors for their contributions to this Special Issue. We also sincerely appreciate the reviewers for their time and dedication in ensuring the quality of this publication. Finally, we are grateful to the staff of the *Micromachines* Editorial Office, particularly Mr. Lebron Tu, for their invaluable assistance.

Conflicts of Interest: The authors declare no conflicts of interest.

References

1. Green, M.A.; Bremner, S.P. Energy conversion approaches and materials for high-efficiency photovoltaics. *Nat. Mater.* **2017**, *16*, 23–34. [CrossRef]
2. Li, W.; Liu, J.; Zhao, D. Mesoporous materials for energy conversion and storage devices. *Nat. Rev. Mater.* **2016**, *1*, 16023. [CrossRef]
3. Gu, C.; Jia, A.-B.; Zhang, Y.-M.; Zhang, S.X.-A. Emerging electrochromic materials and devices for future displays. *Chem. Rev.* **2022**, *122*, 14679. [CrossRef] [PubMed]
4. Liu, Y.; Yu, Z.; Chen, J.; Li, C.; Zhang, Z.; Yan, X.; Liu, X.; Yang, S. Recent development and progress of structural energy devices. *Chin. Chem. Lett.* **2022**, *33*, 1817. [CrossRef]
5. Zhang, X.; Cheng, X.; Zhang, Q. Nanostructured energy materials for electrochemical energy conversion and storage: A review. *J. Energy Chem.* **2016**, *25*, 967–984. [CrossRef]
6. Fratzl, P. Introduction: Sustainable materials. *Chem. Rev.* **2023**, *123*, 1841. [CrossRef]
7. Li, X.; Aftab, S.; Mukhtar, M.; Kabir, F.; Khan, M.F.; Hegazy, H.H.; Akman, E. Exploring nanoscale perovskite materials for next-generation photodetectors: A comprehensive review and future directions. *Nano-Micro Lett.* **2024**, *17*, 28. [CrossRef]
8. Osman, A.I.; Chen, Z.; Elgarahy, A.M.; Farghali, M.; Mohamed, I.M.A.; Priya, A.K.; Hawash, H.B.; Yap, P.-S. Membrane technology for energy saving: Principles, techniques, applications, challenges, and prospects. *Adv. Energy Sustain. Res.* **2024**, *5*, 2400011. [CrossRef]
9. Firoozi, A.A.; Firoozi, A.A.; Oyejobi, D.O.; Avudaiappan, S.; Flores, E.S. Emerging trends in sustainable building materials: Technological innovations, enhanced performance, and future directions. *Results Eng.* **2024**, *24*, 103521. [CrossRef]
10. Elezz, M.A.; Aboeleneen, N.M.; Abd-ElMonem, N.M.; Sorour, F.H. A comprehensive review: Functional nanomaterials for renewable energy: Innovations, applications, and sustainable strategies. *Next Mater.* **2025**, *9*, 101001. [CrossRef]
11. Zhang, Y.; Xia, H.; Yu, J.; Yang, Y.; Li, G. Materials and device engineering perspective: Recent advances in organic photovoltaics. *Adv. Mater.* **2025**, 2504063. [CrossRef]
12. Gu, Y.; Qiu, Z.; Müllen, K. Nanographenes and graphene nanoribbons as multitalents of present and future materials science. *J. Am. Chem. Soc.* **2022**, *144*, 11499. [CrossRef] [PubMed]
13. Jiang, X.; Xue, D.; Bai, Y.; Wang, W.Y.; Liu, J.; Yang, M.; Su, Y. AI4Materials: Transforming the landscape of materials science and enigneering. *Rev. Mater. Res.* **2025**, *1*, 100010. [CrossRef]
14. Yuan, Q.; Gao, W.; Liu, J.L. Achievements in elevating the chemistry enterprise: Advancing energy transition and smart materials. *ACS Energy Lett.* **2025**, *10*, 1366–1373. [CrossRef]
15. Kalair, A.; Abas, N.; Saleem, M.S.; Kalair, A.R.; Khan, N. Role of energy storage systems in energy transition from fossil fuels to renewables. *Energy Storage* **2021**, *3*, e135. [CrossRef]
16. Kittner, N.; Lill, F.; Kammen, D.M. Energy storage deployment and innovation for the clean energy transition. *Nat. Energy* **2017**, *2*, 17125. [CrossRef]
17. Feng, S.; Lazkano, I. Energy storage and clean energy transitions. *Energy Policy* **2025**, *198*, 114447. [CrossRef]
18. Jafarizadeh, H.; Yamini, E.; Zolfaghari, S.M.; Esmaeilion, F.; Assad, M.E.H.; Soltani, M. Navigating challenges in large-scale renewable energy storage: Barriers, solutions, and innovations. *Energy Rep.* **2024**, *12*, 2179–2192. [CrossRef]
19. Khan, A.; Al Rashid, H.; Roy, P.K.; Chowdhury, S.I.; Sathi, S.A. Challenges and the way to improve lithium-ion battery technology for next-generation energy storage. *Energy Environ. Mater.* **2025**, e70088. [CrossRef]
20. Xu, J.; Cai, X.; Cai, S.; Shao, Y.; Hu, C.; Lu, S.; Ding, S. High-energy lithium-ion batteries: Recent progress and a promising future in applications. *Energy Environ. Mater.* **2023**, *6*, e12450. [CrossRef]

21. Choi, J.W.; Aurbach, D. Promise and reality of post-lithium-ion batteries with high energy densities. *Nat. Rev. Mater.* **2016**, *1*, 16013. [CrossRef]
22. Sun, Y.; Liu, N.; Cui, Y. Promises and challenges of nanomaterials for lithium-based rechargeable batteries. *Nat. Energy* **2016**, *1*, 16071. [CrossRef]
23. Li, J.; Wang, Y.; Pei, X.; Zhou, C.; Zhao, Q.; Lu, M.; Han, W.; Wang, L. $ZnMn_2O_4/V_2CT_x$ composites prepared as an anode material via high-temperature calcination method for optimized Li-ion batteries. *Micromachines* **2024**, *15*, 828. [CrossRef] [PubMed]
24. Lee, H.; Yoon, T.; Chae, O.B. Strategies for enhancing the stability of lithium metal anodes in solid-state electrolytes. *Micromachines* **2024**, *15*, 453. [CrossRef] [PubMed]
25. Zhong, Y.; Yang, X.; Guo, R.; Zhai, L.; Wang, X.; Wu, F.; Wu, C.; Bai, Y. Protecting lithium metal anodes in solid-state batteries. *Electrochem. Energy Rev.* **2024**, *7*, 30. [CrossRef]
26. Li, Z.; Zheng, Y.; Liao, C.; Duan, S.; Liu, X.; Chen, G.; Dong, L.; Dong, J.; Ma, C.; Yin, B.; et al. Advanced polymer materials for protecting lithium metal anodes of liquid-state and solid-state lithium batteries. *Adv. Funct. Mater.* **2024**, *34*, 2404427. [CrossRef]
27. Fu, C.; Zhang, X.; Cui, C.; Zhang, X.; Lou, S.; Ma, Y.; Huo, H.; Gao, Y.; Zuo, P.; Yin, G. Molecular bridges stabilize lithium metal anode and solid-state electrolyte interface. *Chem. Eng. J.* **2022**, *432*, 134271. [CrossRef]
28. Ahangari, M.; Xia, F.; Szalai, B.; Zhou, M.; Luo, H. Advancing lithium-ion batteries' electrochemical performance: Ultrathin alumina coating on $Li(Ni_{0.8}Co_{0.1}Mn_{0.1})O_2$ cathode materials. *Micromachines* **2024**, *15*, 894. [CrossRef]
29. Yin, X.; Liu, T.; Yin, X.; Feng, X.; Liu, Y.; Shi, Q.; Zou, X.; Zhao, Y. Free-standing $SnNb_2O_6$@CSN film as flexible anode for high performance sodium-ion batteries. *Chin. Chem. Lett.* **2023**, *34*, 107840. [CrossRef]
30. Dong, X.; Wang, X.; Lu, Z.; Shi, Q.; Yang, Z.; Yu, X.; Feng, W.; Zou, X.; Liu, Y.; Zhao, Y. Construction of Cu-Zn Co-doped layered materials for sodium-ion batteries with high cycle stability. *Chin. Chem. Lett.* **2024**, *35*, 108605. [CrossRef]
31. Ahangari, M.; Zhou, M.; Luo, H. Review of layered transition metal oxide materials for cathodes in sodium-ion batteries. *Micromachines* **2025**, *16*, 137. [CrossRef]
32. Wang, J.; Wang, Z.; Zhang, M.; Huo, X.; Guo, M. Toward next-generation smart windows: An in-depth analysis of dual-band electrochromic materials and devices. *Adv. Opt. Mater.* **2024**, *12*, 2302344. [CrossRef]
33. Zheng, R.; Wang, Y.; Pan, J.; Malik, H.A.; Zhang, H.; Jia, C.; Weng, X.; Xie, J.; Deng, L. Toward easy-to-assemble, large-area smart windows: All-in-one cross-linked electrochromic material and device. *ACS Appl. Mater. Interfaces* **2020**, *12*, 27526–27536. [CrossRef] [PubMed]
34. Wu, S.; Sun, H.; Duan, M.; Mao, H.; Wu, Y.; Zhao, H.; Lin, B. Applications of thermochromic and electrochromic smart windows: Materials to buildings. *Cell Rep. Phys. Sci.* **2023**, *4*, 101370. [CrossRef]
35. Zhang, B.; Wang, J.; Jiang, S.; Yuan, M.; Chen, X. Full spectrum electrochromic WO_3 mechanism and optical modulation via ex situ spectroscopic ellipsometry: Effect of Li+ surface permeation. *Micromachines* **2024**, *15*, 1473. [CrossRef]
36. Wang, Y.; Liu, Y.; Wang, M.; Wu, W.; Tian, M.; Zhu, T. Preparation and performance study of MXene-regulated ethylene glycol-induced WO_3 Film. *Micromachines* **2024**, *15*, 1486. [CrossRef] [PubMed]
37. Torrence, C.E.; Libby, C.S.; Nie, W.; Stein, J.S. Environmental and health risks of perovskite solar modules: Case for better test standards and risk mitigation solutions. *iScience* **2023**, *26*, 105807. [CrossRef]
38. Suo, J.; Pettersson, H.; Yang, B. Sustainable approaches to address lead toxicity in halide perovskite solar cells: A review of lead encapsulation and recycling solutions. *EcoMat* **2025**, *7*, e12511. [CrossRef]
39. Aktas, E.; Rajamanickam, N.; Pascual, J.; Hu, S.; Aldamasy, M.H.; Di Girolamo, D.; Li, W.; Nasti, G.; Martínez-Ferrero, E.; Wakamiya, A.; et al. Challenges and strategies toward long-term stability of lead-free tin-based perovskite solar cells. *Commun. Mater.* **2022**, *3*, 104. [CrossRef]
40. Gao, Z.; Wang, X.; Sun, Z.; Song, P.; Feng, X.; Jin, Z. Vapor-assisted method to deposit compact $(CH_3NH_3)_3Bi_2I_9$ thin films for bismuth-based planar perovskite solar cells. *Micromachines* **2025**, *16*, 218. [CrossRef]
41. Chen, W.; Song, F. Thermally activated delayed fluorescence molecules and their new applications aside from OLEDs. *Chin. Chem. Lett.* **2019**, *30*, 1717–1730. [CrossRef]
42. Jung, S.; Cheung, W.-L.; Li, S.-J.; Wang, M.; Li, W.; Wang, C.; Song, X.; Wei, G.; Song, Q.; Chen, S.S.; et al. Enhancing operational stability of OLEDs based on subatomic modified thermally activated delayed fluorescence compounds. *Nat. Commun.* **2023**, *14*, 6481. [CrossRef] [PubMed]
43. Zhang, W.; Li, Y.; Zhang, G.; Yang, X.; Chang, X.; Xing, G.; Dong, H.; Wang, J.; Wang, D.; Mai, Z.; et al. Advances in host-free white organic light-emitting diodes utilizing thermally activated delayed fluorescence: A comprehensive review. *Micromachines* **2024**, *15*, 703. [CrossRef] [PubMed]
44. Hilbert, M.; López, P. The world's technological capacity to store, communicate, and compute information. *Science* **2011**, *332*, 60–65. [CrossRef] [PubMed]
45. Diddams, S.A.; Hollberg, L.; Mbele, V. Molecular fingerprinting with the resolved modes of a femtosecond laser frequency comb. *Nature* **2007**, *445*, 627–630. [CrossRef]

46. Jetschke, S.; Unger, S.; Schwuchow, A.; Leich, M.; Kirchhof, J. Efficient Yb laser fibers with low photodarkening by optimization of the core composition. *Opt. Express* **2008**, *16*, 15540–15545. [CrossRef]
47. Song, S.; Zhang, M. Broadband near-infrared emission from Bi/Cr Co-doped aluminosilicate glasses. *Micromachines* **2024**, *15*, 1093. [CrossRef]

micromachines

MDPI

Article

Investigation of Halogenated Metallic Phthalocyanine (InPcCl and F_{16}CuPc)-Based Electrodes and Palm Substrate for Organic Solid-State Supercapacitor Fabrication

María Elena Sánchez Vergara [1,2,*], Héctor Iván Sánchez Moore [1] and Luis Alberto Cantera-Cantera [1,3,*]

1 Faculty of Engineering, Universidad Anáhuac México, Av. Universidad Anáhuac 46, Col. Lomas Anáhuac, Huixquilucan 52786, Mexico; hector.sanchez@anahuac.mx
2 Polytechnic University of Cuautitlán Izcalli, Av. Lago de Guadalupe, Colonia Lomas de San Francisco Tepojaco, Cuautitlán Izcalli 54720, Mexico
3 Departamento de Ingeniería en Control y Automatización, ESIME-Z, Instituto Politécnico Nacional, Av. Luis Enrique Erro S/N, Gustavo A. Madero, Zacatenco, Ciudad de Mexico 07738, Mexico
* Correspondence: elena.sanchez@anahuac.mx (M.E.S.V.); lcanterac@ipn.mx (L.A.C.-C.)

Abstract: In this work, we report on the fabrication of a novel Organic Double-Layer Supercapacitor (ODLSC) using recycled palm as the substrate and electrodes based on halogenated indium and copper phthalocyanines. The electrodes were characterized using Reflectance, the Kulbeka–Munk function, and Fluorescence. Finally, their electrical behavior was evaluated, and the results were compared with those obtained for a more conventional supercapacitor fabricated on polyethylene terephthalate substrate and using indium tin oxide film for electrodes. Based on the experimental measurements of the fabricated ODLSC, the parameter identification of the classical equivalent circuit model was carried out using the Least Squares of Orthogonal Distances (LSOD) algorithm. The results indicated that the palm supercapacitor exhibited behavior more like that of traditional supercapacitors, as the root square mean error (RMSE) values in the model approximation of the experimental data were in the order of 10^{-7}. Furthermore, the models obtained allowed a determination of the device's Electrical Impedance Spectroscopy (EIS), revealing that the Palm SC-T1 exhibited capacitive behavior. In contrast, the manufactured Palm SC-T2, PET SC-T1, and PET SC-T2 devices exhibited inductive behavior. All the materials used in this work, such as the substrates, electrodes, separator membranes, and electrolytes, have a high potential to be used in organic supercapacitors.

Keywords: organic double-layer supercapacitor; halogen-substituted phthalocyanine; palm substrate; PET substrate; nylon membrane; electrical behavior

1. Introduction

Supercapacitors (SCs) are becoming increasingly popular in electric vehicles and renewable energy storage applications [1–3]. This increase is due to the great promise they claim to have as energy-storage devices with a long life cycle, high power density, fast charge and discharge cycles, as well as low maintenance costs, all in comparison to other conventional energy sources such as batteries or fuel cells [4–7]. The operation of the energy storage mechanism of SCs is based on the accumulation of charge or reversible redox reactions [8], which also gives excellent potential for use and flexibility, especially in devices that need short energy pulses like wearable electronic devices, electronic paper, or electronic skin [5,9]. Not all SCs are the same; traditional SCs can be categorized into three types depending on their energy storage properties: electrical double-layer capacitors

https://doi.org/10.3390/mi16040455

(EDLC), pseudo capacitors, and hybrid supercapacitors. The EDLCs store charges by using an electrical double-layer on the electrode surface, whereas the pseudo capacitors store charges just in a limited area on the surface, and the hybrid supercapacitors are a combination of the mechanism of energy storage of both supercapacitors previously mentioned, providing intermediate performance [8,10,11].

Mathematical models of supercapacitors are widely used for their design, analysis of electrical behavior, state-of-charge and state-of-health estimation, and the development of controllers. The most common approaches in the literature include electrochemical models, equivalent electrical circuit models, and fractional-order models, which accurately describe the system's electrical variables. Additionally, using neural networks has led to the emergence of intelligent models for predicting the dynamics of energy storage systems based on supercapacitors [1,3,12]. Due to the simplicity of their analysis through linear ordinary differential equations, equivalent electrical circuit models of supercapacitors are widely used to describe their electrical behavior. These models are based on RC networks and are classified into three types: classical, dynamic, and transmission, each with specific applications in the characterization and simulation of supercapacitors [1,12,13]. However, before using any mathematical model, it is essential to address the parameter identification problem, which involves determining the optimal values of the model parameters to minimize the difference between its response and the experimental data obtained from the system for a given input [14–16]. This problem is primarily solved using the Least Squares (LS) method and its variations. In the specific case of the parametric identification of equivalent electrical circuit models of supercapacitors, reference [13] demonstrates that the Least Squares of Orthogonal Distances (LSOD) algorithm offers significant advantages over the traditional LS and Least Absolute Deviations (LAD) methods, particularly in identifying the classical, dynamic, and transmission models of supercapacitors.

It is essential to consider the growing need to create more efficient SCs and greener and more sustainable ones due to pollution and energy consumption, which continues to increase. Organic supercapacitors are a promising alternative to traditional SCs due to their potential for low-cost, lightweight, and optimizable synthesis parameters, their redox potential, and their flexible designs [17,18]. The distinction between organic supercapacitors and traditional SCs is that they depart in terms of the organic materials they use in their structure, such as conducting polymers like polyaniline, polypyrrole, or polythiophenes [19,20], and carbon-based materials like graphene, carbon nanotubes/fibers, or activated carbon that are used for the electrodes [19,21]. These materials offer a high surface area and good electrical conductivity, are inexpensive, easy to obtain from various earth-abundant elements, and are easy to synthesize and control [22]. It is important to emphasize that the proper functioning of organic SCs, charge storage, and mechanical resistance under service conditions depends mainly on the materials and substances used as electrodes, electrolytes, and substrates. Thus, the development of high-performance organic materials is essential in this type of solid-state device, and, in addition, a critical approach that is currently being sought is related to the use of recycled or low environmental impact materials, such as Tetra Pak [23], waste newspaper [24], or cotton [25].

Based on the above, the objective of this work is to manufacture new organic SCs using silver paint coated with indium phthalocyanine (InPcCl) and copper phthalocyanine ($F_{16}CuPc$) halogenated as electrodes and tetrabutylammonium hexafluorophosphate ($TBAPF_6$) as a supporting electrolyte, supplied through a polymeric membrane of Nylon 11. In addition to these innovative materials, a degradable substrate such as recycled palm is proposed. The decreasing ease of mining fossil fuels, the increasing global energy demand, and increased environmental awareness have led to using and producing materials from sustainable and renewable resources. These palm leaves are composed of cellulose, hemicellulose, and lignin, which also have some interesting mechanical properties, like a tensile strength of 42 MPa, a flexural strength of 63 MPa, and an impact strength of 37.2 kJ/m^2; these properties represent a significant benefit in the use of the devices [26]. Cellulose in palm leaves is the main structural component providing rigidity and strength. Since most SCs are made from carbon derivatives and cellulose contains a significant amount of carbon, palm leaves are considered a suitable option. Regarding electrode materials, the metallic phthalocyanines used in this study contain halogenated substituents. This is because halogen-substituted phthalocyanines can be prepared as thin films with high charge carrier mobility, making them suitable for applications in electronic devices. Halogen-substituted phthalocyanines are chemically stable and allow for gas-phase transition, making it possible to obtain film structures with physical vapor deposition. This capability significantly enhances production efficiency for low-cost semiconducting and optoelectronic technology [27]. The high electronegativity of halogens such as F or Cl can withdraw electron density from the phthalocyanine ring, stabilizing the central metal's oxidized states and aiding charge transport. It can increase electrochemical reversibility in energy storage systems requiring multiple redox cycles without degradation. However, to our knowledge, halogen-substituted phthalocyanines have not been widely studied as components of SCs. Hybrid materials with phthalocyanines have been studied mainly, and these can be in the form of nanostructures; some examples are as follows: co-phthalocyanine has been studied as a component of a hybrid SC when it is found in the nanocomposite PcCo@CNTs [5]; ball-type metallophthalocyanines (BTMPc) of Cu, Co, Ni, and Zn have also been studied as electrodes in energy storage systems [6]; and Qi et al. [7] studied tetranitro-substituted cobalt phthalocyanine nanocomposites immobilized on reduced graphene oxide as a supercapacitor electrode hybrid material. Halogenated phthalocyanines such as AlPcCl [27,28], $SiPcCl_2$ [29], InPcCl [27,30], MnPcCl [27,31], and $TiPcCl_2$ [27] have been studied by some authors of this work as organic semiconductors in optoelectronic devices such as OLEDs, photodiodes, or organic solar cells.

Notably, the chloride mono-axially substituted in the InPcCl stabilizes the phthalocyanine's structure by reducing its chemical reactivity. Altering the electron density in the indium atom modifies the delocalization of charges throughout the molecule, affecting its optoelectronic properties [32]. On the other hand, fluoride increases phthalocyanine's thermal and chemical stability due to fluorine atoms in the peripheral rings of the $F_{16}CuPc$ molecule. Introducing fluorine substituents leads to a decrease in the electron density of the macrocycle and an increase in the oxidation potential of the MPc molecule [33]. Also, fluoride allows phthalocyanine to absorb light and transport charges better; $F_{16}CuPc$ has a relatively high mobility of charge carriers and an n-type semiconducting nature [34]. For this reason, in this work, we propose the incorporation of InPcCl and $F_{16}CuPc$ as electrodes in SCs since it is essential to consider that even with all the characteristics and advantages mentioned above, very little research has been conducted on the development of supercapacitors. The main novelty of this work is in demonstrating that efficient solid-state organic SCs can be fabricated using recycled palm leaf as a substrate and metallic phthalocyanines as electrodes. Furthermore, a separating layer made of nylon, which acts as a dispersion membrane for the supporting electrolyte, is also an original proposal for these supercapacitors. To our knowledge, there have been no reports of an Organic Double-Layer Supercapacitor (ODLSC) made with these materials.

2. Materials and Methods

The InPcCl (chloride indium(III) phthalocyanine: $C_{32}H_{16}ClInN_8$), Nylon 11 (polyundecanolactam: $[-NH(CH_2)10CO-]_n$), $F_{16}CuPc$ (copper (II) hexadecafluorophthalocyanine: $C_{32}CuF_{16}N_8$), and $TBA{\cdot}PF_6$ (tetrabutylammonium hexafluorophosphate: $(CH_3H_2CH_2CH_2)_4N(PF_6)$) did not need to be purified before usage and were acquired straight from commercial sources (Sigma-Aldrich, Saint Louis, MO, USA). The two ODLSCs shown in Figure 1 were manufactured, and a commercial palm leaf and PET were used as a substrate. Optimized Ag electrodes with InClPc and F_{16}CuPc coatings were used for the SC with the palm substrate. Initially, the palm was cut, sanded, and cleaned to obtain a smooth surface, upon which conductive silver paint was applied. The paint was dried under ambient conditions and vacuumed to ensure complete solvent removal. Using a vacuum sublimation technique, the InClPc and F_{16}CuPc coatings were deposited on the Ag. The Nylon 11 separator membrane and the TBA·PF_6 electrolyte were also deposited using a high-vacuum thermal sublimation system. This technique provided a high degree of variable control during deposition, including time, sublimation temperature, substrate temperature, and base pressure. The optimization of these variables allowed for the film's deposition with a precise thickness and molecular orientation. For the InClPc and the Nylon 11, an 8.6×10^{-5} Torr vacuum was used; for the TBA·PF_6 and the F_{16}CuPc, an 8.6×10^{-6} Torr vacuum was used. Thicknesses were measured with a quartz crystal microbalance monitor connected to a thickness sensor. The measured thicknesses were InPcCl = 904 Å, Nylon 11 = 186 Å, TBA·PF_6 = 116 Å, and F_{16}CuPc = 470 Å.

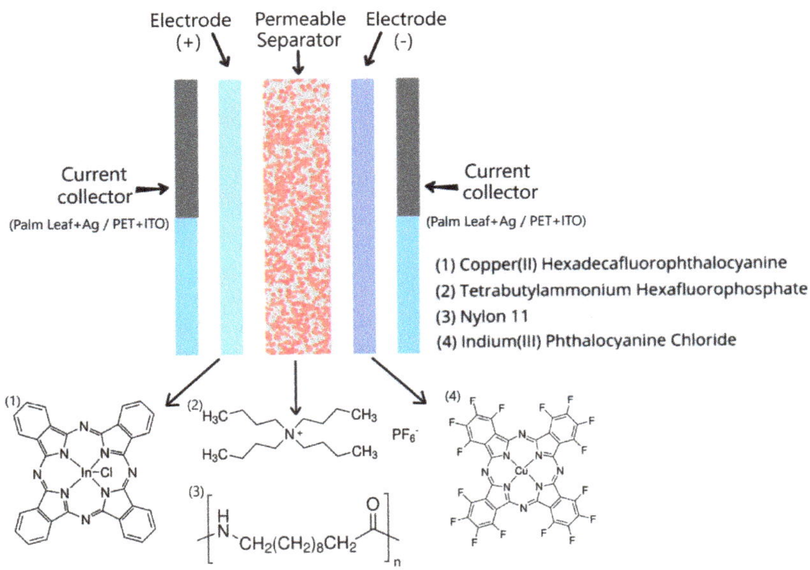

Figure 1. Schematic representation of Organic Double-Layer Supercapacitors.

The ODLSC on PET had important differences concerning the SC manufactured on palm (see Figure 1). For this device, the significant difference was, on the one hand, the change in substrate to commercial polyethylene terephthalate (PET), and one of the electrodes was composed of indium tin oxide film (ITO: $In_2O_3/(SnO_2)_x$) coated with an InClPc film. The second electrode had the same conformation as in the first device, ITO coated with F_{16}CuPc, and the Nylon 11 was included as a structural intermediate layer for the inclusion of the TBA·PF_6. The thicknesses of these materials were the same as those of the palm-mounted device. This was because both SCs were manufactured simultaneously. The devices were treated by thermal relaxation in a Briteg SC-92898 (Instrumentos Científicos, S.A de C.V., México City, México) oven at 140 °C for 12 min. Thermal relaxation was performed to introduce the supporting electrolyte into the nylon membrane and promote charge transport in the supercapacitor. An IR spectroscopy analysis was performed using a spectrometer Nicolet iS5-FT (Thermo Fisher Scientific Inc., Waltham, MA, USA) to verify the main functional groups in phthalocyanines and check whether decomposition occurred during their deposition. IR spectroscopy was carried out on the phthalocyanines used as electrodes, which were previously deposited on silicon, simultaneously using the same vacuum sublimation system and the same operating parameters as were used when deposited on the palm and PET of the ODLSCs. The Reflectance and Kubelka–Munk function were obtained using a Thermo Scientific Evolution 220 UV-Vis spectrophotometer in a wavelength range from 190 to 1100 nm (Thermo Fisher Scientific Inc., Waltham, MA, USA). Fluorescence was evaluated on an FP-8550 spectrofluorometer (Jasco International, Tokyo, Japan). The electrical properties in the ODLSCs were carried out using a sensor station with a Next Robotix lighting controller circuit (Comercializadora K Mox, S.A. de C.V., Mexico City, Mexico) and a Keithley 4200-SCS-PK1 auto-ranging picoammeter (Tektronix Inc., Beaverton, OR, USA).

3. Results and Discussion

3.1. Characterization of the Electrodes

The IR analysis of the electrodes helps to find functional groups on the electrode's surface and determine whether chemical decomposition of the phthalocyanines occurs during their deposition [23]. Furthermore, the InPcCl and F_{16}CuPc films possess unique properties, including strong absorption, emission, and photostability. This is because the molecular orientation and stacking of the phthalocyanine molecules are related to their optical behavior due to the different monoclinic (α-phase) and triclinic (β-phase) forms that they can acquire, which can be monitored by IR spectroscopy. A band can identify the α-phase and the β-phase bands around 720 cm^{-1} and around 778 $cm^{-1,}$ respectively [35,36]. Figure 2 presents the IR spectra for the InPcCl and F_{16}CuPc films, and Table 1 presents the most representative signals of phthalocyanines in film, which are compared to their signals in KBr pellets. The mode at 1608 cm^{-1} in the spectrum of InPcCl attributed to C=C stretching shifts to 1618 cm^{-1} for F_{16}CuPc, the peaks at 1749 $\pm$ 1 cm^{-1} are identified as a C-H stretch, and the C-C benzene stretch is associated with the peaks observed at 1542 and 1475 cm^{-1} for the InPcCl and at 1530 and 1495 cm^{-1} for F_{16}CuPc. The signals at 1328 $\pm$ 6 and 1414 $\pm$ 5 cm^{-1} verify in-plane pyrrole and isoindole stretch, respectively. According to IR spectroscopy, two relevant results are obtained: on the one hand, phthalocyanines are not chemically degraded when deposited as electrodes. This is confirmed by the similarity of the signals in Table 1 for pellets and the film; the small shifts between the values are due to the residual stresses generated in the phthalocyanine molecules during their deposition as films on the substrate. On the other hand, in the electrodes, InPcCl is present in a combination of the α-phase and the β-phase, and F_{16}CuPc is present only in the β-phase.

Table 1. Assignments of the main IR bands for InPcCl and F_{16}CuPc films and KBr pellets.

Sample	C=C Benzene Stretch (cm^{-1})	C-C Benzene Stretch (cm^{-1})	Isoindole Stretch (cm^{-1})	In-Plane Pyrrole Stretch (cm^{-1})	C-H Stretch (cm^{-1})	α-Phase (cm^{-1})	β-Phase (cm^{-1})
InPcCl film	1608	1542, 1475	1410	1334	1748	723	774
InPcCl KBr pellet	1608	1541, 1473	1419	1334	1750	726	771
F_{16}CuPc film	1618	1530, 1495	1419	1323	1750	-	771
F_{16}CuPc KBr pellet	1617	1524, 1505	1417	1319	1748	-	766

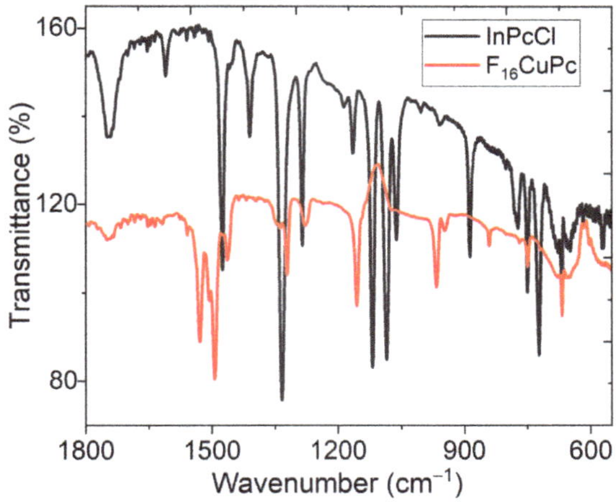

Figure 2. IR spectra of InPcCl and $F_{16}CuPc$ electrodes in film.

InPcCl and $F_{16}CuPc$ are phthalocyanines that absorb visible light radiation and exhibit high optical stability. For this reason, the electrodes are studied in terms of their Reflectance R, since if the ODLSC is used in optical or photovoltaic devices, a low R is required to help avoid energy loss through reflected light, maximizing the device efficiency. Figure 3a,b show the spectra for the electrodes on PET and palm substrates, and it is evident that R changes significantly when the electrode type and substrate are modified. The electrode with InPcCl has the lowest R (Figure 3a) when it is deposited on PET-ITO substrate, while the electrode with $F_{16}CuPc$ has the lowest R (Figure 3b) when it is deposited on palm-Ag substrate. However, the maximum R obtained for each electrode and recorded in Table 2 is low for the two electrodes with $F_{16}CuPc$ and low for the electrode with InPcCl. It is evident that even the $F_{16}CuPc$ electrode on the palm, with 50% maximum Reflectance, has low Reflectance, and the rest of the electrodes, with Reflectance between 24 and 36%, could be used in SCs connected to optical or photovoltaic devices. Additionally, it is crucial to consider that in all spectra, the Q-band is distinguished at wavelength λ, between 400 and 600 nm, corresponding to the visible region of the spectrum and being associated with the electronic transitions $\pi \rightarrow \pi^*$ of the 18 conjugated π electrons of the structure of the phthalocyanines InPcCl and $F_{16}CuPc$. The Q-band refers to the electronic transitions of the highest occupied molecular orbital (HOMO) to the lowest unoccupied molecular orbital (LUMO) of the Pc ring and is very useful, since these transitions increase the electronic conductivity, which is key to improving the performance of $F_{16}CuPc$ and InPcCl electrodes. The presence of the Q-band facilitates the participation of π electrons in reversible redox processes, which could occur in the electrodes and increase their specific capacity. This results in a higher energy density in pseudocapacitive systems than in purely electrostatic supercapacitors. Furthermore, the electron delocalization associated with the Q-band could allow phthalocyanines to form stable complexes with the TBA·PF_6 electrolyte. This favors charge storage mechanisms, especially in pseudocapacitive supercapacitors, where surface redox reactions occur. Although the Q-band of phthalocyanines is known for its role in optical applications, in ODLSCs, it could contribute to the improved electron transport, charge storage capacity, and redox stability of the $F_{16}CuPc$ and InPcCl electrodes [37]. It is important that even on palm, phthalocyanines do not lose their charge transport capacity and their optical behavior, thanks to their Q-band, which maintains its shape and intensity when compared to pristine halogenated phthalocyanine films [27–31]. According to Qi et al. [7] and Wang et al. [38], the presence of phthalocyanines and optical behavior in supercapacitor electrodes can be demonstrated by the characteristic absorption of the Q-band. Another important aspect is the red shift that appears in the spectra when comparing the curves for each substrate. In the case of InPcCl (Figure 3a), the wavelength at maximum Reflectance is 506 nm for PET and shifts to 479 nm for palm. In the case of $F_{16}CuPc$ (Figure 3b), Reflectance is from 518 nm in PET and shifts to 476 nm in palm. These shifts suggest interfacial interactions between the phthalocyanine, ITO, and Ag of PET and palm, respectively [38].

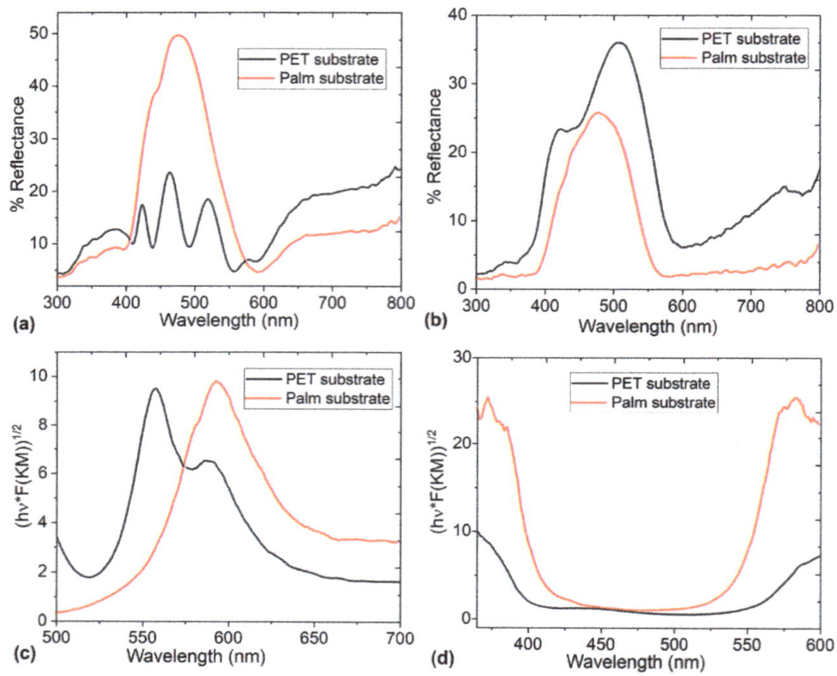

Figure 3. The Reflectance of (**a**) InPcCl and (**b**) $F_{16}CuPc$ electrodes on PET and palm substrates. The Kulbeka–Munk function of (**c**) InPcCl and (**d**) $F_{16}CuPc$ electrodes on PET and palm substrates.

Table 2. Reflectance, Kubelka–Munk gap, and the Stokes shift in the electrodes.

Electrode	Reflectance Maximum (%)	KM Band Gap (eV)
InPcCl on palm	26	1.86
InPcCl on PET	36	1.93
$F_{16}CuPc$ on palm	50	3.04
$F_{16}CuPc$ on PET	24	3.04

As mentioned above, the Q-band provides information on the transport of charges from the HOMO to the LUMO in phthalocyanines, which refers to the optical band gap. This parameter is considered necessary, because if the electrodes have a low band gap, a higher conductivity will be generated, which is related to a lower internal resistance of the electrodes and an improvement in the charge transfer during the charge/discharge cycles. The band gap for the electrodes is calculated using Reflectance measurements by the Kubelka–Munk function F(K-M) [39] and from the Tauc model with Equation (1) [39–41]:

$$h\nu\alpha = A\left(h\nu - E_g\right)^P, \quad (1)$$

where hν is the photon energy (h = Planck's constant y ν = 1/λ); A is a proportionality constant; E_g is the energy band gap; and the exponent p depends on the band structure of the semiconductor, and this value is 2 for indirect transitions, where the participation of phonons is required [39]. The absorption coefficient α is directly proportional to F(K-M), and in the Tauc equation, α can be replaced by this function as follows in Equation (2):

$$(h\nu \times F(KM)) = A\left(h\nu - E_g\right)^p, \quad (2)$$

The procedure consists of plotting the $(h\nu \times F(KM))^{1/2}$ versus λ fitting (Figure 3c,d) the linear portion of this curve with a straight line; the intercept and the use of Equation (3) provide the band gap value (c = light speed). Table 2 presents the band gap values for each electrode on the palm and PET.

$$E_g = \frac{hc}{\lambda}, \quad (3)$$

According to the values shown in Table 2, the electrode with InPcCl presents a much lower band gap than the electrode with $F_{16}CuPc$. This is due to the structure of each phthalocyanine. The indium and its coordination with the chlorine atom introduce electron-donating characteristics, and the electron density is more concentrated around the macrocyclic ring. This raises the energy of the HOMO, narrowing the band gap by bringing it closer to the LUMO. Additionally, the indium atom has a larger ionic radius and induces more π electron delocalization in the phthalocyanine ring, leading to a smaller band gap. On the other hand, $F_{16}CuPc$ has 16 fluorine atoms attached to the phthalocyanine ring; fluorine is an electron-withdrawing group, which lowers the HOMO energy level and stabilizes the LUMO, and this widens the band gap. On the other hand, copper ions stabilize the π system under the influence of the fluorine substituents, and the fluorine atoms pull electron density away from the core, resulting in a higher band gap. The differences in the optical band gaps of the two phthalocyanines and their behavior as electron donors of InPcCl and electron acceptors of $F_{16}CuPc$ are an indication of their adequate capacity as a pair of electrodes to generate the charge transport that must be carried out in the supercapacitor, regardless of whether the substrate is palm-Ag or PET-ITO. On the other hand, it is important to consider that the band gap values obtained for the electrodes are within the range of organic semiconductors [42], which enhances their use not only in SCs but also in optoelectronic devices such as light-emitting diodes, electrochromic devices, and solar cells [43].

The Fluorescence spectra in Figure 4 are obtained for the two supercapacitors at 382 nm excitation. All maximum emissions of wavelengths around 450 nm correspond to blue light wavelengths, and the PET supercapacitor exhibits the highest emission intensity. This high Fluorescence is advantageous for their use with devices such as organic light-emitting diodes (OLEDs). These devices can efficiently emit light when an electric current is applied to them, or when light excites them. Another advantage is that the high Fluorescence can indicate the quality of the phthalocyanine in the electrode, which can facilitate the detection of changes in the electrode state during charging and discharging, improving the monitoring and control of the supercapacitor. The Stokes shift (Δ_{Stokes}), which is the difference in energy or wavelength between the absorbed light and the light emitted by the phthalocyanines at the electrodes after being excited, is 18.5 eV (67 nm) for the supercapacitor on palm and 17.7 eV (70 nm) for the supercapacitor on PET.

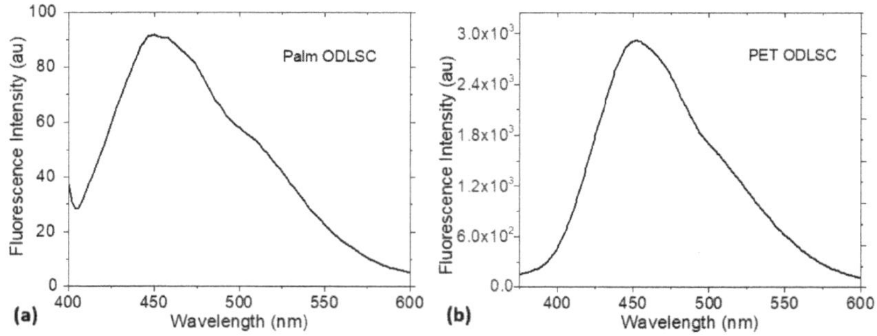

Figure 4. Fluorescence intensity of ODLSs on (**a**) palm substrate and (**b**) PET substrate.

3.2. Parameter Identification of Organic Supercapacitor Devices

The diverse literature, such as [44–46], presents studies and methods for fabricating supercapacitors using organic materials. However, parameter identification is a fundamental technique widely used for characterizing equivalent RC circuit models in inorganic supercapacitors [13]. It enables the analysis of their behavior under different input signals and design control strategies for charging and discharging processes. This section identifies the classical RC model to represent the electrical dynamic of the organic supercapacitor manufactured. The supercapacitor electrical dynamics can be modeled through linear equivalent electric circuit models, such as the classic, dynamic, and transmission circuits, which are compounded by RC networks [13]. These models are represented in linear ordinary differential equations, in which the model parameters must also be identified to ensure that the models accurately describe experimental data. Depending on the equation order, the parameter identification problem becomes more difficult, because more derivatives of the experimental data are necessary. Since the number of RC networks defines the equation order, the classic circuit is the simplest model to describe the electrical dynamics of supercapacitors because it has only one RC network. The classic model of the supercapacitor equivalent electrical circuit is shown in Figure 5.

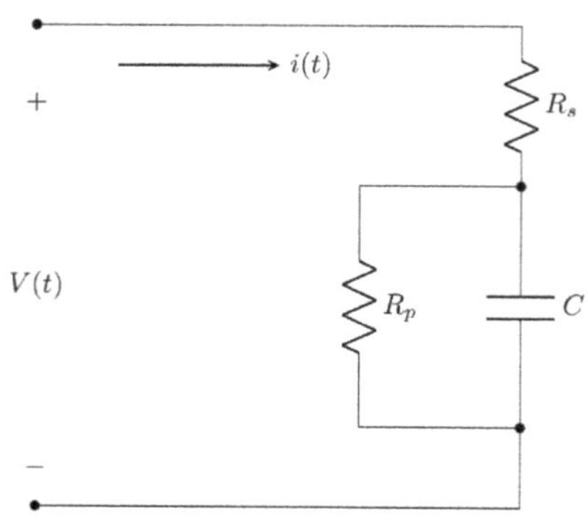

Figure 5. Supercapacitor classic model.

The transfer function of the classic model that relates the supercapacitor output voltage $V(s)$ and supercapacitor input current $I(s)$ from Figure 5 is given by Equation (4).

$$\frac{V(s)}{I(s)} = \frac{b_1 s + b_0}{s + a_0}, \tag{4}$$

with $b_1 = R_s$, $b_0 = \frac{R_s}{R_pC} + \frac{1}{C}$, and $a_0 = \frac{1}{R_pC}$, which are unknown in the models of the manufactured supercapacitors. On the other hand, experimental data are essential to carry out the parameter identification of any mathematical model; in this sense, the experimental data of the manufactured supercapacitors are shown in Figure 6. As shown in Figure 6, the manufactured ODLSCs on palm (Palm SC) and PET (PET SC) are supplied with a voltage from 0 to 1.1 V. Two experiments are conducted on each supercapacitor: one immediately after applying the electrolyte and another 10 min after its application. Figure 6a,c show the V-I curves for the first and second experiments of the Palm SC device, denoted by Palm SC-T1 and Palm SC-T2, respectively. On the other hand, Figure 6b,d show the voltage and current time series for the Palm SC-T1 and Palm SC-T2 experiments, respectively. Regarding the PET SC-T1 and PET SC-T2 experiments, Figure 6e,g present the V-I curves, and Figure 6f,h show the voltage and current time series, respectively. Using the voltage and current time series of the SC devices in Figure 6 and the Least Squares of Orthogonal Distances (LSOD) method [13,47], the estimates of the parameters b_1, b_0, and a_0 from the supercapacitor classic model in Equation (4) and the approximation's root square mean error (RMSE) for all experiments are shown in Table 3. The current responses of the manufactured organic supercapacitor equivalent circuit classic models to the input voltages from 0 to 1.1 V using the parameters from Table 3 are shown in Figure 7.

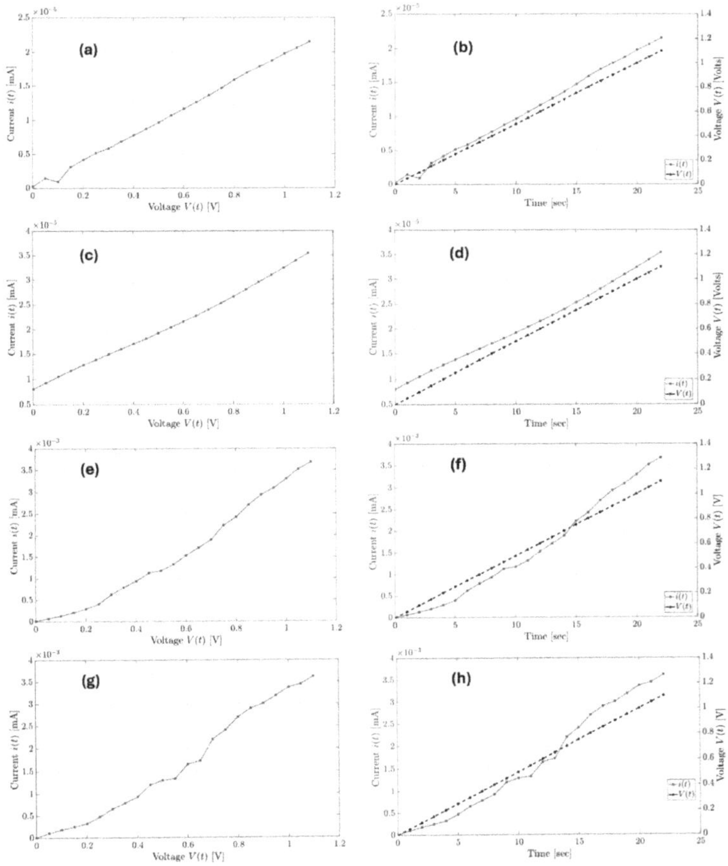

Figure 6. V-I curves of the (**a**) Palm SC-T1, (**c**) Palm SC-T2, (**e**) PET SC-T1, and (**g**) PET SC-T2, and the organic supercapacitor devices' input voltage and output current time series in (**b**) Palm SC-T1, (**d**) Palm SC-T2, (**f**) PET SC-T1, and (**h**) PET SC-T2.

Table 3. Parameters of the classic model's transfer function.

Parameter	Organic Double-Layer Supercapacitor			
	Palm SC-T1	**Palm SC-T2**	**PET SC-T1**	**PET SC-T2**
b_1	45,849.53691	84,316.1838	1201.52793	740.939817
b_0	45,834.70244	84,290.7887	170.674715	97.3210655
a_0	0.88007207	2.07389841	0.74966497	0.42602235
RMSE	3.41×10^{-7}	4.91×10^{-7}	7.02×10^{-5}	1.03×10^{-4}

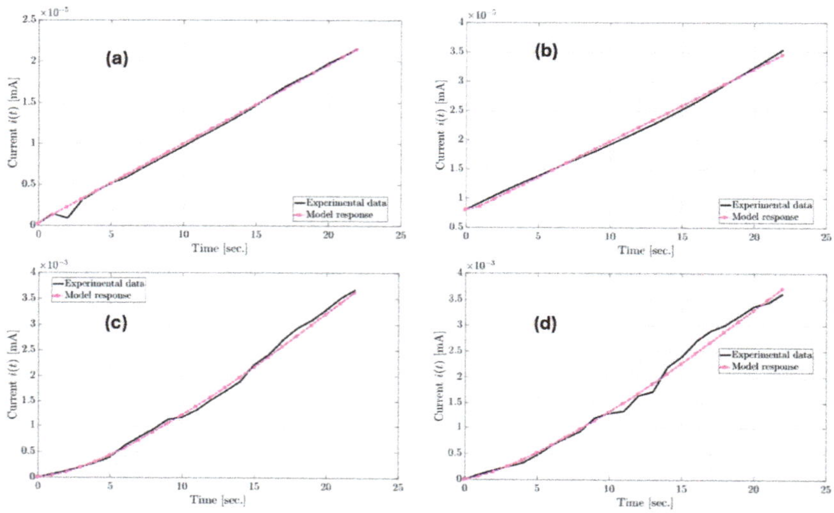

Figure 7. Current responses of the organic supercapacitor equivalent circuit classic models in (**a**) Palm SC-T1, (**b**) Palm SC-T2, (**c**) PET SC-T1, and (**d**) PET SC-T2.

The model responses from Figure 7a,b correspond to the current dynamics of the Palm SC-T1 and Palm SC-T2 devices, respectively. In contrast, the model responses from Figure 7c,d correspond to the current responses of the PET SC-T1 and PET SC-T1 devices. From the RMSE in Table 3, better approximations of the models are carried out with the Palm SC device compared to the approximations of the PET SC device. Still, in general, all approximations are acceptable. On the other hand, it is important to mention that the estimates in Table 3 differ for each manufactured supercapacitor device and test. This means that the performance of the supercapacitor has changed due to the diffusion of the electrolyte inside the device. The estimates presented in Table 3 show the values of the lumped parameters. However, when broken down in terms of the components of the classic equivalent circuit of a supercapacitor, the values detailed in Table 4 are obtained.

Table 4. Parameters of the classic equivalent circuit.

Parameter	Organic Double-Layer Supercapacitor			
	Palm SC-T1	**Palm SC-T2**	**PET SC-T1**	**PET SC-T2**
$C[\mu F]$	182.36	−11.04	−1369.73	−4580
$R_s[k\Omega]$	45.85	84.32	1.202	0.741
$R_p[k\Omega]$	6.23	−43.67	−0.974	−0.512

From the estimates in Table 3, the parameters of the classic equivalent circuit in Table 4 are obtained; however, only the Palm SC-T1 device receives consistent values for the C

capacitor and the R_s and R_p resistances because they are positive. In the other devices, the values of C and R_p are negative. Using the classic equivalent circuit model of the Palm SC-T1 device, the Electrical Impedance Spectroscopy (EIS) of the manufactured supercapacitor from 0.05 to 10,000 rad/s with steps of 0.01 is shown in Figure 8. EIS is a technique used to study the electrical properties of supercapacitors. This technique applies a low-amplitude alternating (AC) voltage signal and measures the current response. One of the standard graphs of EIS analysis is the Nyquist diagram (Figure 8), which represents the imaginary versus real parts of the impedance and helps to interpret resistances, capacitances, and charge transfer processes. Additionally, the 6 mΩ diameter of the semicircle represents the charge transfer resistance (Rct). On the other hand, the EIS graphs for the Palm SC-T2, PET SC-T1, and PET SC-T2 are shown in Figure 9. Figures 8 and 9a show that the only device exhibiting capacitive behavior is the Palm SC-T1, whose EIS graph is in the first quadrant. In contrast, the EIS graphs of the Palm SC-T2, PET SC-T1, and PET SC-T2 devices in Figure 9a,b are in the fourth quadrant, indicating inductive behavior. Therefore, the Palm SC-T1 manufactured device exhibits behavior more consistent with a supercapacitor than the other manufactured devices.

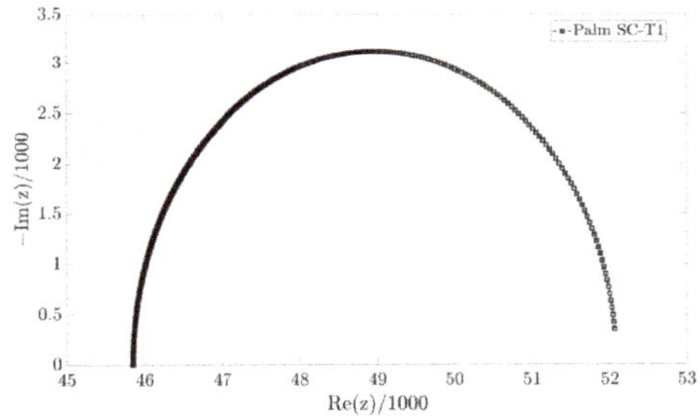

Figure 8. EIS of the classic equivalent circuit model of the Palm SC-T1 device.

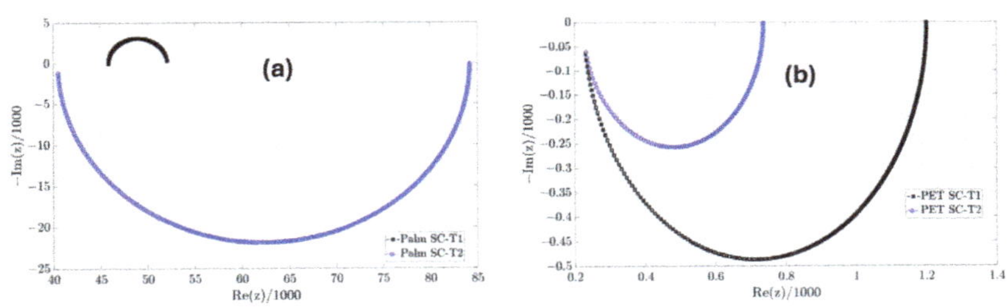

Figure 9. EIS of the classic equivalent circuit model of the (**a**) Palm and (**b**) PET devices.

4. Conclusions

Two Organic Double-Layer Supercapacitors (ODLSCs) were manufactured using InPcCl and $F_{16}CuPc$ electrodes deposited on Ag- and ITO-coated palm and PET substrates, respectively. The electrodes were structurally characterized using IR spectroscopy. The results demonstrated that the high-vacuum sublimation method used to deposit the phthalocyanine electrodes did not alter their chemical composition. The Reflectance of the electrodes was also measured and was found to be less than 50%, which is a good result for SCs used with organic optoelectronic or photovoltaic devices. Additionally, the band gap was calculated using the Kubelka–Munk function. The InPcCl electrode exhibited a band gap ranging from 1.86 to 1.93 eV, while the $F_{16}CuPc$ electrode had a band gap of 3.05 eV. This difference arose from each phthalocyanine's structure and the electron donor or acceptor properties, facilitating charge transport within the supercapacitor. The optical gap values obtained placed InPcCl and F_{16}CuPc as organic semiconductors. The electrodes exhibited Fluorescence at 382 nm excitation, and their maximum emission was found at wavelengths around 450 nm, corresponding to the blue region of the electromagnetic spectrum. Based on the electrical behavior of the fabricated Organic Double-Layer Supercapacitor, the parametric identification of the classical equivalent circuit model was performed for each device. Regarding the approximation of the models to the experimental data, the RMSE values ranged from 10^{-4} to 10^{-7}, indicating a good agreement between the models and the current data. However, in each test performed, the model parameters varied, suggesting that electrolyte diffusion within the supercapacitor affected the current conduction dynamics. On the other hand, when solving the system of equations to determine the equivalent circuit parameters, only the Palm SC-T1 device exhibited consistent values for C, R_s, and R_p. This allowed for the determination of its impedance spectroscopy (EIS), which indicated that this device exhibited behavior more consistent with that of inorganic supercapacitors. In contrast, the other manufactured devices exhibited indictive behavior. The Nyquist diagram for the supercapacitor manufactured on palm suggested that charge transfer processes occurred, and resistive–capacitive behavior was presented. All the new materials studied in this work, such as substrates, electrodes, separator membranes, and electrolytes, have a high potential to be used in organic supercapacitors. Therefore, device optimization and experimentation under different charge–discharge cycles are left for future work.

Author Contributions: Conceptualization, M.E.S.V. and L.A.C.-C.; methodology, M.E.S.V., L.A.C.-C. and H.I.S.M.; software, M.E.S.V. and L.A.C.-C.; validation, M.E.S.V., L.A.C.-C. and H.I.S.M.; formal analysis, M.E.S.V., L.A.C.-C. and H.I.S.M.; investigation, L.A.C.-C. and H.I.S.M.; resources, M.E.S.V.; data curation, M.E.S.V., L.A.C.-C. and H.I.S.M.; writing—original draft preparation, M.E.S.V., L.A.C.-C. and H.I.S.M.; writing—review and editing, M.E.S.V. and L.A.C.-C.; visualization, L.A.C.-C. and M.E.S.V.; supervision, M.E.S.V.; project administration, M.E.S.V.; funding acquisition, M.E.S.V. All authors have read and agreed to the published version of the manuscript.

Funding: M.E.S.V. acknowledges financial support from the Anahuac México University, project number PI0000068.

Data Availability Statement: Data are contained within the article.

Acknowledgments: The authors thank Emilio Iván Sandoval Plata for his technical support. The authors thank Kassandra Sugey Rodríguez Zúñiga for her support in the schemes of this manuscript.

Conflicts of Interest: The authors declare no conflicts of interest.

References

1. Husain, I. *Electric and Hybrid Vehicles: Design Fundamentals*; CRC Press: Boca Raton, FL, USA, 2021; Volume 3, 498p. [CrossR
2. Tran, D.D.; Vafaeipour, M.; El Baghdadi, M.; Barrero, J.; Van Mierlo, O.; Hegazy, O. Thorough state-of-the-art analysis of ele and hybrid vehicle powertrains: Topologies and integrated energy management strategies. *Renew. Sustain. Energy Rev.* **2020**, 109596. [CrossRef]
3. Naseri, F.; Karimi, S.; Farjah, E.; Schaltz, E. Supercapacitor management system: A comprehensive review of modeling, estima balancing, and protection techniques. *Renew. Sustain. Energy Rev.* **2022**, *155*, 111913. [CrossRef]
4. Mu, J.; Zhao, Y.; Guo, Z.; Zhang, Z.; Che, H.; Wang, X.; Zhang, G.; Wang, J.; Mu, L.; Wang, J. Rational design of new, efficient, suitable nickel phthalocyanine reinforced MXene electrodes for supercapacitors. *J. Energy Storage* **2023**, *61*, 106768. [CrossRe
5. Deyab, M.A.; Mohsen, Q.; Slavcheva, E. Co-phthalocyanin/CNTs nanocomposites: Synthesis, characterizations, and applica as an efficient supercapacitor. *J. Mol. Liq.* **2022**, *359*, 119319. [CrossRef]
6. Koçyigit, N.; Arvas, M.B.; Yazar, S.; Açar-Selçuki, N.; Şahin, Y.; Özer, M.; Bekăroğlu, Ö. Synthesis and characterization of n HMPM bridged ball-type metallophthalocyanines, and their DFT studies and electrochemical performance as supercapac electrodes. *Synth. Met.* **2023**, *292*, 117242. [CrossRef]

Qi, F.; Wang, Y.; Xu, J.; Wang, X.; Wang, J.; Shan, H.; Li, M.; Xu, J. Tetranitro-substituted cobalt phthalocyanine immobilized on reduced graphene oxide as supercapacitor electrode material with enhanced capacitance. *Synth. Met.* **2023**, *293*, 117284. [CrossRef]

Arvas, M.B.; Gürsu, H.; Gencten, M.; Sahin, Y. Preparation of different heteroatom doped graphene oxide based electrodes by electrochemical method and their supercapacitor applications. *J. Energy Storage* **2021**, *35*, 102328. [CrossRef]

Artar, E.; Arvas, M.B.; Gorduk, O.; Gorduk, S.; Sahin, Y. Facile synthesis strategy for phthalocyanine-titanium dioxide/multi-walled carbon nanotube/poly(3,4-ethylenedioxythiophene) ternary composite electrodes via one-step electrochemical method for supercapacitor applications. *Synth. Met.* **2023**, *297*, 117401. [CrossRef]

Arvas, M.B.; Yazar, S.; Sahin, Y. An ultra-high power density supercapacitor: Cu(II) phthalocyanine tetrasulfonic acid tetrasodium salt doped polyaniline. *J. Alloys Compd.* **2022**, *919*, 165689. [CrossRef]

Sung, J.; Shin, C. Recent Studies on Supercapacitors with Next-Generation Structures. *Micromachines* **2020**, *11*, 1125. [CrossRef]

Zhang, Z.; Hu, X.; Wang, Z.; Sun, F.; Dorell, D.G. A review of supercapacitor modeling, estimation, and applications: A control/management perspective. *Renew. Sustain. Energy Rev.* **2018**, *81*, 1868–1878. [CrossRef]

Cantera-Cantera, L.A.; Cervantes, I.; Martinez, M. Heteroscedasticity in supercapacitors under nonlinear operating conditions: Analyzing the efficacy of deterministic parameter identification algorithms. *J. Energy Storage* **2025**, *103*, 114189. [CrossRef]

Åström, K.J.; Eykhoff, P. System identification—A survey. *Automatica* **1971**, *7*, 123–162. [CrossRef]

Björck, A. Numerical Methods for Least Squares Problems. *SIAM* **1996**, *51*, 424. [CrossRef]

Bard, Y. *Nonlinear Parameter Estimation*; Academic Press: New York, NY, USA, 1974; Volume 1209, 341p.

Katzuyama, Y.; Takehi, T.; Sokabe, S.; Tanaka, M.; Ishizawa, M.; Abe, H.; Watanabe, M.; Honma, I.; Nakayasu, Y. Series module of quinone-based organic supercapacitor (> 6 V) with practical cell structure. *Sci. Rep.* **2022**, *12*, 3915. [CrossRef]

Zhu, H.; Xu, R.; Wan, T.; Yuan, W.; Shu, K.; Boonprakob, N.; Zhao, C. Nanocomposites of Conducting Polymers and 2D Materials for Flexible Supercapacitors. *Polimers* **2024**, *16*, 756. [CrossRef]

Xiong, Q.; Liu, B.; Liu, Y.; Wang, P.; Cheng, H.; Li, H.; Lu, Z.; Yang, M. In-situ self-templating synthesis of 3D hierarchical porous carbons from oxygen-bridged porous organic polymers for high-performance supercapacitors. *Nano Res.* **2022**, *15*, 7759–7768. [CrossRef]

McAllister, B.T.; Schon, T.B.; DiCarmine, P.M.; Seferos, D.S. A study of fused-ring thieno[3,4-e]pyrazinepolymers as n-type materials for organic supercapacitors. *Polym. Chem.* **2017**, *8*, 5194–5202. [CrossRef]

Wang, K.B.; Xun, Q.; Zhang, Q. Recent progress in metal-organic frameworks as active materials for supercapacitors. *EnergyChem* **2020**, *2*, 100025. [CrossRef]

Sawitri, R.A.; Suryanti, L.; Zuhri, F.U.; Diantoro, M. Dielectric Properties of Dirt Sugarcane Sediment (DSS) Extract-$BaTiO_3$ for Organic Supercapacitors. *IOP Conf. Ser. Mater. Sci. Eng.* **2019**, *515*, 012062. [CrossRef]

Figueroa-González, E.; Oliva, A.I.; Rodriguez-Gonzalez, V.; Gomez-Solis, C.; García, C.R.; Oliva, J. Using recycled Tetrapak and Ag/$BaMoO_4$ nanoarticles to make efficient and flexible solid state supercapacitors. *J. Energy Storage* **2022**, *47*, 103544. [CrossRef]

Adinaveen, T.; Vijaya, J.J.; Sivakumar, R.; Kennedy, L.J. Structural and electrochemical investigation of waste newspaper based electrodes for supercapacitor applications. *Mater. Sci.-Pol.* **2016**, *34*, 302–314. [CrossRef]

Flores-Larrea, L.; Rivera-Mayorga, J.A.; Kshetri, Y.K.; Rodriguez-Gonzales, V.; Garcia, C.R.; Lee, S.W.; Oliva, J. Highly efficient textile supercapacitors fabricated with graphene/NiO:Yb electrodes printed on cotton fabric. *J. Alloys Compd.* **2021**, *886*, 161219. [CrossRef]

Rotich, G.; Dasalegn, A. Mechanical and Water Absorption Properties of Jute/Palm Leaf Fiber-Reinforced Recycled Polypropylene Hybrid Composites. *Int. J. Polym. Sci.* **2022**, *2022*, 4408455. [CrossRef]

Sánchez Vergara, M.E.; Gonzalez Vargas, M.; Sandoval Plata, E.I.; Flores Huerta, A. Chloro-substituted metallo-phthalocyanines ($TiPcCl_2$, MnPcCl, InPcCl, AlPcCl) applied to organic devices. *Heliyon* **2025**, *11*, e42697. [CrossRef]

Sánchez-Vergara, M.E.; Rivera, M. Investigation of optical properties of annealed aluminum phthalocyanine derivatives thin films. *J. Phys. Chem. Solids* **2014**, *75*, 599–605. [CrossRef]

Rodríguez Gómez, A.; Sánchez-Hernández, C.M.; Fleitman-Levin, I.; Arenas-Alatorre, J.; Alonso-Huitrón, J.C.; Sánchez Vergara, M.E. Optical Absorption and Visible Photoluminescence from Thin Films of Silicon Phthalocyanine Derivatives. *Materials* **2014**, *7*, 6585–6603. [CrossRef]

Hamui, L.; Sánchez-Vergara, M.E. Innovative Implementation of an Alternative Tetrathiafulvene Derivative for Flexible Indium Phthalocyanine Chloride-Based Solar Cells. *Micromachines* **2021**, *12*, 633. [CrossRef]

Sánchez Vergara, M.E.; Canseco Juárez, M.J.; Ballinas Indili, R.; Carmona Reyes, G.; Álvarez Bada, J.R.; Álvarez Toledano, C. Studies on the Structure, Optical, and Electrical Properties of Doped Manganese (III) Phthalocyanine Chloride Films for Optoelectronic Device Applications. *Coatings* **2022**, *12*, 246. [CrossRef]

Murphy, A.R.; Fréchet, J.M.J. Organic Semiconducting Oligomers for Use in Thin Film Transistors. *Chem. Rev.* **2007**, *107*, 1066–1096. [CrossRef]

33. Cantera-Cantera, L.A.; Sanchez-Vergara, M.E.; Hamui, L.; Mejia-Prado, I.; Flores-Huerta, A.; Martinez-Plata, T.L. Analysis Flexible Photoconductor, Manufactured with Organic Semiconductor Films. *Micromachines* **2024**, *15*, 446. [CrossRef] [PubM
34. Li, C.F.; Zhang, M.; Liu, S.W.; Chiu, T.L.; Lee, J.H. High Photoelectric Conversion Efficiency of Metal Phthalocyanine/Fulle Heterojunction Photovoltaic Device. *Int. J. Mol. Sci.* **2011**, *12*, 476–505. [CrossRef] [PubMed]
35. Karan, S.; Basak, D.; Mallik, B. Persistence in photoconductivity and optical property of nanostructured copper (II) phthalocya thin films. *Curr. Appl. Phys.* **2010**, *10*, 1117–1122. [CrossRef]
36. Wang, J.B.; Li, W.L.; Chu, B.; Lee, C.S.; Su, Z.S.; Zhang, G.; Wu, S.H.; Yan, F. High speed responsive near infrared photodete focusing on 808 nm radiation using hexadecafluoro–copper–phthalocyanine as the acceptor. *Org. Electron.* **2011**, *12*, 34 [CrossRef]
37. Oni, J.; Ozoemena, K.I. Phthalocyanines in batteries and supercapacitors. *J. Porphyr. Phthalocyanines* **2012**, *16*, 754–760. [Cross
38. Wang, Y.; Li, M.; Ramachandran, R.; Shan, H.; Chen, Q.; Luo, A.; Wang, F.; Xu, Z.-X. Peripheral octamethyl-substituted nicke phthalocyanine-decorated carbon-nanotube electrodes for high-performance all-solid-state flexible symmetric supercapacito *Energy Chem.* **2023**, *76*, 214–225. [CrossRef]
39. Sharma, A.; Jain, A.; Saxena, S. The structure-activity relationship of some hexacoordinated dimethyltin (IV) complexe fluorinated β-diketone/β-diketones and sterically congested heterocyclic β-diketones. *Appl. Organomet. Chem.* **2015**, *29*, 499– [CrossRef]
40. Kayunkid, N.; Tammarugwattana, N.; Mano, K.; Rangkasikorn, A.; Nukeaw, J. Growth and characterizations of tin-do nickel-phthalocyanine thin film prepared by thermal co-evaporation as a novel nanomaterial. *Surf. Coat. Technol.* **2016**, 101–105. [CrossRef]
41. Pirriera, M.D.; Puigdollers, J.; Voz, C.; Stella, M.; Bertomeu, J.; Alcubilla, R. Optoelectronic properties of CuPc thin films depos at different substrate temperatures. *J. Phys. D Appl. Phys.* **2009**, *42*, 145102. [CrossRef]
42. Scharber, M.C.; Sariciftci, N.S. Low Band Gap Conjugated Semiconducting Polymers. *Adv. Mater. Technol.* **2021**, *6*, 200C [CrossRef]
43. Roncali, J. Molecular Engineering of the Band Gap of π-Conjugated Systems: Facing Technological Applications. *Macromol. R Commun.* **2007**, *28*, 1761–1775. [CrossRef]
44. Shalini, S.; Naveen, T.B.; Durgalakshmi, D.; Balakumar, S.; Rakkesh, R.A. Progress in flexible supercapacitors for weara electronics using graphene-based organic frameworks. *J. Energy Storage* **2024**, *86*, 111260. [CrossRef]
45. Muthu, D.; Dharman, R.K.; Muthu, S.E.; Oh, T.H. Recent developments in metal-organic framework-derived transition m oxide@ carbon nanostructure and carbon nanostructure for supercapacitor applications. *J. Energy Storage* **2025**, *119*, 116 [CrossRef]
46. Hou, J.; Ning, Y.; Guo, K.; Jiao, W.; Chen, C.; Zhang, B.; Wu, X.; Zhao, J.; Lin, D.; Sun, S. Application of metal-organic framew materials in supercapacitors. *J. Energy Storage* **2025**, *113*, 115535. [CrossRef]
47. Cantera-Cantera, L.A.; Garrido, R.; Luna, L.; Vargas-Jarillo, C.; Asiain, E. Identification of Linear Time-Invariant Systems: A L Squares of Orthogonal Distances Approach. *Mathematics* **2023**, *11*, 1238. [CrossRef]

micromachines

le

por-Assisted Method to Deposit Compact $(CH_3NH_3)_3Bi_2I_9$ in Films for Bismuth-Based Planar Perovskite Solar Cells

o Gao [1,2], Xinjie Wang [1,2], Zhen Sun [1,2], Ping Song [1,2], Xiyuan Feng [3,*] and Zhixin Jin [1,2,*]

1 State Key Laboratory of Metastable Materials Science and Technology, Yanshan University, Qinhuangdao 066004, China
2 School of Science, Yanshan University, Qinhuangdao 066004, China
3 School of Microelectronics, Northwestern Polytechnical University, Xi'an 710129, China
* Correspondence: fengxy@nwpu.edu.cn (X.F.); jinzx@ysu.edu.cn (Z.J.)

Abstract: Bismuth-based perovskite derivatives, $(CH_3NH_3)_3Bi_2I_9$ (MBI), are promising non-toxic light-absorbing materials widely used in various photoelectric devices because of their excellent stability. However, MBI-based perovskite solar cells (PSCs) are limited by poor film quality, and the performance of such a device is far behind that of lead-based PSCs. In this work, the crystal structure and morphological properties of MBI films were compared across different preparation methods. The two-step vapor-assisted method can prepare continuous dense MBI films because MBI crystal nucleation is induced by the BiI_3 seed layer. The MBI film grown by this method is better for the production of excellent PSCs compared to the film prepared by the solution method. The best photovoltaic device based on the MBI film could obtain a power conversion efficiency of 1.13%. An MBI device is stored in the glove box for 60 days, and the device's performance is maintained at 99%. These results indicate that the vapor-assisted deposition of MBI films can be an effective method to improve the performance of bismuth-based planar PSCs.

Keywords: $(CH_3NH_3)_3Bi_2I_9$ thin films; vapor-assisted method; seed layer

1. Introduction

Lead-based perovskite solar cells (PSCs) have achieved a great breakthrough in power conversion efficiency (PCE), and the certified efficiency currently exceeds 26% [1]. However, the natural defects of lead-based perovskites, such as those related to stability and toxicity, are still the obstacles restricting their industrialization. As a result, several attempts have been made to develop lead-free perovskite materials for photovoltaic devices. For instance, multiple environmentally benign cations, including Sn^{2+} and Bi^{3+}, have been explored as replacements for Pb [2–5]. To date, the PCE of Sn-based PSCs has exceeded 15% [6]. Nevertheless, Sn-based perovskite films face challenges due to the accelerated oxidation of Sn^{2+} to Sn^{4+}, which compromises both the device efficiency and long-term stability [2].

Bismuth-based perovskite derivatives have emerged as promising alternatives to lead-based perovskite, owing to their outstanding photoelectric properties and remarkable stability, which position them as ideal candidates for photovoltaic applications [7,8]. Among them, MBI has been subjected to a lot of research work in solar cells due to its excellent stability and simple preparation process, although its PCE remains significantly lower [9–11]. At present, the maximum efficiency of MBI solar cells based on mesoporous structures is 3.17%, which is still far behind that of lead and tin-based PSC [12]. One of the main reasons for the low PCE is that the solution fabrication route cannot prepare highly compact MBI films [7,12]. At present, the common solutions of spin coating, two-step

https://doi.org/10.3390/mi16020218

soaking and solvent evaporation have not effectively solved this problem [9,13–21]. The primary cause is likely due to the significantly faster crystal growth rate of MBI, along with a broader range of solvents dissolving both BiI_3 and MBI compared to their lead-based counterparts [22].

Past studies have demonstrated that the crystallographic structure and morphology of MBI films are essential for the performance of MBI PSCs [10,23]. However, few studies have been focused on the relationship between the MBI growth method and the film morphology. In this work, MBI films were prepared by various methods, including the spin coating method (SC), two-step soaking method (TSS) and two-step vapor-assisted (TSVA) method. Using BiI_3 as the seed layer, the crystal growth direction of MBI can be changed by the TSVA method. It was important that the morphology of MBI films was changed from a hexagonal lamellar structure to a continuous dense film. Compared with the traditional SC method, the PCE of Bi-based planar PSCs prepared by the TSVA method could be increased by 4 times, reaching 1.13%. In addition, at room temperature, the PSC retained 99% of its initial performance when stored in a nitrogen atmosphere for two months.

2. Results

Figure 1 illustrates the preparation of thin films for MBI by different methods. Under the SC method, BiI_3 and MAI were dissolved in DMF, MBI films were spin-coated on the SnO_2 substrate, and finally the whole was annealed and crystallized. Under the TSS method, the BiI_3 film is first spun onto a conductive substrate and then annealed at different temperatures in rapid sequence. Afterwards, the BiI_3 films are immersed in MAI solution for a certain time and then annealed to form MBI films. At last, the preparation process of the BiI_3 film by the TSVA method is the same as that under the TSS method. The BiI_3 was put in a petri dish, which was placed on a hot table in reverse; the BiI_3 film was facing down, the hot table was placed on the MAI, and the hot table was heated to a certain temperature, after which MBI films were obtained by vapor deposition. For further details on the film fabrication process, please refer to the Supplementary Section.

The morphology of the smooth, dense, and large grains in perovskite solar cells (PSCs) plays a significant role in enhancing power conversion efficiency (PCE) through several interconnected mechanisms. Firstly, large grains reduce the number of grain boundaries and defects, which are known to be sites for carrier recombination. This reduction in recombination losses allows for more efficient charge transport and collection. Additionally, the larger grain size facilitates improved carrier mobility by minimizing scattering events, thereby enhancing the overall electrical conductivity. Furthermore, a smooth surface morphology contributes to superior light absorption by reducing reflection, thereby increasing the number of photons absorbed and converted into charge carriers. Moreover, films with larger grains typically exhibit better crystallinity, which not only improves the mobility of charge carriers but also enhances the stability of the perovskite material, reducing performance degradation over time. Lastly, the reduction in the number of grain boundaries in larger grains minimizes the impact of ionic migration, which is more pronounced in smaller-grained films and can negatively affect both device stability and efficiency [10,13]. Collectively, these factors underscore the importance of a smooth, dense, and large-grain morphology in optimizing the PCE of PSCs.

To gain insights into the relationship between film morphology and preparation method, scanning electron microscopy (SEM) was applied. Significant differences in the morphologies of MBI films prepared using different methods were observed (Figure 2). Photographic images of the prepared samples are shown in Figure S1. The MBI polycrystalline film prepared by the conventional SC method exhibited poor morphology, characterized by large grains and noticeable gaps (Figure 2a). The edges of the MBI sheets reached lengths greater than 500 nm. The layer was generally perpendicular to the substrate, so the thickness of the MBI film was about 800 nm (Figure S2). With a change in the MBI growth method, the morphology of the MBI changed accordingly. Figure 2b shows the TSS method, whereby a granular film with small grain sizes was produced, but the film was irregular. Fortunately, the morphology of MBI films prepared by TSVA can be significantly changed. No isolated hexagonal, cross-ridged crystalline grains were noticed in the TSVA-based MBI films. Similar to that observed in lead-based perovskite thin films, the MBI films have a smooth and dense large grain morphology. The grain sizes of the samples were in the range of 500 to 800 nm, with larger grains, exceeding 1 μm, being readily fabricated. Our experiments revealed that the films prepared by the two-step method were more uniform and flat than those prepared by the one-step method. The TSS-based and TSVA-based films showed particle packed structures, while those prepared by the one-step SC method exhibited a vertical hexagonal lamellar structure, consistent with the findings reported in the literature [18,24]. This may be because the two-step method grew the structure on the basis of BiI_3 as the crystal seed layer, which could regulate the crystal growth direction. In order to verify the causes of the changes in the morphology of MBI films, MBI films were prepared by a two-step spin coating method, which consisted of spin coating BiI_3 film and MAI films successively, and SEM images were displayed in Figure S3. The film prepared by this method can also form a uniform small granular film. Therefore, the reason for the change in morphology is that, in the two-step preparation process, the BiI_3 in the first step is used as the crystal seed layer, which can regulate the crystal growth direction during the formation of the MBI film [14,23].

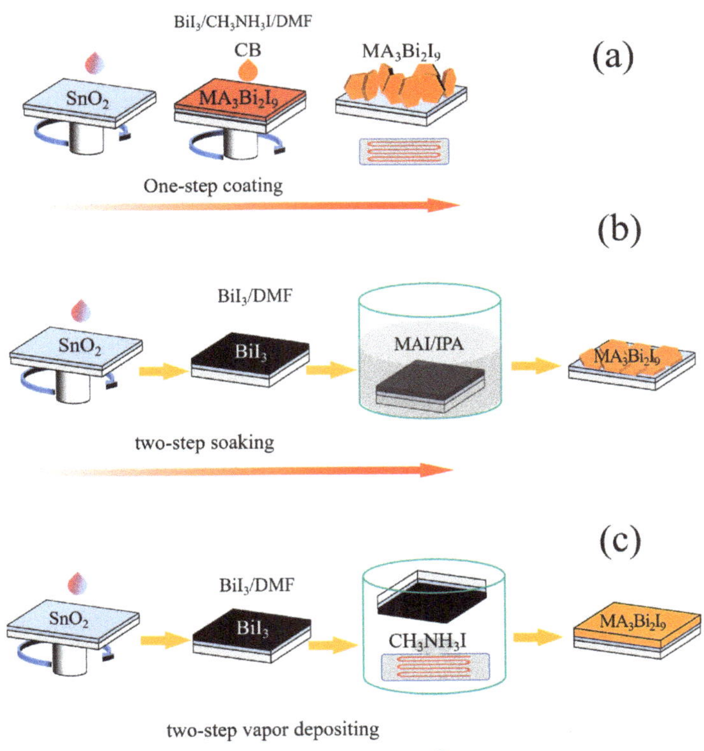

Figure 1. (**a**) Preparation procedure of the MBI thin film by the SC method. (**b**) TSS method. (**c**) TSVA method.

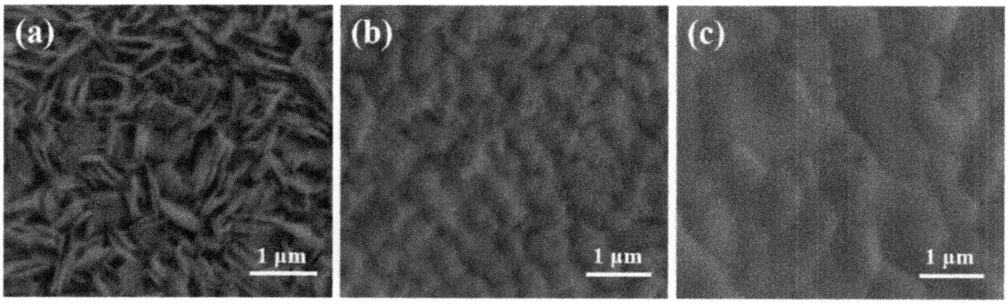

Figure 2. The surface morphologies of MBI films were prepared by different methods. (**a**) SC method, (**b**) TSS, (**c**) TSVA method.

The BiI_3 films were annealed at 60 °C and 100 °C, respectively. Figure S4a shows the X-ray diffraction (XRD) of BiI_3 films before and after annealing. The peaks of 12°, 24.5°, 27°, 41°, and 43.6° matched with the crystal faces of (003), (021), (113), (300), and (303) of BiI_3, respectively [23,25]. The peaks marked with an asterisk (*) in the figure correspond to the diffraction peaks of the FTO substrate. We found that with the increase in temperature, the (300) crystal plane gradually strengthened, and the (003) crystal plane gradually weakened. This indicates that BiI_3 changes from longitudinal to transverse growth. In the two-step process, the crystal orientation of the seed layer in the first step is a critical factor in the quality of the final product. Hence, BiI_3's crystal orientation is the root cause of the vertical or horizontal growth of MBI films [5,23]. Figure S4b presents the ultraviolet–visible (UV-vis) absorption curve of the BiI_3 films. The BiI_3 film exhibited a wide absorption in the visible range of 350–680 nm, agreeing well with the reported result [26]. Compared with the unannealed film, the absorption edge of the annealed sample was redshifted, which was caused by the crystallization of the intermediate product ($Bi(DMF)_8$–Bi_3I_{12}) prepared by the solution method [27]. When the annealing temperature reached 100 °C, the absorption was obviously enhanced, indicating that the crystallinity of BiI_3 films was greatly improved. Figure S4c–e present SEM images of BiI_3 films treated at different annealing temperatures. As the annealing temperature increased, the grain dimension of the BiI_3 films grew steadily, and a large number of cavities were created on the surface. With the evaporation of DMF from the film, precursor molecules aggregated with each other to form groups, resulting in depletion regions during film growth. The formation of these depleted regions contributed to the development of cavities on the surface of the film [27]. The multihole structure of the BiI_3 film significantly enhanced the surface area, thereby promoting an effective interfacial reaction with MAI vapor [28].

The XRD patterns of MBI films grown via different methods are shown in Figure 3a. The main peaks at 2θ = 12.2°, 25.3°and 42.2° correspond to the (101), (006) and (400) crystal faces of the MBI crystal, respectively [7]. By changing the preparation method, the crystal orientations of MBI films changed from a single orientation of layered films to a multi-crystalline orientation of compact flat films. The preponderant crystal direction of the MBI film prepared by the SC method was (101), with (006) and (400) crystallographic directions, in which 006 was the dominant crystal direction. Hence, a vertical layered structure was formed, which also corresponded to the previous SEM. When the preparation methods were changed to TSS and TSVA, the crystal orientation changed obviously. In particular, the crystal orientations of TSVA, MBI included (100), (101), (006), (203), (204), and (400) planes, with no obvious preferred orientation. Therefore, nucleation sites of the BiI_3 seed layer can grow synchronously in all directions [14]. Therefore, homogeneous and dense MBI films can be obtained by adjusting the crystallization direction. Meanwhile, no residue of BiI_3 was observed in the film produced by the two-step method, indicating that BiI_3 was completely transformed into MBI.

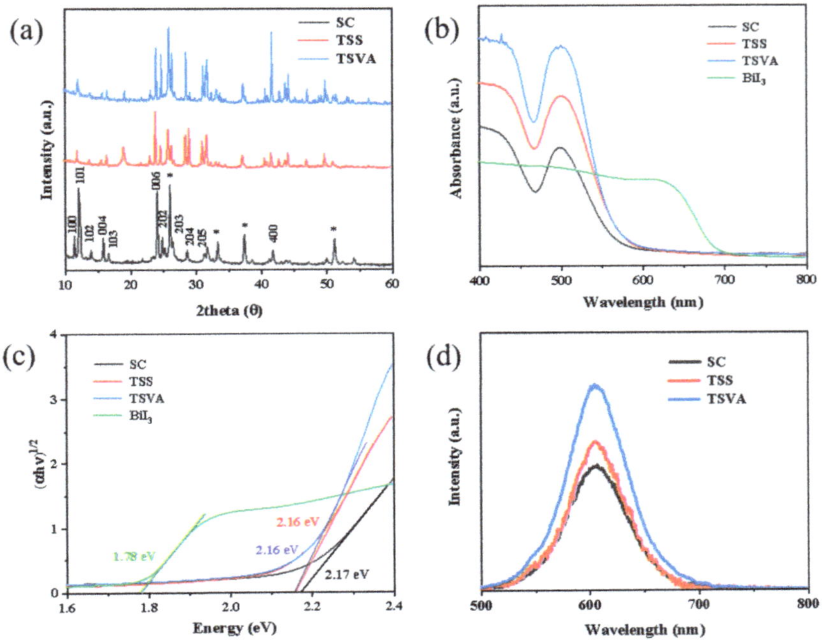

Figure 3. (**a**) XRD patterns. The "*" symbols represent the signals of the FTO substrates. (**b**) UV–visible spectra, (**c**) Tauc plot versus the light energy, (**d**) photoluminescence spectra of MBI films grown via different methods.

Figure 3b shows the UV-vis results of several films. The BiI_3 film exhibited broad absorption across the visible wavelength range of 350–700 nm, consistent with previous observations. Upon transformation to MBI, the spectrum of the MBI film displayed a distinct exciton absorption peak at 505 nm, which was attributed to the electron transition of Bi^{3+} in the isolated $Bi_2I_9{}^{3-}$ cluster from the ground S_0 state to the excited $3P_1$ state as the energy of light increased [29]. The MBI films prepared by TSS and TSVA showed significantly higher absorption than the films prepared by SC. The absence of a characteristic BiI_3 absorption band in the MBI film suggests the complete conversion of BiI_3. Figure 3c shows the band gap diagram, indicating that the preparation method has no effect on the optical band gap of the MBI film, and the band gap of the prepared film is 2.13 eV. This is basically consistent with the previous values in the literature [23,30]. A Tauc plot was employed to analyze the linear absorption edge by plotting $(\alpha h\nu)^{1}/^{2}$ as a function of $h\nu$, revealing a band gap of 1.8 eV for the evaporated BiI_3 film, consistent with the reported value [31,32].

Photoluminescence (PL) measurements were conducted to further investigate the optical properties of the MBI films. Figure 3d demonstrates the PL spectra of the MBI grown via different methods. The strongest peak at 605 nm (2.06 eV) belonged to the band edge exciton radiation luminescence [22]. The PL intensity of MBI films grown by two-step method is obviously higher than that of MBI films grown by the one-step method. This is attributed to the enhanced crystallinity and morphology of the MBI perovskite film, which suppressed radiative recombination. The strength of the films prepared by the two-step method is better than that of those prepared by the one-step method, which indicates that the crystal quality of bismuth-based perovskite can be improved by the two-step method. The film prepared by TSVA showed the highest PL intensity in these methods, indicating that the film grown by this method had the highest crystal quality. The PL attenuation spectra of MBI films obtained by different methods are shown in Figure S5. The average lifetimes without a charge transport layer fitted via a single-exponential function were 5.83, 6.78 and 7.52 ns for MBI (SC), MBI (TSS), and MBI (TSVA), respectively. The increased lifetime indicates that carriers have a longer diffusion length. Therefore, the photovoltaic device prepared by the two-step MBI film is expected to improve FF and PCE.

MBI-based PSCs were fabricated using compact films, with the configuration FTO/SnO_2/MBI/Spiro-OMeTAD/Au (Figure 4a). Figure 4b showed the band locations of all functional layers of the device. UPS measurements (Figure S6) indicate a Fermi level positioned 0.6 eV above the valence band of MBI, highlighting its intrinsic semiconducting nature, which helps reduce carrier recombination. The observation of low radiative recombination in the higher-energy region with a bandgap of more than 2.13 eV was crucial to achieving high open-circuit voltages (V_{oc}) and efficient photovoltaic conversion. The improved band arrangement of MBI films is anticipated to enhance the collection and transport of charge carriers at the MBI/Spiro-OMeTAD interface, leading to enhanced photoelectric performance [33,34]. The performances of the MBI PSCs are displayed in Figure 5 and Table 1. The best TSVA-based device has demonstrated significant improvement over the other groups in three *J*-*V* parameters. The best device could achieve a short-circuit current (J_{sc}) of 2.03 mA/cm^2 while maintaining a high V_{oc} of 0.78 V. Additionally, the fill factor (FF) was substantially elevated, contributing to the overall performance enhancement. It is found that the FF of the TSVA-based device exceeded 0.6 in terms of device performance. The enhanced FF is related to the optimization of the crystallinity and morphology of the MBI film, and the improved MBI layer can play a more effective role, given the presence of a charge transport layer. Therefore, the high-quality MBI film prepared by the TSVA route can overcome the bottleneck of high FFs. Figure S7 shows that all PSCs display a small hysteresis behavior.

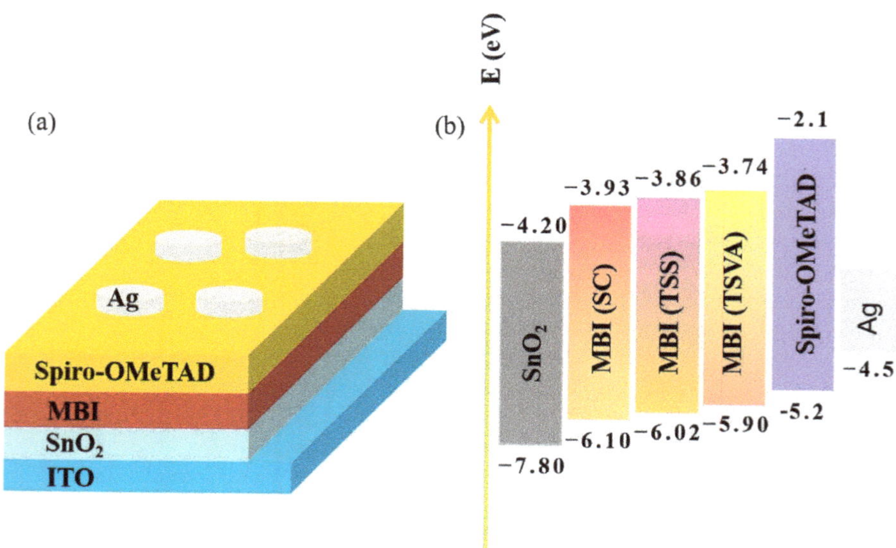

Figure 4. (**a**) Configuration of the planar MBI photovoltaic device; (**b**) schematic band diagram of MBI film grown via different methods.

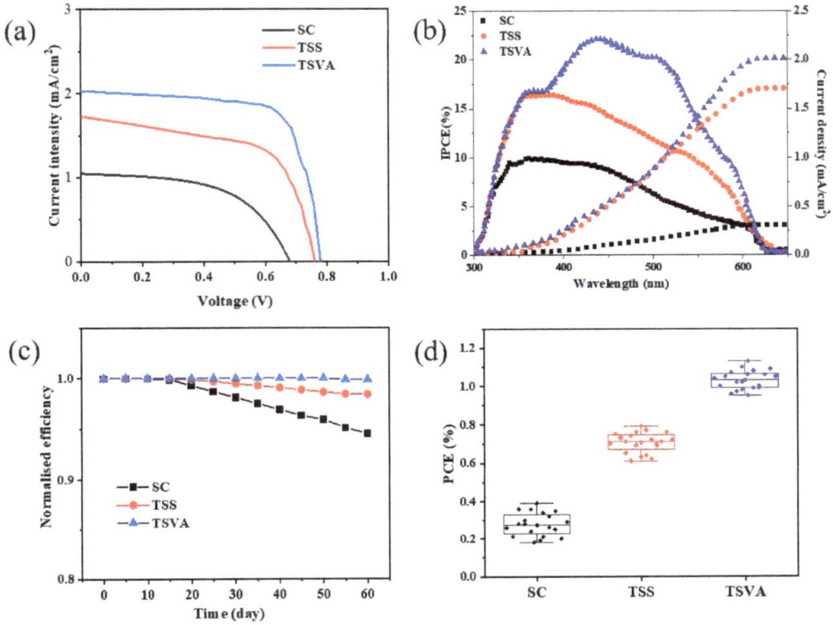

Figure 5. (**a**) *J-V* curves of MBI devices with different growth methods. (**b**) IPCE spectra and the integrated current for the different devices. (**c**) Stability of MBI devices with different preparation methods in a glove box. (**d**) PCE box statistics of 20 photovoltaic devices prepared by different methods.

Table 1. Photovoltaic performance at 100 W/cm^2 of MBI devices made via several methods.

Sample	V_{oc} (V)	J_{sc} (mA/cm^2)	*FF* (%)	η (%)
TSVA	0.78	2.03	71.4	1.13
TSS	0.75	1.73	60.4	0.79
SC	0.68	1.05	55.2	0.39

In accordance with the optical absorption of MBI films prepared using different growth methods, the incident photon-to-current conversion efficiency (IPCE) showed a good correlation. The IPCE spectrum exhibited a maximum value of approximately 23% at 458 nm, and the corresponding integrated current was consistent with that of the *J-V* curve. The IPCE curve had a cutoff point at 650 nm and did not have a high platform across the entire visible range. The IPCE of the film prepared by vapor deposition was 2 times higher than that fabricated with the one-step method, and 10% higher than that prepared by the immersion method, which may be due to the higher quality of the film prepared by the vapor deposition method.

In addition to the toxicity of perovskite materials, long-term stability is another key factor limiting the industrialization of PSCs. Bismuth, as a non-toxic element, offers better chemical stability in MBI compared to Pb-based perovskites. The non-encapsulated devices were evaluated over a period of 60 days, showing only a minimal decrease in PCE of 0.1%, demonstrating a significant improvement in stability compared to lead-based perovskites. The long-term stability and its performance over time are depicted in Figure 5c. In addition, the best device was stored in the glove box for 60 days with virtually no attenuation in device performance, maintaining 99.9%. To characterize the reproducibility of MBI PSCs, Figure 5d exhibits the statistical distribution of PCE for the three kinds of MBI PSCs. The photovoltaic device-based MBI grown by TSVA showed a higher average PCE compared to that of solar cells constructed via other methods. Compared with the PSC prepared by the SC method, MBI prepared by the TSVA method showed a narrower PCE distribution range and better reproducibility.

The electrical properties of the devices were characterized by Mott–Schottky testing. Figure 6a presents the capacitance–voltage (C-V) characteristics of all devices. The built-in voltages (V_{bi}) of the SC, TSS and TSVA solar cells showed values of 0.69 V, 0.75 V, and 0.78 V, respectively. The larger V_{bi} indicates that the driving force of photogenerated carrier separation increased, resulting in the inhibition of electron–hole recombination [34]. Regarding the MBI films, the large V_{bi} could be attributed to the wide bandgap. In addition, in order to contrast the charge transport properties of the devices, the charge space limited current (SCLC) method can be used to study the trap density of the films. Figure 6b shows the SCLC curves of three hole-only devices with MBI films. First, the SCLC curve indicates the Ohmic region at low bias voltage. When the bias voltage surpasses the trap filled limit voltage (V_{TFL}), the SCLC curve enters the trap-filled region, indicating that the trap state of the device was completely filled. When the bias voltage continued to increase, the device entered the Child's region [35–37]. The trap state densities in the MBI films were calculated using the following equation [38]:

$$N_t = \frac{2\varepsilon_0 \varepsilon_r V_{TFL}}{qL^2}$$

where q is the elementary charge; L denotes the thickness of the MBI layer; ε_0 and ε_r represent the vacuum dielectric constant and relative dielectric constant, respectively. When MBI was grown using the TSVA method, the V_{TFL} of hole-only devices was intended to be 0.12 V, which is lower than that of the MBI films prepared by the SC and TSS methods. This indicates that the trap density in MBI films prepared by this method was significantly reduced. Hence, the trap densities of the TSVA MBI films were calculated to be 1.65×10^{15} cm^{-3}.

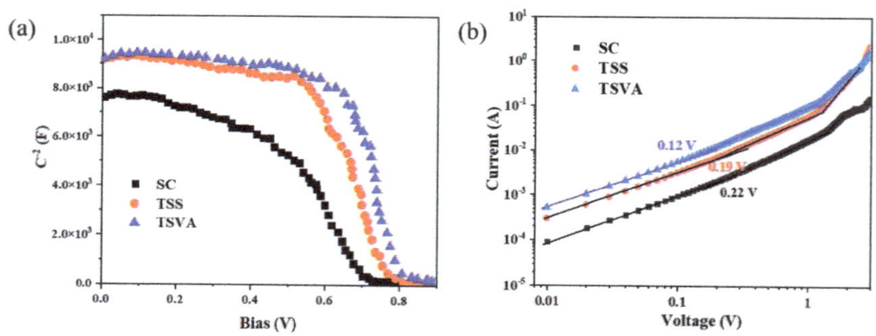

Figure 6. (**a**) C-V curves of different devices yielded by Mott–Schottky testing; (**b**) SCLC curves for hole-only devices prepared via different methods.

The charge mobility (μ) was extracted from the Child region using the Mott–Gurney equation [16]:

$$\mu = \frac{8JL^3}{9\varepsilon_0 \varepsilon_r V^2}$$

where J/V^2 denotes the slope of J as a function of V^2 in the Child region. The TSVA MBI film exhibited a charge mobility of 8.5×10^{-7} cm^{2}V^{-1}s^{-1}. The trap density and mobility values of MBI films prepared by other methods are shown in Table S2. The results of low trap state density and large charge mobility fully account for the improved J_{sc} and V_{oc} in MBI solar cells prepared by the TSVA method.

3. Conclusions

In summary, different methods (SC, TSS, TSVA) have been applied to prepare the MBI thin films, among which the SC method can produce a hexagonal sheet morphology, while the TSVA method can obtain a uniform and dense surface morphology. The experimental data show that the BiI_3 seed layer plays a key role in controlling the final morphology and crystallinity of MBI. By controlling the annealing temperature, the crystal direction of the BiI_3 seed layer changes from 003 to 300, which is conductive to the growth of flat films. During the conversion of BiI_3 to MBI, the TSVA method can inhibit the preferred orientation of MBI and make the seed layer nucleate uniformly. Further, the preparation of photovoltaic devices with dense MBI as the absorption layer can achieve better performance due to their lower carrier defect density and higher carrier mobility. As a result, the TSVA device has a PCE of 1.13%, and after 2 months of storage in a nitrogen atmosphere at room temperature, the PCE remains at 99% of the initial PCE. Hence, the TSVA method may provide a new way to prepare high-quality MBI films.

Supplementary Materials: The following supporting information can be downloaded at: https://www.mdpi.com/article/10.3390/mi16020218/s1, Figure S1: Phototographic images of the prepared samples; Figure S2: SEM of the surface and cross-section of MBI films prepared by spin-coating; Figure S3: The surface SEM of MBI films was prepared by continuous two-step spin-coating; Figure S4: (a) XRD of BiI_3 films annealed at different temperatures. The "*" symbols represent the signals of the FTO substrates. (b) UV–visible absorption spectra of the BiI_3 annealed at different temperatures. (c–f) SEM images of BiI_3, BiI_3-60, BiI_3-100 and BiI_3-150 films; Figure S5: PL decay spectra of MA3Bi2I9 film prepared by different growth methods; Figure S6: UPS spectra and the secondary electron cutoff for the MBI layers produced by different growth methods; Figure S7: Forward and reverse scans of *J-V* curves of Bi-based PSC devices prepared by different growth methods; Table S1: Table of photovoltaic parameters of Bi-based PSCs prepared by different methods; Table S2: The carrier mobility and trap density of different preparation methods.

Author Contributions: Conceptualization, Z.J.; methodology, Z.G.; validation, X.W.; investigation, X.F.; resources, Z.S. and P.S.; writing—original draft preparation, Z.G. and X.W.; writing—review and editing, Z.J. and Z.G.; supervision, X.F.; project administration, Z.J. and X.F.; funding acquisition, Z.J., X.F., Z.S. and P.S. All authors have read and agreed to the published version of the manuscript.

Funding: This work was supported by National Natural Science Foundation of China (Grant Nos. 12204401, 52101233, 52371200), the innovation capability improvement Project of Hebei province (No. 22567605H), and the project supported by the Fundamental Research Funds for the Central Universities of Ministry of Education of China (No. D5000240188).

Data Availability Statement: Data is contained within the article or supplementary material. The original contributions presented in this study are included in the article/supplementary material. Further inquiries can be directed to the corresponding author.

Conflicts of Interest: The authors declare no conflicts of interest.

ferences

Park, J.; Kim, J.; Yun, H.S.; Paik, M.J.; Noh, E.; Mun, H.J.; Kim, M.G.; Shin, T.J.; Seok, S., II. Controlled growth of Perovskite Layers with Volatile Alkylammonium Chlorides. *Nature* **2023**, *616*, 724–730. [CrossRef] [PubMed]
Jin, Z.; Yu, B.-B.; Liao, M.; Liu, D.; Xiu, J.; Zhang, Z.; Lifshitz, E.; Tang, J.; Song, H.; He, Z. Enhanced efficiency and stability in Sn-based perovskite solar cells with secondary crystallization growth. *J. Energy Chem.* **2021**, *54*, 414–421. [CrossRef]
Na, H.; Alam, S.; Ham, G.; Baek, D.; Yoon, T.O.; Lee, G.; Li, M.Q.; Lee, M.; Cha, H.; Lee, J.; et al. Interlayer molecular doping to enhance efficiency in tin perovskite solar cells. *ACS Energy Lett.* **2024**, *9*, 4306–4315. [CrossRef]
Li, Z.; Yang, M.; Li, L.; Tu, Z.; Song, L.; Ding, B.; Wang, R.; Xu, Y. Density functional theory analysis of structural and electronic properties of hexagonal hybrid perovskite $(CH_3NH_3)_3Bi_2I_9$. *Phys. B Condens. Matter* **2022**, *630*, 413695. [CrossRef]
Schileo, G.; Grancini, G. Lead or no lead? Availability, toxicity, sustainability and environmental impact of lead-free perovskite solar cells. *J. Mater. Chem. C* **2021**, *9*, 67–76. [CrossRef]

6. Wang, L.; Bi, H.; Liu, J.Q.; Wei, Y.Y.; Zhang, Z.; Chen, M.M.; Baranwal, A.K.; Kapil, G.; Kitamura, T.; Yang, S.Z.; et al. Excee 15% performance with energy level tuning in tin-based perovskite solar cells. *ACS Energy Lett.* **2024**, *9*, 6238–6244. [CrossR
7. Zhang, X.Q.; Wu, G.; Gu, Z.W.; Guo, B.; Liu, W.Q.; Yang, S.D.; Ye, T.; Chen, C.; Tu, W.W.; Chen, H.Z. Active-layer evolu and efficiency improvement of $(CH_3NH_3)_3Bi_2I_9$-based solar cell on TiO_2-deposited ITO substrate. *Nano Res.* **2016**, *9*, 2921– [CrossRef]
8. Park, B.W.; Philippe, B.; Zhang, X.; Rensmo, H.; Boschloo, G.; Johansson, E.M.J. Bismuth Based Hybrid Perovskites A_3 (A:Methylammonium or Cesium) for Solar Cell Application. *Adv. Mater.* **2015**, *27*, 6806–6813. [CrossRef]
9. Singh, T.; Kulkarni, A.; Ikegami, M.; Miyasaka, T. Effect of electron transporting layer on bismuth-based lead-free perov $(CH_3NH_3)_3Bi_2I_9$ for photovoltaic applications. *ACS Appl. Mater. Interfaces* **2016**, *8*, 14542–14547. [CrossRef] [PubMed]
10. Wang, S.; Ma, Z.; Liu, B.; Wu, W.; Zhu, Y.; Ma, R.; Wang, C. High-performance perovskite solar cells with large grain-size obta by using the Lewis acid-base adduct of thiourea. *Sol. Rrl.* **2018**, *2*, 1800034. [CrossRef]
11. Oz, S.; Hebig, J.C.; Jung, E.; Singh, T.; Lepcha, A.; Olthof, S.; Jan, F.; Gao, Y.; German, R.; van Loosdrecht, P.H.; et al. roDimensional $(CH_3NH_3)_3Bi_2I_9$ Perovskite for Optoelectronic Applications. *Sol. Energy Mater. Sol. Cells* **2016**, *158*, 195- [CrossRef]
12. Jain, S.M.; Phuyal, D.; Davies, M.L.; Li, M.; Philippe, B.; De Castro, C.; Qiu, Z.; Kim, J.; Watson, T.; Tsoi, W.C.; et al. An effe approach of vapour assisted morphological tailoring for reducing metal defect sites in lead-free, $(CH_3NH_3)_3Bi_2I_9$ bismuth-b perovskite solar cells for improved performance and long-term stability. *Nano Energy* **2018**, *49*, 614–624. [CrossRef]
13. Sajid, S.; Alzahmi, S.; Salem, I.B.; Obaidat, I.M. Perovskite-surface-confined grain growth for high-performance perovskite s cells. *Nanomaterials* **2022**, *12*, 3352. [CrossRef]
14. Kulkarni, A.; Singh, T.; Ikegami, M.; Miyasaka, T. Photovoltaic Enhancement of Bismuth Halide Hybrid Perovskite by N-Me Pyrrolidone-Assisted Morphology Conversion. *RSC Adv.* **2017**, *7*, 9456–9460. [CrossRef]
15. Vuong, V.H.; Pammi, S.; Pasupuleti, K.S.; Hu, W.; Tran, V.D.; Jung, J.S.; Kim, M.D.; Pecunia, V.; Yoon, S.G. Engineering chem vapor deposition for lead-free perovskite-inspired $MA_3Bi_2I_9$ self-powered photodetectors with high performance and stab *Adv. Opt. Mater.* **2021**, *9*, 2100192. [CrossRef]
16. Abulikemu, M.; Ould-Chikh, S.; Miao, X.; Alarousu, E.; Murali, B.; Ngongang Ndjawa, G.O.; Barbe, J.; El Labban, A.; Amas A.; Del Gobbo, S. Optoelectronic and Photovoltaic Properties of The AirStable Organohalide Semiconductor $(CH_3NH_3)_3Bi_2$ *Mater. Chem. A* **2016**, *4*, 12504–12515. [CrossRef]
17. Mali, S.S.; Kim, H.; Kim, D.H.; Kook Hong, C. Anti-Solvent Assisted Crystallization Processed Methylammonium Bismuth Io Cuboids towards Highly Stable Lead-Free Perovskite Solar Cells. *ChemistrySelect* **2017**, *2*, 1578–1585. [CrossRef]
18. Hoye, R.L.Z.; Brandt, R.E.; Osherov, A.; Stevanovic, V.; Stranks, S.D.; Wilson, M.W.B.; Kim, H.; Akey, A.J.; Perkins, J.D.; Kur R.C.; et al. Methylammonium Bismuth Iodide as A LeadFree. Stable Hybrid. *Org.—Inorg. Sol. Absorber. Chem. Eur. J.* **2016** 2605–2610.
19. Ran, C.X.; Wu, Z.X.; Xi, J.; Yuan, F.; Dong, H.; Lei, T.; He, X.; Hou, X. Construction of compact Methylammonium Bismuth Io Film Promoting Lead-Free inverted Planar Heterojunction Organohalide Solar cells with Open-Circuit Voltage over 0.8V. *J. F Chem. Lett.* **2017**, *8*, 394–800. [CrossRef] [PubMed]
20. Momblona, C.; Kanda, H.; Sutanto, A.A.; Mensi, M.; Roldán-Carmona, C.; Nazeeruddin, M.K. Co-evaporation as an opti technique towards compact methylammonium bismuth iodide layers. *Sci. Rep.* **2020**, *10*, 10640. [CrossRef]
21. Achoi, M.; Noman, M.; Kato, S.; Kishi, N.; Soga, T. Pinhole-free methylammonium bismuth iodide perovskite solar cells all-solution-processed multi-step spin coating. *J. Electron. Mater.* **2022**, *51*, 577–585. [CrossRef]
22. Lyu, M.; Yun, J.-H.; Cai, M.; Jiao, Y.; Bernhardt, P.V.; Zhang, M.; Wang, Q.; Du, A.; Wang, H.; Liu, G.; et al. Organic—Inorg Bismuth(III)-Based Material: A Lead-Free, Air-Stable and SolutionProcessable Light-Absorber beyond Organolead Perovsk *Nano Res.* **2016**, *9*, 692–702. [CrossRef]
23. Okano, T.; Suzuki, Y. Gas-assisted coating of Bi-based $(CH_3NH_3)_3Bi_2I_9$ active layer in perovsktie solar cells. *Mater. Lett.* **2017**, 77–79. [CrossRef]
24. Zhang, Z.; Li, X.; Xia, X.; Wang, Z.; Huang, Z.; Lei, B.; Gao, Y. High-quality $(CH_3NH_3)_3Bi_2I_9$ film-based solar cells: Push efficiency up to 1.64%. *J. Phys. Chem. Lett.* **2017**, *8*, 4300–4307. [CrossRef]
25. Li, J.H.; Han, H.F.; Li, B.C.; Zhao, C.X.; Xu, J.; Yao, J.X. Solvent evaporation induced preferential crystal orientation BiI_3 film the high efficiency $MA_3Bi_2I_9$ perovskite solar cells. *J. Alloys Compd.* **2022**, *909*, 164725. [CrossRef]
26. Brandt, R.E.; Kurchin, R.C.; Hoye, R.L.; Poindexter, J.R.; Wilson, M.W.; Sulekar, S.; Lenahan, F.; Yen, P.X.; Stevanovic, V.; Nino, et al. Investigation of bismuth triiodide (BiI_3) for photovoltaic applications. *J. Phys. Chem. Lett.* **2015**, *6*, 4297–4302. [CrossRe
27. Ma, S.; Yang, Y.; Liu, C.; Cai, M.; Ding, Y.; Tan, Z.; Shi, P.; Dai, S.; Alsaedi, A.; Hayat, T. Vertically oriented BiI_3 temp featured BiI_3/polymer heterojunction for high photocurrent and long-term stable solar cells. *ACS Appl. Mater. Interfaces* **2019** 32509–32516. [CrossRef]
28. Hamdeh, U.H.; Nelson, R.D.; Ryan, B.J.; Bhattacharjee, U.; Petrich, J.W.; Panthani, M.G. Solution-processed BiI_3 thin films photovoltaic applications: Improved carrier collection via solvent annealing. *Chem. Mater.* **2016**, *28*, 6567–6574. [CrossRef]

Günzler, A.; Bermu, E.; Muscarella, L.A.; Ochoa, M.; Ochoa-Martínez, E.; Ehrler, B.; Saliba, M.; Steiner, U. Shaping perovskites: In situ crystallization mechanism of rapid thermally annealed, prepatterned perovskite films. *ACS Appl. Mater. Interfaces* **2021**, *13*, 6854–6863. [CrossRef] [PubMed]

Jia, X.; Hu, Z.; Xu, J.; Huang, L.; Zhang, J.; Zhang, J.; Zhu, Y. Improved perovskite morphology and crystallinity using porous PbI_2 layers for efficient planar heterojunction solar cells. *Appl. Phys. Lett.* **2017**, *111*, 243902. [CrossRef]

Shin, S.S.; Baena, J.P.C.; Kurchin, R.C.; Polizzotti, A.; Yoo, J.J.; Wieghold, S.; Bawendi, M.G.; Buonassiss, T. Solvent-engineering Method to Deposit compact Bismuth-based thin films: Mechanism and Application to Photovoltaics. *Chem. Mater.* **2018**, *30*, 336–343. [CrossRef]

Wang, H.J.; Tian, J.H.; Jiang, K.J.; Zhang, Y.; Fan, H.C.; Huang, J.H.; Yang, L.M.; Guan, B.; Song, Y.L. Fabrication of methylammonium bismuth iodide through interdiffusion of solution-processed BiI_3/CH_3NH_3I stacking layers. *RSC Adv.* **2017**, *7*, 43826. [CrossRef]

Kawai, S.T.; Ishii, A.; Kitamura, T.; Shimanuki, S.; Iwata, M.; Ishibashi, Y. Optical absorption in band-edge region of $(CH_3NH_3)_3Bi_2I_9$ single crystals. *J. Phys. Soc. Jpn.* **1996**, *65*, 1464–1468. [CrossRef]

Emara, J.; Schnier, T.; Pourdavoud, N.; Riedl, T.; Meerholz, K.; Olthof, S. Impact of film stoichiometry on the ionization energy and electronic structure of $CH_3NH_3PbI_3$ perovskites. *Adv. Mater.* **2016**, *28*, 553–559. [CrossRef] [PubMed]

Bai, D.; Zhang, J.; Jin, Z.; Bian, H.; Wang, K.; Wang, H.; Liang, L.; Wang, Q.; Liu, S. Interstitial Mn^{2+}-driven high-aspect-ratio grain growth for low-trap-density microcrystalline film for record efficiency inorganic $CsPbI_2Br$ solar cells. *ACS Energy Lett.* **2018**, *3*, 970–978. [CrossRef]

Zhang, C.; Wang, H.; Li, H.; Zhuang, Q.; Gong, C.; Hu, X.; Cai, W.; Zhao, S.; Chen, J.; Zang, Z. Simultaneous passivation of bulk and interface defects through synergistic effect of anion and cation toward efficient and stable planar perovskite solar cells. *J. Energy Chem.* **2021**, *63*, 452–460. [CrossRef]

Hu, X.; Wang, H.; Ying, Y.; Wang, M.; Zhang, C.; Ding, Y.; Li, H.; Li, W.; Zhao, S.; Zang, Z. Methylammonium chloride as an interface modificator for planar-structure perovskite solar cells with a high open circuit voltage of 1.19 V. *J. Power Sources* **2020**, *480*, 229073. [CrossRef]

Bube, R.H. Trap density determination by space-charge-limited currents. *J. Appl. Phys.* **1962**, *33*, 1733–1737. [CrossRef]

micromachines

Article

Preparation and Performance Study of MXene-Regulated Ethylene Glycol-Induced WO_3 Film

Yuqi Wang [1,†], Yong Liu [1,†], Minmin Wang [2], Wenjun Wu [1], Maofei Tian [1] and Tao Zhu [1,*]

1 College of Materials and Metallurgy, Guizhou University, Guiyang 550025, China
2 School of Energy and Power Engineering and State Key Laboratory of Coal and CBM Co-Mining, North University of China, Taiyuan 030051, China
* Correspondence: ivzhutao@126.com
† Yuqi Wang and Yong Liu contributed equally to this work.

Abstract: This study introduces the development of a W-$M_{1.0}$ electrochromic film, characterized by a "coral"-like TiO_2@WO_3 heterostructure, synthesized via a hydrothermal process leveraging the inherent instability of MXene. The film showcases exceptional electrochromic performance, with a coloring response time of 2.8 s, a bleaching response time of 4.6 s, and a high coloring efficiency of 137.02 cm^2C^{-1}. It also demonstrates a superior light modulation ability of 73.83% at 1033 nm. Notably, the W-$M_{1.0}$ film exhibits remarkable cyclic stability, retaining over 90% of its initial light modulation capacity after 4000 cycles, outperforming many existing electrochromic materials. The film's enhanced performance is credited to its coral-like structure, which boosts the specific surface area and promotes ion transport, and the TiO_2@WO_3 heterojunctions, which enhance charge transfer and stabilize the material. Devices fabricated with the W-$M_{1.0}$ film as the cathode and a PB film as the anode exhibit a seamless transition from dark blue to colorless, underscoring their potential for smart window and dynamic glass applications.

Keywords: two-dimensional material; tungsten oxide; heterojunction; electrochromic

1. Introduction

Electrochromism is a phenomenon where electrochromic (EC) material undergoes electrolyte ion insertion/extraction under an applied electric field, leading to repeated changes in the chemical valence state and band structure, which in turn cause reversible alterations in its optical properties [1,2]. EC materials offer a multitude of advantages, such as intelligent color adjustment capabilities, high optical contrast, ease of control, low energy consumption, and robust stability [3–6]. Consequently, they have found broad applications in various domains, including smart windows, anti-glare rearview mirrors, electronic displays, memory devices, and wearable technology [7–12].

Inorganic materials have garnered significant attention due to their exceptional stability and well-established synthesis techniques. Tungsten oxide (WO_3), the first inorganic material identified to exhibit electrochromic properties, has been extensively studied for optimizing its chromic behavior. It is well known that the electrochromic properties of WO_3 are highly dependent on factors such as surface morphology, structure, composition, and crystallinity, all of which are closely linked to the film preparation process. Consequently, researchers focus on enhancing the performance of WO_3 films by refining these preparation methods, including metal doping, designing nanostructures, and material compounding, which have proven effective in improving EC properties. However, these methods often lead to the formation of a dense WO_3 membrane structure, which impedes the insertion and extraction of electrolyte ions and slows down ion migration. Additionally, this dense architecture decreases the reaction interface, limiting electrochemical reactivity and negatively impacting the response time and efficiency of electrochromic properties. During electrochromic processes, the materials expansion and contraction can create local stress concentrations, raising the likelihood of cracking and spalling, ultimately affecting cycle stability [13,14]. Therefore, simplifying the process to regulate the structure of WO_3 films is crucial for advancing their practical applications.

The hydrothermal synthesis of WO_3 films has emerged as a favored approach due to its low energy consumption, straightforward equipment requirements, and environmentally benign characteristics. This method facilitates film formation, with the introduction of a seed layer or the employment of self-seeding agents being instrumental in enhancing the deposition process. However, the incorporation of a seed layer can sometimes result in suboptimal interfacial bonding, which may compromise the film's cyclic stability. In this context, the utilization of self-seeding agents for film formation presents considerable advantages, particularly in the fabrication of electrochromic WO_3 films [15–17]. Studies have successfully demonstrated the fabrication of WO_3 films using self-seeding agents, such as the creation of coral-like nanostructured films with enhanced cyclic stability through the use of glycerol as a self-seeding agent and ammonium sulfate as a capping agent [18]. Furthermore, ethylene glycol has been effectively utilized as a self-seeding agent to produce nest-like nanostructured WO_3 films, which exhibit rapid response rates [15].

Herein, a high-performance W-M_x EC film (denotes X mL addition of MXene dispersion in the reaction) was successfully prepared using a simple hydrothermal method. Ethylene glycol was employed as a self-seeding agent to induce WO_3 nucleation on the FTO substrate, followed by the introduction of a two-dimensional MXene material as a capping agent. Due to the unique two-dimensional sheet structure and unstable chemical feature of MXene, it is oxidized into TiO_2 during the preparation process [19–21]. Therefore, this method is not only effective but also allows specific morphology to be controlled and also involves doping WO_3 films through the introduction of TiO_2 to enrich their energy band structure. As anticipated, the W-M_x films exhibited porous, coral-like nanostructures that provide ion transfer channels and abundant electrochemically active sites, and the TiO_2-doped heterojunction film enriched the band structure that would promote the electronic transfer process. These favorable factors can effectively promote the EC process and lead to enhancements in EC performance, like response time, coloration efficiency, modulation capability, and cyclic stability.

2. Experimental Section

2.1. Materials

Sodium tungstate dihydrate ($Na_2WO_4{\cdot}2H_2O$), hydrochloric acid (HCl), hydrogen peroxide (H_2O_2), ethylene glycol ($C_2H_6O_2$), propylene carbonate ($C_4H_6O_3$), lithium fluoride (LiF), and lithium perchlorate ($LiClO_4$) were procured from Aladdin Reagent Co., Ltd. (Shanghai, China), and used as received without further purification. Titanium aluminum carbide powder ($TiAlC_2$) was sourced from Jilin Yiyi Technology Co., Ltd. (Changchun, China) and employed without additional purification. Fluorine-doped tin oxide (FTO) conductive glass substrates were obtained from Liaoning Preferred New Energy Technology Co., Ltd. (Yingkou, China), and were cut into dimensions of 0.8×4 cm^2 prior to use. The substrates underwent ultrasonic cleaning in a sequence of acetone, isopropyl alcohol, deionized water, and ethanol for 30 min each, followed by immersion in ethanol for storage. Before experimental use, the glass substrates were dried using a nitrogen gun and subjected to ultraviolet ozone cleaning for 20 min to ensure surface sterility and removal of organic contaminants, thus preparing the substrates for subsequent experimental procedures.

2.2. Preparation of Precursor Solution

W-M_X films were synthesized via a one-step hydrothermal process. Initially, 3.29 g Na_2WO_4 $2H_2O$ was dissolved in 30 mL of deionized water under continuous stirring to form a transparent solution. Subsequently, 3 M hydrochloric acid was added dropwise until no further precipitate formation occurred. The solution was then centrifuged at 5000 rpm to pellet the precipitate, which was washed with deionized water until neutral. The precipitate was redissolved in 30 mL hydrogen peroxide and diluted to 100 mL with deionized water at

60 °C to obtain the hydrothermal precursor solution. The etching of $TiAlC_2$ powder resulted in the formation of accordion-like MXene sheets, as shown in Figure S1. The preparation of the MXene dispersion has been detailed in our previous publication [22].

2.3. Preparation of W-Mx Films

In the preparation of W-Mx films, 2 mL precursor solution, 0.3 mL 3M hydrochloric acid, 5 mL deionized water, and 1 mL ethylene glycol were combined with a specific volume of MXene dispersion liquid for ultrasonic homogenization. The resulting mixture was then transferred to the liner of a 25 mL Teflon-lined stainless-steel hydrothermal reaction vessel. A pristine fluorine-doped tin oxide (FTO) glass substrate, with the conductive surface facing downwards, was positioned within the liner. The liner was subsequently sealed within a high-pressure, explosion-proof reaction vessel, and the temperature of the blast drying oven was set to 120 °C for a duration of 2.5 h to facilitate the hydrothermal reaction, as shown in Figure 1a. The incorporation of varying quantities of the end-capping agent MXene served to modulate the film properties, thereby yielding the desired W-M_x films.

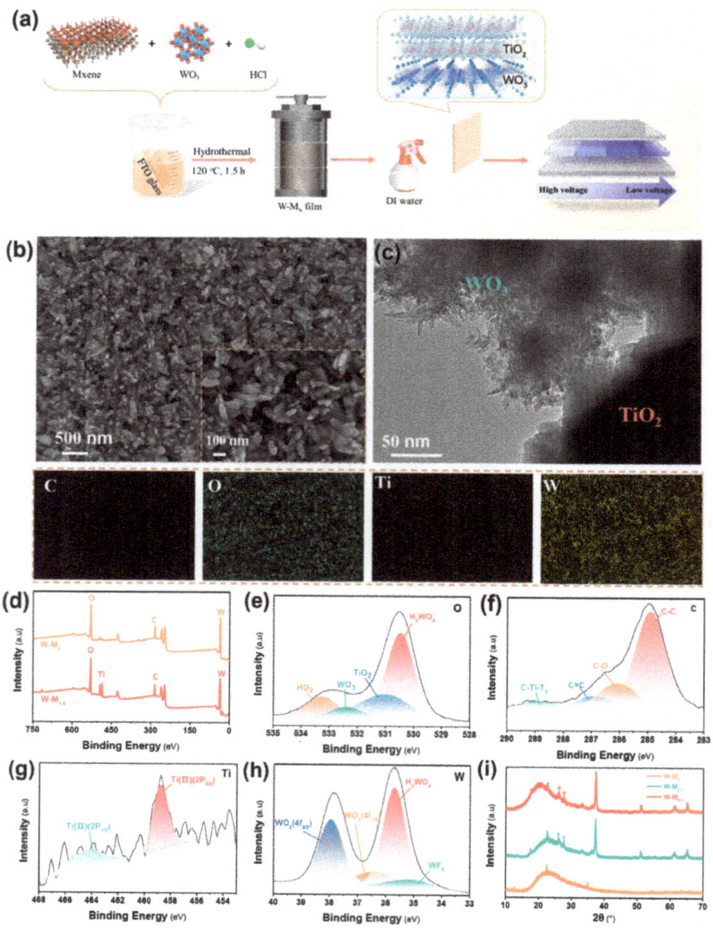

Figure 1. (**a**) Process flow diagram for the hydrothermal one-step synthesis of W-M_x films and a schematic of their color change under varying voltages. (**b**,**c**) SEM images and EDS-mapping and TEM images of the W-$M_{1.0}$ films, respectively. (**d**–**i**) XPS and XRD characterization of the W-M_x films.

2.4. Fabrication of Electrochromic Device

Encapsulation of W-$M_{1.0}$ EC devices was performed utilizing a methodical approach. PB (Prussian Blue) film and W-$M_{1.0}$ film were selected as anode and cathode coloring materials, respectively, while PC/$LiClO_4$ (1 M) was used as the electrolyte. The fabrication of W-$M_{1.0}$ devices was followed by a series of performance tests to evaluate their functionality and efficiency.

2.5. Characterization

A comprehensive characterization of the composition and morphology of the W-M_x films was conducted using various analytical techniques. Scanning electron microscopy (SEM, Gemini 300, ZEISS, Oberkochen, Germany) and transmission electron microscopy (TEM, Tecnai F20, FEI, Hillsboro, OR, USA) were employed to examine the films' microstructure and surface topography. Elemental analysis of the W-M_x films was performed using X-ray photoelectron spectroscopy (XPS, K-Alpha, Thermo Scientific, Waltham, MA, USA) and energy-dispersive X-ray spectroscopy (EDS) mapping, which was integrated with the SEM system (Gemini 300, ZEISS, Oberkochen, Germany). The crystallographic properties of the films were investigated through X-ray diffraction (XRD, X'Pert PRO MPD, PANalytical, Almelo, The Netherlands). The films' and devices' tuning ability, response time, coloration efficiency, and cycling stability were evaluated using a UV-Vis spectrophotometer (UV6300S, Mapada, Shanghai, China) in tandem with the electrochemical workstation (CHI 760E, Chenhua, Shanghai, China). The color coordinates (L*, a*, b*) of the W-M_x films were determined with a portable colorimeter (YS4510, 3nh, Shenzhen, China), providing quantitative data on the films' color characteristics. These comprehensive analyses were essential for a thorough understanding of the W-M_x films' properties and their potential applications in electrochromic devices.

3. Results and Discussion

3.1. Structural Characterization of W-M_X Films

To ascertain the morphological influence of MXene on W-M_x films, a comprehensive characterization was conducted using scanning electron microscopy (SEM), transmission electron microscopy (TEM), and X-ray photoelectron spectroscopy (XPS). Under identical process parameters, the W-M_x films synthesized via a one-step hydrothermal method are depicted in Figure 1a. In stark contrast to the nanofloral structure of the W-M_0 membrane, the W-$M_{0.5}$ and W-$M_{1.0}$ membranes display a coral-like morphology (Figures 1b, S2 and S3). During the formation of the W-M_0 film, ethylene glycol induces the generation of sufficient nuclei, enabling WO_3 crystals to grow unrestrained within the confined space. Upon contact, these crystals coalesce to form nanoflower-like clusters on the FTO surface. The dense structure of the W-M_0 membrane results in a reduced specific surface area, consequently limiting the number of active sites available for electrochemical reactions.

The addition of MXene significantly enhances the control over the morphology of W-M_x films. In the W-$M_{0.5}$ films, the presence of MXene as a capping agent restricts the growth of WO_3 crystals within the confined space, preventing the formation of large nanoflower-like structures. Consequently, the W-$M_{0.5}$ film exhibits a dispersed coral-like nanostructure. As the MXene content increases, the coral-like WO_3 structure in the W-$M_{1.0}$ membrane becomes even more dispersed. However, with further increments in MXene content, the growth of WO_3 is increasingly inhibited, leading to a lack of distinct W-M_x film formation on the FTO surface.

The morphological analysis of the W-$M_{1.0}$ films was conducted using TEM at multiple magnifications, providing a detailed examination of their structural features. The films were observed to possess a lamellar structure that is indicative of two-dimensional (2D) characteristics. This is attributed to the retention of the original 2D structure of MXene during its oxidative transformation into TiO_2, as shown in Figure 1c. As a result, the W-$M_{1.0}$ film exhibits a microstructural profile that is reminiscent of certain 2D materials, highlighting the influence of MXene on the film's morphology. High-resolution TEM images, as presented in Figure S4, further elucidate the structural details of the W-$M_{1.0}$ film. These images reveal the presence of lattice fringes measuring 0.34 nm in regions of high crystallinity and 0.24 nm in regions of low crystallinity. The lattice fringe at 0.34 nm corresponds to WO_3, while the 0.24 nm lattice fringe corresponds to TiO_2 with an anatase structure, in agreement with the literature [21,23]. The latter is a consequence of the oxidation of MXene by hydrogen peroxide (H_2O_2) during the hydrothermal reaction process, which is a critical step in the formation of the film [24–26].

To elucidate the role and distribution of MXene within the W-M_X film formation process, energy-dispersive X-ray spectroscopy mapping (EDS-Mapping) was employed to analyze the elemental composition on the surface of the W-M_x films. The W-M_0 film, which consists solely of WO_3, exhibits a nanoflower structure with the distribution of W and O elements closely matching this morphology, as shown in Figure S5. The EDS-Mapping results for the W-$M_{1.0}$ films, as shown in Figure 1b, reveal the concurrent presence of W, O, C, and Ti elements. The elements W and O are predominantly derived from WO_3, while C is contributed by both MXene and ethylene glycol. The Ti element, which is a signature component of MXene, is uniformly dispersed throughout the W-$M_{1.0}$ film, indicating the thorough incorporation of MXene within the film's structure. These findings suggest that ethylene glycol plays a crucial role in the hydrothermal reaction by facilitating the nucleation and growth of WO_3 grains on the FTO surface through the formation of "bridge bonds". MXene, on the other hand, plays a dual role in film formation and structural regulation, ultimately existing within the film as anatase TiO_2 post-reaction.

The elements in W-M_x films were characterized by XPS, revealing peaks at 36.1, 530.1, and 285.1 eV for W, O, and C, respectively. The W and O elements originate from WO_3, while C and O elements from the crystallization agent ethylene glycol are only present in MXene, which also contains Ti elements [27,28]. Compared to the W-M_0 film, the W-$M_{1.0}$ film exhibited a higher detection of Ti elements (Figure 1d) [29,30]. This confirms that MXene is instrumental in regulating the morphology and composition of W-$M_{1.0}$ films during their formation.

The XPS W4f characteristic spectrum of the W-$M_{1.0}$ film shows peaks at 37.98, 36.38, 35.68, and 35.08 eV, corresponding to the characteristic peaks of $WO_3(4f_{7/2})$, $WO_3(4f_{5/2})$, H_2WO_4, and WF_6, respectively, as shown in Figure 1h. The XPS O 1s characteristic spectrum, with peaks at 533.28, 532.38, 531.08, and 530.48 eV, corresponds to H_2O, WO_3, TiO_2, and H_2WO_4, respectively, as shown in Figure 1e. The XPS C 1s characteristic spectrum, with peaks at 288.88, 286.98, and 284.98 eV, corresponds to C-Ti-T_x, C=O, C-O, and C-C bonds, respectively, as shown in Figure 1f. The XPS Ti 2p spectrum, with peaks at 464.38 and 458.78 eV, corresponds to the Ti (II)($2p_{1/2}$) and Ti(II)($2p_{3/2}$) characteristic peaks, respectively, as shown in Figure 1g. XPS analysis confirms the presence of TiO_2 in the anatase structure in the W-$M_{1.0}$ film following the one-step hydrothermal reaction with MXene.

As shown in Figure 1i, the XRD patterns of W-M_x and W-$M_{1.0}$ films display characteristic peaks for FTO (indium tin oxide) at 37.28°, 50.96°, 61°, and 65.07° [31]. Additionally, TiO_2 characteristic peaks are observed at 25.93° and 27.54° [32,33]. In contrast, W-M_0 films do not show prominent FTO or TiO_2 peaks. The dense structure and significant thickness of the films make it challenging to detect FTO signals in XRD tests (Figure S6), with the results primarily highlighting tungsten oxide characteristic peaks. This confirms that MXene was oxidized to anatase TiO_2 under acidic conditions during the preparation of W-M_x films, contributing to the adjustment of morphology and construction of the films.

3.2. Electrochemical Properties of the W-M_x Film

The cyclic voltammetry curves of W-M_x films were studied, as shown in Figure 2a. During testing, W^{6+} in the W-M_x film undergoes continuous reduction to W^{5+}. The current peaks when the reduction rate of W^{6+} is at its highest before gradually declining. As the scanning speed increases, the electrochemical reactions occurring on the electrode surface can reach a stable state more rapidly, leading to a larger current response and consequently increasing the area of the curve, as shown in Figure 2c–e. Notably, the reduction peak of the W-$M_{1.0}$ film is shifted compared to that of the W-M_0 film. The shift can be attributed to the introduction of MXene, which causes distortion of the WO_3 lattice and alters the characteristics of the film interface. These changes lead to modifications in the energy barrier of the charge transfer process, thereby impacting the reduction reaction potential.

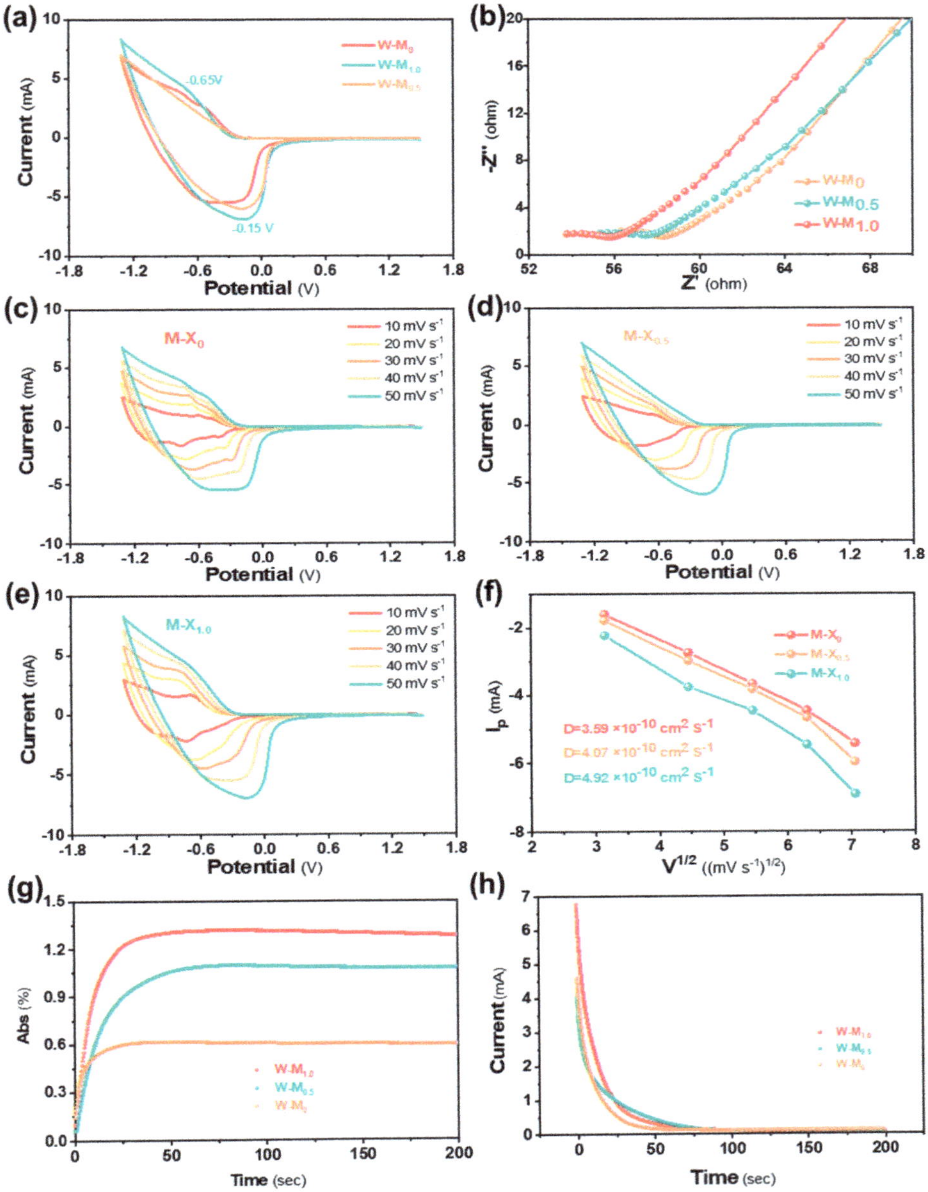

Figure 2. (**a**) The CV curves of W-M_X films. (**b**) The Nyquist plots of the W-M_x films. (**c**–**e**) The CV curves of W-M_X films at different scanning rates. (**f**) Current density changes with the square root of the scanning rates. (**g**) The relationship between absorption and time change in W-M_x films. (**h**) The relationship between current and time change in W-M_x films.

The electrochemical impedance of W-M_0 and W-$M_{1.0}$ films was analyzed via Nyquist curves in Figure 2b. Both films exhibit similar arc radii in the high-frequency region, indicating consistent system resistance, electrode capacitance, and polarization resistance. However, the low-frequency region shows significant slope differences due to varying ion diffusion impedances. The loose structure of W-$M_{1.0}$ film facilitates ion transport, while the TiO_2@WO_3 heterostructure enhances charge transfer. Conversely, the dense structure of W-M_0 nanoparticles and single energy bands hinders ion transport.

When a coloration voltage is applied to W-M_x films, the transmittance and current of the film change over time, as shown in Figure 2g,h, with the absorption increasing rapidly at the initial stage. When a bias voltage is applied, the current profile exhibits a rapid response followed by a sharp decrease, indicating an efficient electrochemical reaction process controlled by ion diffusion. The electrochromic process of W-M_x films is controlled by the diffusion of electrolyte ions during the application of the colored voltage. In this process, Li^+ ions are repeatedly intercalated and deintercalated in the film, altering the valence state of the W element and causing a reversible redox reaction in the material system. The TiO_2@WO_3 heterostructure in the W-$M_{1.0}$ film deforms at the grain boundaries, leading to spatial migration. The diffusion coefficient of Li^+ in the W-M_0, W-$M_{0.5}$, and W-$M_{1.0}$ films can be calculated using the Randles-Sevcik Equation (1) [34]

$$i = 2.69 \times 10^5 \times n^{3/2} \times C_0 \times A \times D^{1/2} \times v^{1/2} \tag{1}$$

where i is the peak current (mA), D is the diffusion coefficient ($cm^2{\cdot}s^{-1}$), A is the working electrode area (cm^2), n is the number of electrons transferred, C_0 is the concentration of active ions ($mol{\cdot}cm^{-3}$), and v is the scan rate in cyclic voltammetry ($mV{\cdot}s^{-1}$).

The ion diffusion coefficients of the W-M_0, W-$M_{0.5}$, and W-$M_{1.0}$ films are 3.59×10^{-10}, 4.07×10^{-10}, and 4.92×10^{-10} $cm^2\ S^{-1}$, respectively. The different chemical compositions and lattice structures of TiO_2 and WO_3 in the heterostructure induce lattice distortion, enhancing the electrochemical reactivity of the materials. Additionally, the varied energy band arrangement reduces the charge transfer barrier, making it easier for charges to transfer at the interface. As the MXene content increases, the lattice distortion and energy band changes become more pronounced, leading to an increase in the ion diffusion coefficient of the W-M_x films.

3.3. Electrochromic Properties of W-M_x Films

The addition of MXene significantly enhanced the electrochemical properties of the W-M_x films, leading to further exploration of their electrochromic behavior. As shown in Figure 3a, the optical transmission curves of the W-M_x films show distinct differences when the applied voltage varies from −1.6 V to 1.4 V, particularly in the near-infrared region, where the films exhibit excellent modulation capability (ΔT). Notably, the W-$M_{1.0}$ film displays the largest modulation range, while the W-M_0 film shows the smallest. At a voltage of −1.6 V, the W-$M_{1.0}$ film is in a colored state, with a transmittance of only 6.69% at 1033 nm. When the voltage is increased to 1.4 V, the W-$M_{1.0}$ film fades, and its transmittance rises to 94.30%, resulting in a ΔT value of 87.61% at 1033 nm. In comparison, the ΔT values for the W-$M_{0.5}$ and W-M_0 films at 1033 nm are 77.92% and 44.87%, respectively, as shown in Figure 3b.

The CIELAB color coordinates (L*, a*, b*) of the W-$M_{1.0}$ film in both its colored and faded states were quantitatively determined using a portable color difference meter. In the colored state (−1.6 V), the color coordinate values are (50.48, −3.70, −18.04), while in the faded state (1.4 V), the values shift to (74.53, 0.57, 0.19). The corresponding color coordinates and optical photographs for these different states are shown in Figure S7, and the switching process is shown in SMOV1.

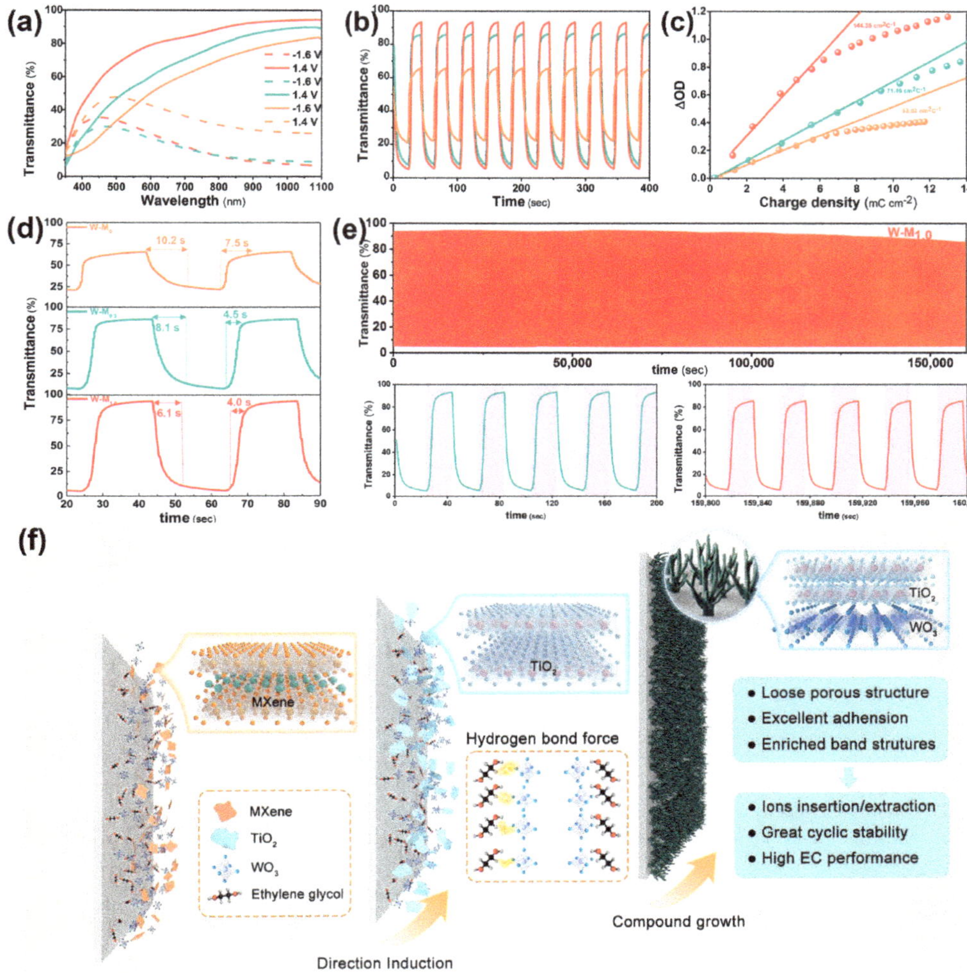

Figure 3. EC performance of W-M_x films: (**a**) Transmittance curves of W-M_0 (yellow), W-$M_{0.5}$ (green), and W-$M_{1.0}$ (red) films at different voltages. (**b**) Optical modulation of W-M_x films at 1033 nm. (**c**) Coloration efficiency of W-M_x film. (**d**) Response time of W-M_x films. (**e**) Long-term stability of W-$M_{1.0}$ films. (**f**) Schematic illustration of the fabrication process of the TiO_2@WO_3 heterostructure.

Response time is a critical factor in assessing the performance of electrochromic materials, typically defined as the time required to achieve 90% of the maximum modulation capacity. As shown in Figure 3d, the W-$M_{1.0}$ film exhibited a coloring time of 6.1 s and a fading time of 4.0 s. In comparison, the W-$M_{0.5}$ film had a coloring time of 8.1 s and a fading time of 4.5 s, while the W-M_0 film showed a coloring time of 10.2 s and a fading time of 7.5 s.

Coloring efficiency (CE) is a crucial parameter for assessing the speed of optical modulation in electrochromic materials. It is defined as the ratio of the change in optical absorption at a specific wavelength to the amount of charge gained or lost per unit area. This can be calculated using the following Equation (2) [35]:

$$CE = \frac{\Delta OD}{Q_d} = \frac{\log(T/T_c)}{Q_d} \tag{2}$$

where T represents the transmittance at different time points, T_c is the transmittance of the colored state, and Q_d is the amount of charge injected or withdrawn per unit area. The ΔOD and Q_d at different time points were plotted in a scatter plot and subjected to linear fitting to determine the coloring efficiency of the W-M_0, W-$M_{0.5}$, and W-$M_{1.0}$ films, which were calculated to be 53.03, 71.46, and 144.35 cm^2 C^{-1}, respectively, as shown in Figure 3c.

The results indicate that with increased MXene content, the response time of the films is reduced and the coloring efficiency is improved. The enhancement is attributed to the loose coral-like structure of the W-$M_{1.0}$ film, which facilitates electrolyte ion transport, in contrast to the dense nanoflower structure of the W-M_0 film that impedes ion transport. The coral-like morphology of the W-$M_{1.0}$ film offers more reaction sites for the electrochromic process, thereby accelerating the reaction rate. Additionally, the TiO_2 formed from MXene introduction integrates effectively with WO_3, enhancing electron transfer during the electrochemical reaction and thus increasing the electrochromic reaction rate, as shown in Figure 3f.

Cyclic stability is a crucial index for evaluating their performance. To assess electrochromic film stability, step voltages of −1.6 V and 1.4 V were applied to W-M_x films for several cycles, with each step lasting 20 s. The change in transmittance at a wavelength of 1033 nm was recorded in real-time, as shown in Figure 3e and Figure S8. Initially, the optical regulation ability of the W-M_0 film slightly decreased. This decrease occurred during the early cycles as the film established its electrolyte ion transport channels, causing minor structural collapse. Once these channels were properly formed, the film's optical regulation ability stabilized. However, after 2000 cycles (80,000 s), the ΔT value of the W-M_0 film diminished by 19.21%, indicating a relatively quick decline in its optical regulation ability and poorer cyclic stability.

In contrast, the W-$M_{1.0}$ film demonstrated the least attenuation of regulatory capacity during cyclic testing. As shown in Figure 3f, this can be attributed to the heterostructure formed between TiO_2 and WO_3 that creates a better interface that enhances charge transfer while reducing interface defects, thereby improving stability. The dispersed coral-like structure helps to increase the specific surface area of the film, contributing to higher ion transport rates and larger electrochemical reaction interfaces. Concurrently, the potential damage to the film's structure, which can be caused by lattice expansion and contraction during the ion embedding and deembedding processes, is mitigated. Additionally, the enriched band structure of the TiO_2/WO_3 heterojunction improves the conductivity and high carrier concentration to promote the electronic transfer to facilitate electrochemical reaction rates and enhance the EC performance. Consequently, the W-$M_{1.0}$ film exhibited superior cycle stability, with only a 9.89% reduction in its photomodulation capacity after 4000 cycles (160,000 s). The regulation capacity of the W-$M_{1.0}$ film reached 87.61% at a wavelength of 1033 nm, showcasing better cyclic stability compared to most previously reported [36–39].

In the field of electrochromic materials, many materials perform well in certain aspects, but achieving a balance between optical contrast, response speed, and optical transmittance remains challenging. According to the comparison data in Table S1, while some modified WO_3 films and other electrochromic materials excel in optical contrast or response speed, their coloration efficiency and cycling stability are generally inferior to those of the W-$M_{1.0}$ film, often significantly lower, which limits the further development of these materials [40–42]. In contrast, the W-$M_{1.0}$ film demonstrates fast response, remarkable coloration efficiency, and excellent cycling stability, offering unique advantages for electrochromic applications. For instance, it can achieve rapid color switching within a short period and offers a wide range of optical modulation, making it highly attractive for applications such as smart windows, optical blinds, and display devices. Additionally, the exceptional cycling stability of W-$M_{1.0}$ ensures that it can maintain stable electrochromic performance over extended periods of use, making it an ideal material for wearable devices, portable electronics, e-readers, and flexible displays. In these devices, the electrochromic effect not only provides an energy-efficient alternative to traditional displays but also meets complex optical modulation requirements.

3.4. Fabrication of Electrochromic Device

To further assess the electrochromic performance and practical application potential of the W-$M_{1.0}$ film, PB film was chosen as the anode electrochromic material and W-$M_{1.0}$ as the cathode. PC/$LiClO_4$ (1 M) was used as the electrolyte, and devices measuring 6.0×4.0 cm^2 were fabricated. When a voltage of −3.2 V was applied, the EC device based on the W-$M_{1.0}$ film displayed a dark blue color, while at 2.5 V, it turned colorless, as shown in Figure 4a. As the voltage was gradually increased from −3.2 V to 2.5 V, the color of the electrochromic device transitioned smoothly, accompanied by a decrease in absorption rate, as shown in Figure 4b. The device achieved a significant optical contrast of 73.83% at 1033 nm within the voltage range of −3.2 to 2.5 V, as shown in Figure 4c and SMOV2. The response times were measured under a −3.2/2.5 V step voltage, revealing a coloration time of 2.8 s and a bleached time of 4.6 s, as shown in Figure 4d. When a constant colored voltage of −3.2 V was applied, the device transitioned from colorless to dark blue, with changes in optical transmittance recorded at 1033 nm, as shown in Figure 4e. The calculated coloration efficiency of the device was 137.02 cm^{2}C^{-1}. These findings demonstrate that the W-$M_{1.0}$ device exhibits excellent electrochromic properties, underscoring its potential for practical applications.

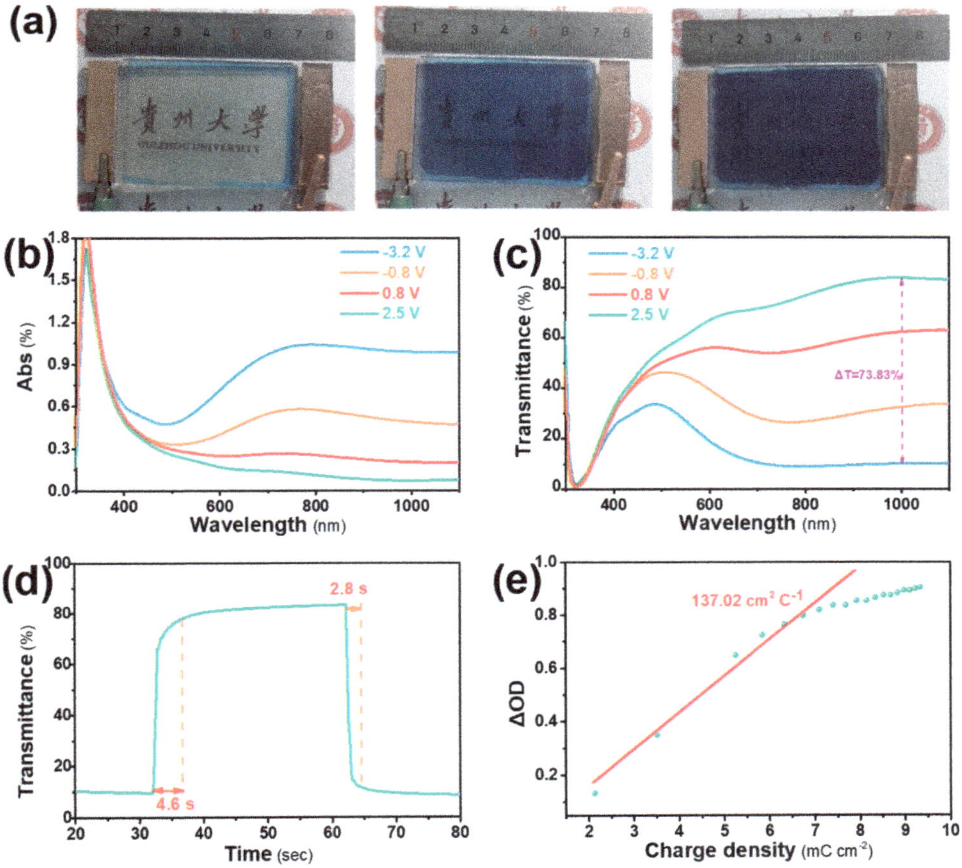

Figure 4. The ECD performances of W-$M_{1.0}$ film are (**a**) optical images, (**b**) absorption curves, (**c**) transmittance curves, (**d**) response time, and (**e**) coloration efficiency.

4. Conclusions

In summary, the optimal process for preparing WO_3 EC membranes via hydrothermal methods involves using ethylene glycol as a self-seeding agent, two-dimensional MXene as a capping agent, and producing W-M_x films under acidic conditions. During the preparation, MXene is oxidized into a rutile-type TiO_2 structure, which regulates the overall film structure and forms a porous coral-like heterostructure through mutual doping with WO_3. The interfacial bonding, unique band structure, and lattice distortion within the heterostructure enhance the electrochromic properties of the W-M_x films. The preparation process for the W-$M_{1.0}$ film is straightforward, featuring a low redox voltage window, a fast ion migration rate (4.92×10^{-10} cm^2/S), short response times (colored time of 6.1 s, bleached time of 4.0 s), high coloration efficiency (144.35 cm^2/C), and excellent cyclic stability (ΔT decays only 9.89% after 4000 cycles at 1033 nm). Furthermore, the EC device based on W-$M_{1.0}$ film, when packaged and evaluated, demonstrated good modulation capability (73.83% at 1033 nm), short response times (colored time of 2.8 s, bleached time of 4.6 s), and high coloration efficiency (137.02 cm^2/C). Thus, the EC film produced by this method holds significant practical application value.

Supplementary Materials: The following supporting information can be downloaded at: https://www.mdpi.com/article/10.3390/mi15121486/s1, Figure S1. TiC_2T_x sheet after etching. Figure S2. SEM image of the W-M_0 film. Figure S3. SEM image of the W-$M_{0.5}$ film. Figure S4. HR-TEM images of the W-$M_{1.0}$ films. Figure S5. EDS-Mapping of the W-M_0 film. Figure S6. The thickness of (a) W-M_0 film, (b) W-$M_{0.5}$ film and W-$M_{1.0}$ film. Figure S7. Optical photos and color values of W-$M_{1.0}$ film under different voltages. Figure S8. Long-term stability of (a) W-M_0 films and (b) W-$M_{0.5}$ films. Table S1. Comparison of t_c, t_b, $\Delta T\%$, CE and cycling stability of W-$M_{1.0}$ films with other electrochromic materials.

Author Contributions: All authors have significantly contributed to the research presented in this manuscript. Conceptualization, Y.W. and Y.L.; methodology, Y.W. and T.Z.; validation, M.W., M.T., and W.W.; writing—original draft preparation, Y.W. and Y.L.; supervision, T.Z. and M.T.; project administration, T.Z. All authors have read and agreed to the published version of the manuscript.

Funding: This work was supported by the National Natural Science Foundation of China (grant No. 52163022), Guizhou Science and Technology Plan general Program (No. ZK2022-155), Guizhou Provincial Education Department Youth Science and Technology Talent Development project (No. KY2022-136), Natural Science Special Foundation of Guizhou University (No. 202112), and Guizhou Graduate Research Fund project (YJSKYJJ 2021-004).

Data Availability Statement: The original contributions presented in the study are included in the article; further inquiries can be directed to the corresponding authors.

Conflicts of Interest: The authors declare no conflict of interest.

References

1. Deb, S.K. A novel electrophotographic system. *Appl. Opt.* **1969**, *8*, 192–195. [CrossRef] [PubMed]
2. Sven, M.; Marco, S.; Martin, D.; Abdelbast, G.; Karim, Z.; Uwe, P.; Peer, L. Large-area electrochromic devices on flexible poly substrates with high optical contrast and enhanced cycling stability. *J. Adv. Mater. Technol.* **2021**, *6*, 2000836.
3. Zhu, L.; Nuo Peh, C.K.; Zhu, T.; Lim, Y.F.; Ho, G.W. Bifunctional 2D-on-2D MoO_3 nanobelt/$Ni(OH)_2$ nanosheets supercapacitor-driven electrochromic energy storage. *J. Mater. Chem. A* **2017**, *5*, 8343–8351. [CrossRef]
4. Nguyen, T.D.; Yeo, L.P.; Ong, A.J.; Wang, Z.; Mandler, D.; Magdassi, S.; Tok, A.I.Y. Electrochromic smart glass coating functional nano-frameworks for effective building energy conservation. *Mater. Today Energy* **2020**, *18*, 100496. [CrossRef]
5. Neo, W.T.; Ye, Q.; Chua, S.J.; Xu, J. Conjugated polymer-based electrochromics: Materials, device fabrication and applica prospects. *J. Mater. Chem. C* **2016**, *4*, 7364–7376. [CrossRef]
6. Pan, M.; Zhou, Q.; Liu, J.; He, Q.; Gong, C.; Tang, Q.; Shen, W. Electrochromic materials containing pyridinium salt and benz moieties with dual-colored and long-life performance. *Sol. Energy Mater. Sol. Cells* **2022**, *240*, 111712. [CrossRef]
7. Niu, J.; Wang, Y.; Zou, X.; Tan, Y.; Jia, C.; Weng, X.; Deng, L. Infrared electrochromic materials, devices and applications. *A Mater. Today* **2021**, *24*, 101073. [CrossRef]
8. Yang, G.; Zhang, Y.M.; Cai, Y.; Yang, B.; Gu, C.; Zhang, S.X.A. Advances in nanomaterials for electrochromic devices. *Chem. Rev.* **2020**, *49*, 8687–8720. [CrossRef]

Lahav, M.; Van Der Boom, M.E. Polypyridyl metallo-organic assemblies for electrochromic applications. *Adv. Mater.* **2018**, *30*, 1706641. [CrossRef]
Wu, C.; Shao, Z.; Zhai, W.; Zhang, X.; Zhang, C.; Zhu, C.; Yu, Y.; Liu, W. Niobium tungsten oxides for electrochromic devices with long-term stability. *ACS Nano* **2022**, *16*, 2621–2628. [CrossRef]
Hao, T.; Wang, S.; Xu, H.; Zhang, X.; Xue, J.; Liu, S.; Song, Y.; Li, Y.; Zhao, J. Stretchable electrochromic devices based on embedded WO_3@AgNW Core-Shell nanowire elastic conductors. *Chem. Eng. J.* **2021**, *426*, 130840. [CrossRef]
Mohanadas, D.; Sulaiman, Y. Recent advances in development of electroactive composite materials for electrochromic and supercapacitor applications. *J. Power Sources* **2022**, *523*, 231029. [CrossRef]
Zhao, L.; Chen, Z.; Peng, Y.; Yang, L.; Ai, J.; Zhou, J.; Miao, L. High-performance complementary electrochromic energy storage device based on tungsten trioxide and manganese dioxide films. *Sustain. Mater. Techno.* **2022**, *32*, e00445. [CrossRef]
Ahmad, K.; Kim, H. Fabrication of nickel-doped tungsten trioxide thin film-based highly stable flexible electrochromic devices for smart window applications. *ACS Sustain. Chem. Eng.* **2023**, *11*, 10746–10754. [CrossRef]
Li, H.; Shi, G.; Wang, H.; Zhang, Q.; Li, Y. Self-seeded growth of nest-like hydrated tungsten trioxide film directly on FTO substrate for highly enhanced electrochromic performance. *J. Mater. Chem. A* **2014**, *2*, 11305–11310. [CrossRef]
Pan, J.; Zheng, R.; Wang, Y.; Ye, X.; Wan, Z.; Jia, C.; Weng, X.; Xie, J.; Deng, L. A high-performance electrochromic device assembled with hexagonal WO_3 and NiO/PB composite nanosheet electrodes towards energy storage smart window. *Sol. Energy Mater. Sol. Cells* **2020**, *207*, 110337. [CrossRef]
Kadam, A.V. Propylene glycol-assisted seed layer-free hydrothermal synthesis of nanostructured WO_3 thin films for electrochromic applications. *J. Appl. Electrochem.* **2016**, *47*, 335–342. [CrossRef]
Pan, J.; Wang, Y.; Zheng, R.; Wang, M.; Wan, Z.; Jia, C.; Weng, X.; Xie, J.; Deng, L. Directly grown high-performance WO_3 films by a novel one-step hydrothermal method with significantly improved stability for electrochromic applications. *J. Mater. Chem. A* **2019**, *7*, 13956–13967. [CrossRef]
A Soomro, R.; Zhang, P.; Fan, B.; Wei, Y.; Xu, B. Progression in the oxidation stability of MXenes. *Nano-Micro Lett.* **2023**, *15*, 2–18. [CrossRef]
Lotfi, R.; Naguib, M.; Yilmaz, D.E.; Nanda, J.; Duin, A.C.T. A comparative study on the oxidation of two-dimensional Ti_3C_2 MXene structures in different environments. *J. Mater. Chem. A* **2018**, *6*, 12733–12743. [CrossRef]
Zhao, X.; Vashisth, A.; Blivin, J.W.; Tan, Z.; Holta, D.E.; Kotasthane, V.; Shah, S.A.; Habib, T.; Liu, S.; Lutkenhaus, J.L.; et al. pH, Nanosheet concentration, and antioxidant affect the oxidation of $Ti_3C_2T_x$ and Ti_2CT_x MXene dispersions. *Adv. Mater. Interfaces* **2020**, *7*, 2000845. [CrossRef]
Liu, Y.; Yuan, J.; Wang, Y.; Zheng, R.; Liu, Q.; Shang, X.; Shao, J.; Wan, Z.; Luo, J.; Jia, C. Approach to significantly enhancing the electrochromic performance of PANi by in situ electrodeposition of the PANi@MXene composite film. *ACS Appl. Mater. Interfaces* **2023**, *15*, 58940–58954. [CrossRef] [PubMed]
Kim, J.; Inamdar, A.I.; Jo, Y.; Cho, S.; Aqueet, A.T.; Hou, A.B.; Cha, S.; Kim, T.G.; Kim, H.; Im, H. Nanofilament array embedded tungsten oxide for highly efficient electrochromic supercapacitor electrodes. *J. Mater. Chem. A* **2020**, *8*, 13459–13469. [CrossRef]
Zhai, M.; Liu, Y.; Huang, J.; Wang, Y.; Chen, K.; Fu, Y.; Li, H. Efficient suspension plasma spray fabrication of black titanium dioxide coatings with visible light absorption performances. *Ceram. Int.* **2019**, *45*, 930–935. [CrossRef]
Rehani, D.; Saxena, M.; Balal, M.; Barman, S.R.; Dhakate, S.R.; Sharma, S.N. Role of Eu and Fe in TiO_2 for magneto-opto-electronic applications. *Appl. Phys. A* **2022**, *128*, 2–14. [CrossRef]
Du, J.; Zhang, Z.; Yue, C.; Nie, Z.; Tan, H.; Tang, Z.; Li, N.; Xu, L.; Xu, J. Two-dimensional oxygen vacancy-doped tungsten oxide hydrate nanosheets for high-performance electrochromic device. *Mater. Today Chem.* **2022**, *26*, 101089. [CrossRef]
Ranjbar, M.; Mahdavi, S.M.; Iraji, A. Pulsed laser deposition of W-V-O composite films: Preparation, characterization and gasochromic studies. *Sol. Energy Mater. Sol. Cells* **2008**, *92*, 878–883. [CrossRef]
Kunyapat, T.; Xu, F.; Neate, N.; Wang, N.; Sanctis, A.D.; Russo, S.; Zhang, S.; Xia, Y.; Zhu, Y. Ce-doped bundled ultrafine diameter tungsten oxide nanowires with enhanced electrochromic performance. *Nanoscale* **2018**, *10*, 4718–4726. [CrossRef]
Mustafa, M.N.; Abdah, M.A.A.M.; Numan, A.; Sulaiman, Y.; Walvekar, R.; Khalid, M. Development of high-performance MXene/nickel cobalt phosphate nanocomposite for electrochromic energy storage system using response surface methodology. *J. Energy Storage* **2023**, *68*, 107880. [CrossRef]
Murugesan, R.A.; Raja, K.C.N. Capacitance performance of $Ti_3C_2T_x$ MXene nanosheets on alkaline and neutral electrolytes. *Mater. Res. Bull.* **2023**, *163*, 112217. [CrossRef]
Wu, W.; Fang, H.; Ma, H.; Wu, L.; Zhang, W.; Wang, H. Correction to: Boosting transport kinetics of Ions and electrons simultaneously by $Ti_3C_2T_x$ (MXene) addition for enhanced electrochromic performance. *Nano-Micro Lett.* **2021**, *13*, 20. [CrossRef] [PubMed]
Qin, Z.M.; Wang, P.C.; Yang, R.; Chen, H.B. Fast pyrolysis of silicones at low temperatures catalyzed by anatase titanium dioxide. *Polym. Degrad. Stab.* **2022**, *182*, 109387. [CrossRef]
Patel, R.K.; Chawla, A.K.; Manna, S.; Pandey, J.K. Defluoridation of water using anatase titanium dioxide nano-powder loaded 3D printed model devise. *J. Water Process Eng.* **2021**, *40*, 101785. [CrossRef]
Huang, Y.; Wang, B.; Lyu, P.; Zhao, S.; Wu, X.; Zhang, S.; Li, R.; Jiang, Q.; Wang, F.; Zhao, Y.; et al. Oxygen-deficient tungsten oxide nanoflowers for dynamically tunable near-infrared light transmittance of smart windows. *Nano Res.* **2023**, *16*, 12165–12172. [CrossRef]

35. Lin, C.; Chen, Y.; Chen, E. Preparation and characterization of electrochromic tungsten oxide-titania composite thin films w different tungsten/titanium ratios. *Thin Solid Film.* **2014**, *556*, 48–53. [CrossRef]
36. Zhou, L.; Wei, P.; Fang, H.; Wu, W.; Wu, L.; Wang, H. Self-doped tungsten oxide films induced by in situ carbothermal reduc for high performance electrochromic devices. *J. Mater. Chem. C* **2020**, *8*, 13999–14006. [CrossRef]
37. Huo, X.; Zhang, H.; Shen, W.; Miao, X.; Zhang, M.; Guo, M. Bifunctional aligned hexagonal/amorphous tungsten oxide core/s nanorod arrays with enhanced electrochromic and pseudocapacitive performance. *J. Mater. Chem. A* **2019**, *7*, 16867–16 [CrossRef]
38. Weng, S.; Zhang, C.; Wang, Q.; Xu, G.; Zhou, J.; Song, K.; Rogachev, A.A.; Yarmolenko, M.A.; Cao, H.; Zhang, H. In situ forma of solid electrolyte interphase for improved cyclability of electrochromic tungsten oxide thin films. *Surf. Interfaces* **2024**, *46*, 103 [CrossRef]
39. Morankar, P.J.; Amate, R.U.; Teli, A.M.; Beknalkar, S.A.; Chavan, G.T.; Ahir, N.A.; Jeon, C.W. Nanogranular advancement molybdenum-doped tungsten oxide for superior electrochromic energy storage. *J. Energy Storage* **2024**, *84*, 110978. [CrossRe
40. Lei, Y.; Feng, K.; Zeng, A.; Yang, H.; Zhang, L.; Liu, Z. Sol-gel deposited ZnO substrate for the modulation of electrodepos PEDOT nanostructures and enhancement of electrochromic stability. *Appl. Surf. Sci.* **2025**, *681*, 161480. [CrossRef]
41. Li, X.; Chu, J.; Cheng, Y.; Yang, F.; Xiong, S. Novel prussian blue@Carbon-dots hybrid thin film: The impact of carbon-dots material structure and electrochromic performance. *Electrochim. Acta* **2020**, *355*, 136659. [CrossRef]
42. Yuan, J.; Xia, L.; Wu, Y.; Liu, Z.; Mishra, Y.K.; He, L.; Xiong, J. High durable TiO_2 electrochromic films by Ni doping. *J. Mater. Mater. Electron.* **2024**, *35*, 961. [CrossRef]

micromachines

cle

ıll Spectrum Electrochromic WO_3 Mechanism and Optical odulation via Ex Situ Spectroscopic Ellipsometry: Effect of Li^+ ırface Permeation

ue Zhang [1], Jintao Wang [2], Shuhui Jiang [3], Meng Yuan [1,4,*] and Xinyu Chen [1,*]

1 School of Physics, Changchun University of Science and Technology, Changchun 130012, China; zhangbuyue729@163.com
2 School of Infromation Engineering, Yantai Institute of Technology, Yantai 261499, China
3 School of Materials Science and Engineering, Jilin Architecture University, Changchun 130119, China
4 Key Laboratory of Functional Materials Physics and Chemistry of the Ministry of Education, Jilin Normal University, Changchun 130012, China
* Correspondence: author: ymzy9424@163.com (M.Y.); chenxinyucust@163.com (X.C.)

Abstract: Tungsten oxide (WO_3) electrochromic devices are obtaining increasing interest due to their color change and thermal regulation. However, most previous work focuses on the absorption or transmission spectra of materials, rather than the optical parameters evolution in full spectrum in the electrochromic processes. Herein, we developed a systematic protocol of ex situ methods to clarify the evolutions of subtle structure changes, Raman vibration modes, and optical parameters of WO_3 thin films in electrochromic processes as stimulated by dosage-dependent Li^+ insertion. We obtained the below information by ex situ spectroscopic ellipsometry. (1) Layer-by-layer Li^+ embedding mechanism demonstrated by individual film thickness analysis. (2) The details of its optical leap in the Brillouin zone in the full spectral. (3) The optical constants varied with the Li^+ insertion in the ultraviolet, visible, and near-infrared bands, demonstrating the potential for applications in chip fabrication, deep-sea exploration, and optical measurements. (4) Simulated angular modulation laws of WO_3 films for full spectra in different Li^+ insertion states. This ex situ method to study the optical properties of electrochromic devices are important for monitoring phase transition kinetics, the analysis of optical leaps, and the study of ion diffusion mechanisms and the stoichiometry-dependent changes in optical constants over the full spectral. This work shows that electrochromic films in Li^+ surface permeation can be applied in the field of zoom lenses, optical phase modulators, and other precision optical components. Our work provides a new solution for the development of zoom lenses and a new application scenario for the application of electrochromic devices.

Keywords: WO_3; electrochromic; protocol of ex situ methods; ex situ spectroscopic; ellipsometry

1. Introduction

In recent years, adaptive lenses with zoomable capability have attracted attention for their wide range of applications in correction of ophthalmic lens [1,2], zoom lens [3,4], aberration compensation [5,6], biomedical imaging [7–9], optical tweezers [10,11], beam steering [12,13], fiber coupling [14,15], and so on. These zoom lenses are usually realized based on three techniques [16]. The first is based on an optical system constructed with multiple sets of optical elements, where the focal length is changed by a nonlinear movement of a spherical or cylindrical lens along the optical axis [17]. The second is the liquid crystal lens [18]; the main working principle is to adjust the refractive index of liquid crystal to realize the change in focal length. The third is the liquid lens [19], the main way is through the external force changes the shape of the lens to realize the focal length change. Replacing the moving optics with a stationary adaptive lens can simplify the system, reduce the size and power consumption, and eliminate vibration caused by conventional mechanic movement [20].

In recent years, tungsten oxide (WO_3) electrochromic devices have attracted much attention in the direction of color change and thermal regulation; it has broad application prospects in smart windows [21], low-power displays [22], antiglare rearview mirrors [23], smart textiles [24], energy storage systems [25], heat-shielding systems [26], sensing [27], catalysis and military camouflage, etc. These properties are based on the properties defined by changes in the absorption or transmission spectra of WO_3 films. The ion insertion process changes the crystal structure of the WO_3 films and causes a change in the optical constants [28]. This indicated the possibility of electrochromic technology applied in the field of adaptive lenses. In this paper, we have used WO_3, a classical electrochromic system, as a template to capture the refractive index changes in the full spectral of the film during lithium-ion (Li^+) insertion using a series of ex situ measurement techniques. The refractive angle modulation law of the Li^+ insertion dosage-dependent at full spectrum is analyzed.

In this work, we prepared γ-WO_3 by sol–gel spin-coating, and the film gradually shifted from transparent to dark blue with the layer-by-layer embedding of Li^+. The samples were tested by ex situ X-ray diffraction (XRD), ex situ Raman spectroscopy, and ex situ spectroscopic ellipsometry during the coloration process. The results of ex situ XRD and ex situ Raman spectroscopy demonstrate that the crystal structure of the samples gradually transforms the $\gamma \rightarrow \beta \rightarrow \alpha$-phase during the coloration process. We obtained the optical constants in the full spectral of the samples in different lithium-ion insertion dosage states by ex situ spectroscopic ellipsometry. The Li^+ insertion into the γ-WO_3 samples is a layer-by-layer insertion mechanism, as can be seen from the results of the thickness variation in each film layer. And we fitted a quantitative relationship between the Li^+ insertion dosage (x) and the Li_xWO_3 layer thickness. The detail of all optical leap points in the Brillouin zone corresponding to the range of 1–5 eV is also found. The change in extinction coefficient (k) with Li^+ insertion shows that its band gap decreases from 3.22 eV to 3.14 eV. The results of its refractive index (n) in the ultraviolet, visible, and near-infrared bands show obvious peak motion and peak shape changes, which demonstrates the difference in its n adjusting ability for specific wavelengths of light and shows the prospect of its application in the fields of chip processing, undersea exploration, and optical inspection. Our work provides a new solution for the development of precision optical components such as zoom lenses, as well as a new application scenario for the application of electrochromic devices.

2. Results

2.1. Ex Situ XRD Monitored Structure Evolution in Electrochromic Process

The as-fabricated WO_3 film is composed of uniformly distributed small particles that are difficult to discern with a diameter smaller than 10 nm, according to the SEM in Figure 1a. The as-coated WO_3 layer shows an average thickness of 380 nm uniformly and compactly on the top surface of the ITO layer in Figure 1b. The as-prepared WO_3 film crystallized into a typical monoclinic crystalline system (γ-phase) as indicated in JCPDS card No. of 18–0376.

With the insertion of lithium, the peaks of WO_3 at 23.1° and 24.3° are shifted toward a small angle due to the monoclinic structure of WO_3 toward an increasing symmetry [29]. It has been widely reported that for WO_3 with temperature, in bulk form, phase transformation occurs in the following sequence: monoclinic II (ε-WO_3, $T < -43$ °C) $\rightarrow$ triclinic (δ-WO_3, -43 °C $< T <$ 17 °C) $\rightarrow$ monoclinic I (γ-WO_3, 17 °C $< T <$ 330 °C) $\rightarrow$orthorhombic (β-WO_3, 330 °C $< T <$ 740 °C) $\rightarrow$ tetragonal (α-WO_3, $T >$ 740 °C) [30] For the lithium insertion process in this work, the structure undergoes a similar change: $\gamma \rightarrow \beta \rightarrow \alpha$; we listed the details of the specific peaks in Table 1.

In Figure 2, the XRD pattern of WO_3 embedded with Li^+ was plotted with details of the peak position change near 24–25°. Contour plots make it more obvious to observe changes in the peak position, peak value, and half-peak width of the plots. By comparison with the JCPDS card No. of 18–0376, the peak at 24.54° corresponds to the (200) crystal plane. As the value of x increases, the peak position is shifted towards a small angle. It decreases from 24.54 to 24.32°. According to Bragg's Law [31], the insertion of Li^+ increased the sample cell volume from 423.6 to 434.4, an increase of 2.5%. This results in compressive stress after the sample has been colored [32–34].

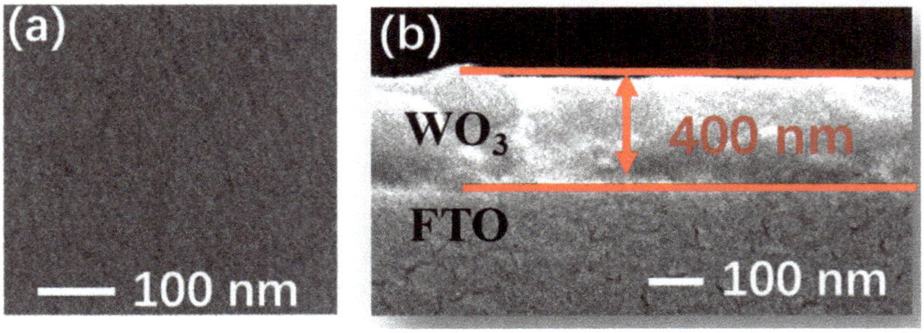

Figure 1. (**a**) Planar and (**b**) cross-sectional photographs of field emission scanning electron microscopy of WO_3 thin film. The as-fabricated WO_3 film is composed of uniformly distributed small particles that are difficult to discern with a diameter smaller than 10 nm. The as-coated WO_3 layer shows an average thickness of 380 nm uniformly and compactly on the top surface of the ITO layer.

Table 1. The XRD pattern details the specific peak of Li_xWO_3.

x (Li)	0	0.10	0.15	0.20	0.25	0.30	0.35
1	23.16	23.10	23.06	23.14	23.12	23.0	23.0
2	23.66	23.5	23.54	23.56	23.5	23.48	23.5
3	24.34	24.3	24.22	24.14	24.18	24.14	24.14
4	26.52	26.56	26.52	26.54	26.52	26.5	26.58
5		——	——	——	——	——	28.26
6	28.76	28.82	28.66	28.8	28.72	28.66	28.72
7	33.48	33.66	33.58	33.62	33.6	33.56	33.62
8	——	——	33.94	33.96	34.02	33.95	33.86
9	34.2	34.18	34.22	34.24	34.24	34.26	34.28
10	37.8	37.78	37.76	37.0	37.8	37.78	37.78

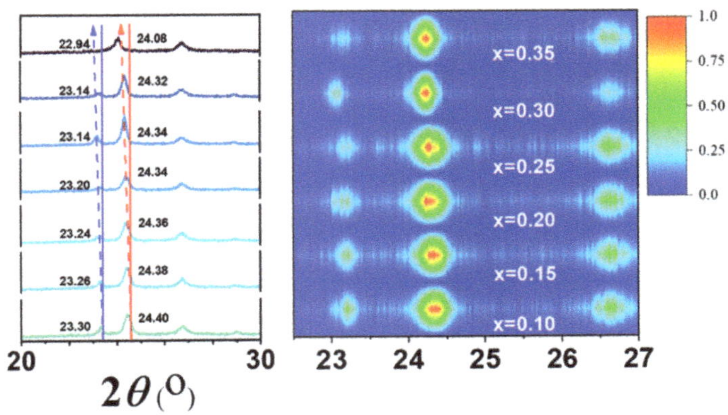

Figure 2. The XRD results of the Li_xWO_3 films ($0.10 \leq x \leq 0.35$). To show the structural changes more clearly, we drew a detailed view of the main diffraction peaks of the sample and the contour of the film. Blue lines represent (002) crystal surfaces and red lines represent (200) crystal surfaces.

2.2. *Ex Situ Raman Spectroscopy*

In the monoclinic (I) phase, space group $P2_1/n(C_{2h}^5)$, each unit cell contains eight formula units, and the total representation corresponds to 96 normal modes, but these 48 are

Raman active: Γ(Raman) = $24A_g + 24B_g$. As shown in Table 2, the A_g modes have a Raman tensor with non-zero diagonal elements and the xz components (labeled as y on the unique axis). The tensor for the B_g modes contains non-zero xy and yz components [35]. Raman spectrum of γ-WO_3 shows eight vibration modes. Raman peaks at 713.2 and 805.2 cm^{-1} are attributed to O-W-O and W=O stretching vibrations. The peaks at 567.2 cm^{-1} and 625.0 cm^{-1} can be seen by comparing the FTO substrate. In the Raman spectrum of the FTO, the two peaks were characteristic peaks of SnO_2 [36,37]. The O-W-O bending pattern of the bridging oxide ion peaks appears at 271.0 cm^{-1} and 328.3 cm^{-1} [38]. The peak at 181.1 cm^{-1} and 132.4 cm^{-1} corresponds to the $(W_2O_2)_n$ chain [39,40].

Table 2. Raman peak position and associated vibrational modes of γ-WO_3.

Raman Peak Position (cm^{-1})	Peak Assignments
132.4, 181.1	Lattice vibration mode
271.0, 328.3	δ (O-W-O), bending vibration modes
567.2, 625.0	attributed to FTO
713.2, 805.2	ν (O-W-O/W=O), stretching vibration modes

It causes a strong decrease in the total Raman intensity due to optical absorption after coloring WO_3 [41]. Additionally, the structural changes associated with the variable lithium content also induce strong changes in the intensity ratio between the various bands. Figure 3 shows the Raman spectrum of Li_xWO_3 films with various amounts of lithium insertion and have ten vibration modes. The ten peaks are listed in Table 3. By comparing the Raman spectra of γ-WO_3 and Li_xWO_3, we found three new modes for A_3, A_6, and A_7 in addition to the seven modes that were affected by lithium insertion in the peak positions. The new modes A_6 and A_7 have been identified due to vibrations of the O-W^{5+}-O and W^{5+}=O bonds, respectively [42–44]. When Li^+ and electrons are inserted into γ-WO_3 films, the electrons reduce W^{6+} ions to W^{5+} and form O-W^{5+}-O and W^{5+}=O bonds. These bonds are weaker than the bonds involving W^{6+} ions, so their corresponding Raman peaks appear at lower energies than those for the O-W^{6+}-O and bonds [45]. It has also been established that the intensities of A_6 and A_7 directly correlate with the coloration of γ-WO_3 films. According to previous reports [46,47], the A_9 mode can be classified as O-W^{6+}-O stretching vibrations, and the A_{10} mode can be classified as W^{6+}=O stretching vibrations [48]. In fact, the 713.2 cm^{-1} mode decreases strongly with respect to the 815.9 cm^{-1} mode upon lithium insertion and disappears for ion concentrations lower than the maximum attainable. This behavior corresponds well to the observed quenching of the same mode in a temperature range below the orthorhombic → tetragonal transition in the case of temperature-driven evolution in pure WO_3 [47]. Here, we focus on the A_{10} mode. The blue shift in the near 810 cm^{-1} peaks decreases with increasing lithiation seems to be in good agreement with their analysis [46]. This implies that the content of O-W^{6+}-O is reduced and transformed to O-W^{5+}-O, while the shift in the peak phase in the short wavelength direction implies that the insertion of Li reduces the amplitude of the O-W-O bond and produces a deformation in the structural length. However, the sudden red shift in the vibrational modes in $Li_{0.10}WO_3$ film and $Li_{0.35}WO_3$ film may be due to the luminescence contribution caused by the change in optical properties during the phase transition [35]. Combined with the peak position results of XRD, there is an increase in theta angle at (120) (−202) and (202), so the W^{6+}=O bonds at this point should correspond to these three crystal planes. After Li^+ insertion, the attenuation of the vibrational mode of γ-WO_3 at 713 cm^{-1} is obvious, and then the vibrational mode at 805 cm^{-1} is shifted, suggesting that the W^{6+}→W^{5+} reaction occurs preferentially in the interior O-W-O, followed by W=O at the surface [49,50].

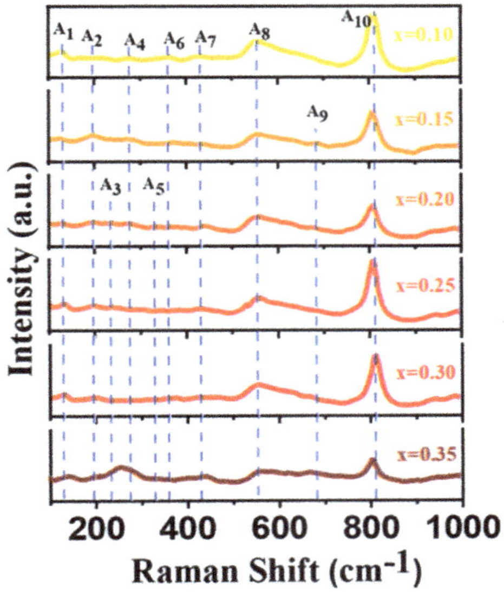

Figure 3. The Raman spectra of the Li_xWO_3 ($0.10 \leq x \leq 0.35$).

Table 3. Raman peak position for the Li_xWO_3 films.

x (Li)	0.10	0.15	0.20	0.25	0.30	0.35
A_1 (cm^{-1})	130.1	130.1	132.4	137.1	137.1	141.7
A_2 (cm^{-1})	197.3	199.6	201.9	201.9	204.3	204.3
A_3 (cm^{-1})	——	——	241.2	241.2	248.1	254.9
A_4 (cm^{-1})	277.9	282.5	275.6	307.7	323.7	325.9
A_5 (cm^{-1})	——	——	335.1	348.8	364.7	351.0
A_6 (cm^{-1})	369.6	376.1	373.7	387.4	378.3	398.7
A_7 (cm^{-1})	430.4	439.4	446.2	430.4	437.2	439.4
A_8 (cm^{-1})	567.2	567.2	567.2	567.2	567.2	567.2
A_9 (cm^{-1})	——	691.3	678.05	673.6	667.0	667.0
A_{10} (cm^{-1})	813.7	811.5	809.4	809.4	809.4	802.8

The sign "—" denotes that no peak could be read for the possible vibration mode in the specific Raman spectrum for the materials.

2.3. Ex Situ Spectroscopic Ellipsometry

The optical parameters and their evolution in the lithiation processes of Li_xWO_3 films were measured and calculated by spectroscopic ellipsometry and DeltaPsi 2 software. As described in the experiment section, the film is deposited on the FTO substrate. To analyze the optical properties of the materials, the optical parameters of the substrate should be determined previously. Commercial FTO substrate is assumed to be a multilayer stack of glass and F-doped SnO_2 (Figure 4a). A three-phase model of roughness layer/SnO_2-F bulk layer/glass substrate was built to simulate the optical constants of the FTO substrate. The optical constants for each layer were imported from the Application Library of the software DeltaPsi 2. The corresponding simulated spectra of the FTO substrate agreed well for the entire spectral range with the χ^2 value of 1.77. The ellipsometric data, Ψ and Δ, measured (solid line) and simulated (dotted line) for FTO, show in Figure 4b.

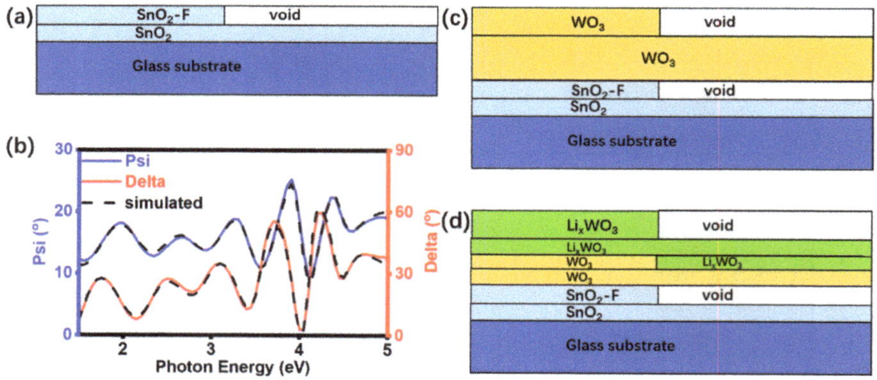

Figure 4. The model of (**a**) the FTO substrate and (**b**) the fit result of it. The model of (**c**) γ-WO_3 films on the FTO substrate and (**d**) Li_xWO_3 films, respectively.

Spectroscopic ellipsometry results with different Li^+ insertion dose-dependent were measured to illustrate the optical parameters evolution due to the phase transition pre and post the color change. Figure 4c shows the optical modeling of the γ-WO_3 film, and Figure 4d shows the optical modeling of the Li_xWO_3 film.

2.4. Layer-by-Layer Diffusion Mechanism of Li^+ in γ-WO_3

In Figure 5a, we can consider it as a doped film of γ-WO_3 and Li_xWO_3 for different x values of the films; therefore, calculating the thickness and ratio of the two materials of the γ-WO_3/Li_xWO_3 layer can be used to indicate the Li^+ insertion dose-dependent of the film. Due to the presence of roughness layer thickness, we added the thickness of the roughness layer in amounts of 50% to each of the two layers. Table 4 shows the fitting thickness results of Li_xWO_3 films by the software DeltaPsi 2. We fitted the relationship between x and the ratio of the Li_xWO_3 layer to the thickness of the film, as shown in Figure 5b. The relational equation was obtained by polynomial fitting: $y = 118.69x^2 - 11.53x + 1.12$ ($0 < x \leq 0.35$). When x = 0, there is no Li^+ insertion, and the thickness of Li_xWO_3 is also zero. This case applies only to the coloration of crystalline WO_3 films in lithium hexafluorophosphate electrolyte. Since the degree of coloration is affected by the crystal structure, specific surface area, cation radius, etc., the establishment of this system requires more research subsequently.

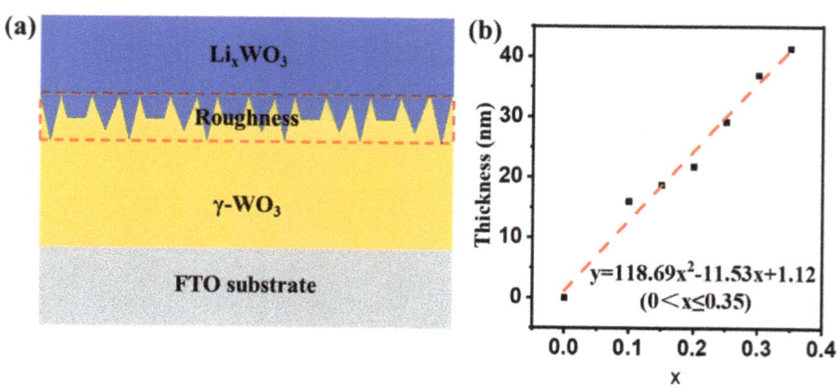

Figure 5. (**a**) Schematic structure of thin film after lithiation. (**b**) Relationship between Li^+ insertion value x and film thickness.

Table 4. Fitting thickness results of Li_xWO_3 films by DeltaPsi 2.

x (Li)	0	0.10	0.15	0.20	0.25	0.30	0.35
Li_xWO_3 (nm)	——	16.01	18.74	21.82	29.14	36.94	41.35
γ-WO_3 (nm)	380	364	361.42	358.22	351.83	343.06	338.66

We added the thickness of the roughness layer in amounts of 50% each to each of the two layers.

Through the analysis above we can realize that the thickness of the Li_xWO_3 layer progressively deepens with the layer-by-layer diffusion of Li^+ in it. This indicates that the γ-WO_3 films prepared by the spin-coating method are denser, and the diffusion mechanism of Li^+ in them is uniform diffusion. Although there is also a rough layer present, there is no particularly obvious ion-preferential diffusion channel. These films are more suitable for applications in the preparation of precision optical components than those prepared by conventional magnetron sputtering [51].

2.5. Details of Optical Leaps in the Full Spectrum

$$\frac{d^2(E^2\varepsilon)}{dE^2} = \begin{cases} n(n-1)A_m e^{i\phi_m}(E - E_{cpm} + i\Gamma_m)^{n-2} & n \neq 0 \\ A_m e^{j\phi_m}(E - E_{cpm} + i\Gamma_m)^{-2} & n = 0 \end{cases} \quad (1)$$

To analyze the optical transitions of Li_xWO_3 films, we computed $d^2(E^2\varepsilon)/dE^2$ in Figure 6. The $d^2(E^2\varepsilon)/dE^2$ is shown in Equation (1). The positions as well as the shapes of these critical points (CPs) can be totally determined by the CPs analysis method [52]. The transition energies were 2.5 eV, 3.05 eV, 3.65 eV, and 4.55 eV, respectively. We assigned the origins of the four interband transitions according to the results from density functional calculations and previous analysis of the respective materials. The E_{cp1} transition arises from a direct gap transition between the highest valence band (HVB) and the lowest conduction band (LCB) at the Γ point in the Brillouin zone (BZ). Another interband transition labeled E_{cp2} occurred at the X point in the BZ. The E_{cp3} = 3.65 eV transition arises from an indirect gap transition between the HVB at the R point and the LCB at the X point in the BZ. The E_{cp4} = 4.55 eV transition arises from a direct gap transition between the HVB and the LCB at the R point in the BZ.

For Li_xWO_3 thin film, five distinct transitions labeled E_{cp1}, E_{cp2}, E_{cp3}, E_{cp4}, and E_{cp5} appear in the spectra (n = 1/2). By comparing the data in Table 5, we can know that the CPs of Li_xWO_3 films are identical. The transition energies for the Li_xWO_3 thin film were 1.2 eV, 1.95 eV, 3.1 eV, 4.2 eV, and 4.45 eV, respectively. The E_{cp1} = 1.2 eV transition arises from a direct gap transition between the highest valence band (HVB) and the lowest conduction band (LCB) at the Γ point in the Brillouin zone (BZ). Another interband transition labeled E_{cp2} occurred at the X point in the BZ. At the same time, the contribution from the indirect transition of the HVB at the M point and the LCB at the Γ point in the BZ is also included. The E_{cp3} = 3.1 eV transition arises from an indirect gap transition between the HVB at the X point and the LCB at the M point in the BZ. The E_{cp4} = 4.2 eV transition arises from a direct gap transition between the HVB and the LCB at the M point in the BZ. The E_{cp5} = 4.45 eV transition arises from a direct gap transition between the VB and the CB at the R point in the BZ. In all cases, the LCB and HVB are formed by hybridization of W 5*d* and O 2*p* states, but the CB is mainly dominated by W *d* states while the VB has an O *p* character [53]. According to the Burstein–Moss shift effect, when Li is inserted, the sample shows metallicity and the spectral change drastically shifts the Fermi level to CB and slightly reduces the size of the band gap. This effect is typical of a simplex *n*-type semiconductor, where the semiconductor becomes metallic when the doping level is very high. In metallic materials, the Fermi level is in the CB, and this part becomes a partially filled conduction band (PFCB); the band below this one is the highest fully occupied band (B_{-1}), and the band above is the lowest unoccupied band (B_1) [54]. The W 5*d* states dominate both the PFCB and the B_1, whereas the B_{-1} presents an O 2*p* character number of states. The Li 2*s* states lay beyond the Fermi level higher than the W 5*d* states.

The 2*s* level of atomic Li is higher in energy than the 5*d* level of atomic W, and thus, one would expect that the W 5*d* states lay closer to the Fermi level. The position of the Li 2*s* states also helps us elucidate the role played by the Li atom. Since these states lie outside the Fermi energy level, the corresponding valence electrons enter the PFCB (more precisely, the t_{2g} conduction band), thus leaving the Li atom in the ionized state [55]. This suggests that the insertion of Li provides the system with roving electrons, which is consistent with the metallicity reported experimentally [56]. Thus, the metallic transition reoccurs due to the injection of electrons into the system (electron doping) [57]. This is the main reason why the Tauc–Lorentz and Drude models were selected for analytical modeling.

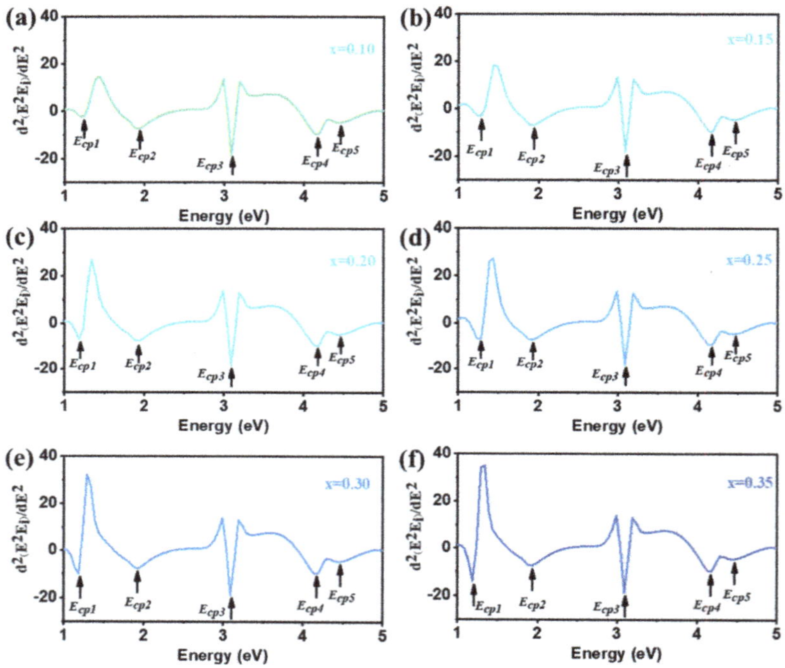

Figure 6. Using the Second derivatives of the pseudo dielectric function numerically calculated spectra of Li_xWO_3 (blue) thin films at 300K. The E_{cp1}, E_{cp2}, E_{cp3}, E_{cp4}, and E_{cp5} transition features are indicated by arrows. (**a**–**f**) is the calculated spectra of Li_xWO_3 for x = 0.10, x = 0.15, x = 0.20, x = 0.25, x = 0.30, x = 0.35, respectively.

Table 5. Trip thresholds for γ-WO_3 (yellow) and Li_xWO_3 (blue) films calculated from the second derivative of the pseudo-dielectric constant.

x (Li)	0	0.10	0.15	0.20	0.25	0.30	0.35
E_{cp1}	——	1.2	1.2	1.2	1.2	1.2	1.2
E_{cp2}	2.5	1.95	1.95	1.95	1.95	1.95	1.95
E_{cp3}	3.05	3.1	3.1	3.1	3.1	3.1	3.1
E_{cp4}	3.65	4.2	4.2	4.2	4.2	4.2	4.2
E_{cp5}	4.55	4.45	4.45	4.45	4.45	4.45	4.45

The analysis of the Brillouin zone point locations and details is crucial for the design of optical materials and devices and is important for optimizing the structure of devices such as optical fibers and optical waveguide devices.

2.6. Full Spectrum Properties

In Figure 6, the second-order derivative curves of the imaginary part of the dielectric constant are essentially the same, except for the peak intensity of the peak near 1.3 eV, where a significant difference can be observed. As x increases, the intensity of the peak here increases as well. This is because the absorption of the film in the near-infrared band gradually increases as the value of x increases. We have tested the infrared absorption of the sample below and compared its absorption intensity in the infrared band. And it is consistent with its trend.

To observe the change in optical constants during the film color change, we tested the absorption for different values of x, as shown in Figure 7a. It is shown in previous studies that tungsten exists in WO_3 electrochromic films in tetra-, penta-, and hexavalent forms, and the ratio of each valence state is related to the structure of the material and the color-changing conditions. It is recognized that the hexavalent states of tungsten (d^0) exhibit charge transfer UV absorption only, while the rest of the valences W^{5+} (d^1) and W^{4+} (d^2)show d-d transitions in the visible region [58]. In the absorption spectrum, we can observe two distinct peaks, one near 420 nm and one between 600 and 700 nm. Based on the above findings, the absorption spectra of the films were designated as follows: (a) The UV absorption of the film at 420 nm is attributed to the absorption of W (VI) (d^0). (b) The broad band centered at 650–800 nm is attributed to W (V) (d^1). (c) The presence of significant absorption in the near-infrared band is due to the contribution of W (IV) (d^2). The peak of 650–800 nm is not easily identifiable in the absorption spectrum, which is caused by the change in the light source in the spectrometer. However, we found a similar peak in the simulated spectrum later. In the transmission spectrum of Figure 7b, the transmittance decreases with increasing x-value, while the color becomes darker. Direct allowed optical transitions were obtained following the expression $\alpha = \frac{A}{hv}(hv - E_g)^{\frac{1}{2}}$, where A is the photon energy-independent constant, hv is the photon energy, and E_g is the optical bandgap [59]. Optical bandgaps decrease from 3.22 eV to 3.14 eV were estimated of the hybrid Li_xWO_3 and γ-WO_3 films in the illustrations of Figure 7a, respectively, by extrapolating the square of α plots to the energy intercept and are in good agreement with previously reported values [60]. It is influenced by both γ-WO_3 and Li_xWO_3 films together.

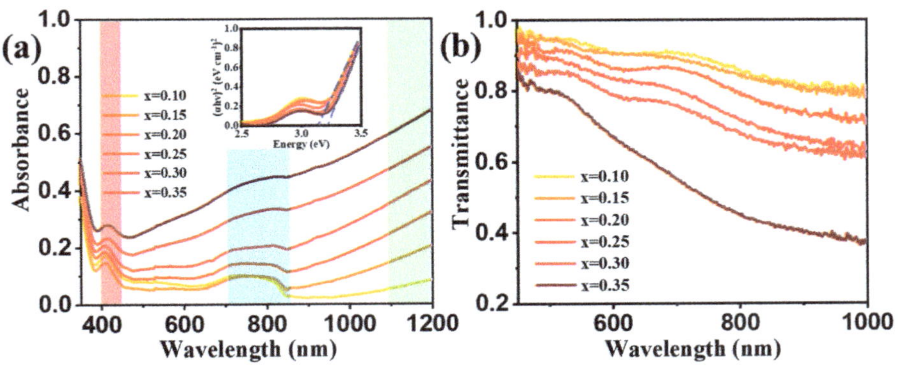

Figure 7. (**a**) The absorption and (**b**) transmission spectra for different values of x. The illustration is the optical bandgaps by simulated. Absorption peaks in the red area is attributed to the absorption of W (VI) (d^0). Absorption peaks in the blue area is attributed to the absorption of W (V) (d^1). Absorption peaks in the green area is attributed to the absorption of W (IV) (d^2).

Due to the high light modulation ability in the visible light band of Li^+ insertion in WO_3 and the strong absorption in the near-infrared band, it has attracted a great deal of attention in the direction of color-changing and heat-modulated glasses in recent years.

Meanwhile, the high transmittance in the initial state for color change is also the main reason for its application in the field of smart windows and low power displays [61–63].

2.7. Optical Constants for Different Li^+ Insertions

In the electrochromic process of Li^+ insertion in crystalline WO_3, the resulting Li_xWO_3 films are the same substance and do not change with the value of x. It is only with the different amount of Li^+ insertion that it participates in the reaction in different proportions, leading to the color change in the films. The optical constants of the resulting Li_xWO_3 films are only affected by the ratio of reacted Li_xWO_3 to unreacted WO_3 and by the roughness layer we mentioned above.

We also obtained the entire optical and dielectric constants of the film for different values of x. As a result of the fits, the imaginary part of the dielectric function ($\varepsilon_2 = 2$ nk), which represents optical absorption, is composed of two parts. Figure 8 shows the complex optical constants of the Li_xWO_3 film for different x values (the refractive index, *n*, and extinction coefficient, k). The yellow line is the *n* and k for γ-WO_3 with the value of 0.1 to 0.35 in layer 4. Due to the diffuse reflective influence of the roughness layer, there will be a slight difference from γ-WO_3, but the overall trend is the same. The refractive index is relatively high in the vicinity of 3–5 eV, i.e., its refractive index is higher in the near UV-blue band at 248–413 nm. There are also smaller shoulder peaks near 1.8 and 2.8 eV, suggesting that it also has a relatively high refractive index for peaks near 450 nm and 690 nm.

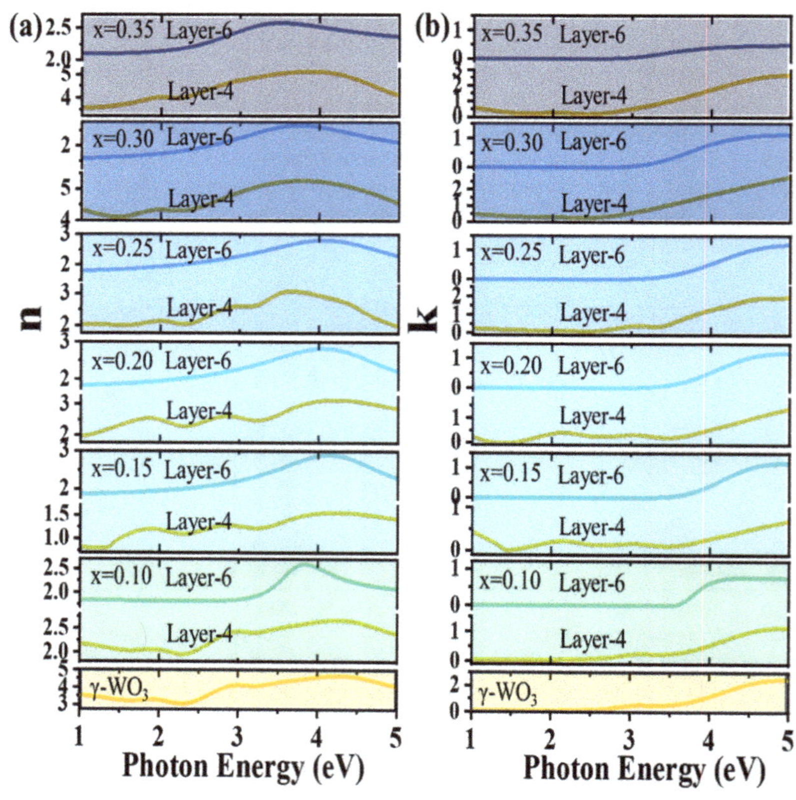

Figure 8. (**a**) The refractive index and (**b**) extinction coefficient of the Li_xWO_3 film for different x values. As a comparison, the values of the γ-WO_3 layer are also given. In fitting Li_xWO_3 layers, the optical constants of the γ-WO_3 layer are fixed.

We have obtained "pseudo" complex optical constants in this work. As the value of x increases, n gradually decreases, and the corresponding photon energy at the non-zero of k decreases. The shift in k represents a shift in the band gap as x increases, decreasing from 3.65 eV to 2.75 eV. However, the roughness layer mainly affects the real part of the complex optical constants rather than the imaginary part. The pseudo-extinction coefficient is reasonably equal to the intrinsic extinction coefficient of almost all spectral features containing inter-band transitions. The extinction coefficient of the film shows a near-monotonic increase with increasing photon energy. The refractive indices of the samples have more pronounced variations around 1.2 eV, 1.8 eV, 3.0 eV, and 3.8 eV, indicating more pronounced modulation of the refractive indices of the light sources at 980 nm, 688 nm, 405 nm, and 325 nm wavelengths. Therefore, it can also be used to prepare optical components such as optical phase modulators, filters, polarizers, etc. It also has the potential to be applied in the scenarios of chip fabrication [64,65], medical imaging [66], and medical therapy [67]. The ex situ spectroscopic ellipsometry provides us with the optical constants of the full spectrum of the samples in various states, which provides more accurate information for the study of the properties and applications of the materials.

We confirmed the validity of the spectroscopic ellipsometry analysis by comparing the experimental and simulated absorption spectra [68]. Figure 9 shows the absorption coefficient α spectra and the transmittance intensity of the thin film samples obtained from the experiment, while the dashed line shows the spectrum calculated by SE modeling. The simulated spectra were calculated by fitting the values of the extinction coefficients obtained. The absorption coefficients of the thin films are obtained from $\alpha = \frac{4\pi k}{\lambda}$, where α is the absorption coefficient, k is the extinction coefficient, and λ is the wavelength of light in a vacuum [69–71]. Spectra obtained via the method is matched well. The peak values are slightly deviated, which may be a result of the roughness between γ-WO_3 and Li_xWO_3.

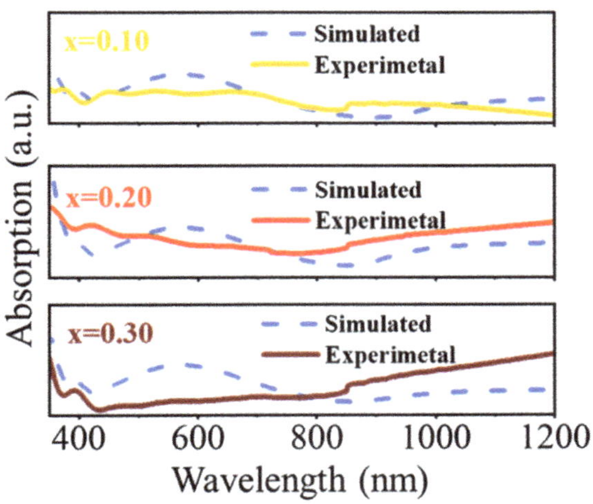

Figure 9. Simulated and experimental absorption coefficient spectra of thin films on the FTO substrate in the visible region with x = 0.10, 0.20, and 0.30.

3. Discussion

In summary, we developed a systematic protocol of ex situ methods to clarify the evolutions of subtle structure changes, Raman vibration modes, and optical parameters of WO_3 thin films in electrochromic processes as stimulated by dosage-dependent Li^+ insertion. The crystal structure of the samples gradually transforms the $\gamma \rightarrow \beta \rightarrow \alpha$-phase during the coloration process. We proved the layer-by-layer Li^+ insertion mechanism by analyzing the film thickness of each layer. The relationship between the Li^+ insertion dosage and the thickness of Li_xWO_3 films was obtained: $y = 118.69x^2 - 11.53x + 1.12$

$(0 < x \leq 0.35)$. The detail of all optical leap points in the Brillouin zone corresponding to the range of 1–5 eV is also found. The change in extinction coefficient (k) with Li^+ insertion shows that its band gap decreases from 3.22 eV to 3.14 eV. The refractive index of each film layer was also used to calculate the relationship between the variation in optical phase angle with wavelength at different Li^+ insertion dosages. This work provides new ideas for controllable preparation and smart applications of smart optical films.

4. Methods

4.1. Chemicals and Materials

The tungsten powder (W, 99.80%) and hydrogen peroxide (H_2O_2, 30%) were purchased from Sinopharm Chemical Reagent Co., Ltd. (Shanghai, China). Lithium hexafluorophosphate ($LiPF_6$) electrolyte purchased from Dongguan Koluder Innovation Technology Co., Ltd. (Dongguan, China). Specification of the electrolyte: 1.0 mol $LiPF_6$ in ethylene carbonate (EC): dimethyl carbonate (DMC): ethyl methyl carbonate (EMC) = 1:1:1. Analytical purity acetone, isopropyl alcohol, and ethanol were purchased from Beijing Chemical Industry (Beijing, China). Fluorine-doped tin oxide (FTO) substrates (size 2 × 2 cm^2) with a surface square resistance of 15 Ω/□ were purchased from Advanced Election Technology Co., Ltd. (Liaoning, China).

4.2. Films Preparation

The WO_3 film was fabricated via the spin-coating method: (1) substrate pretreatment. FTO substrates were cleaned in the solvents of acetone, isopropyl alcohol, ethanol, and deionized water for 15 min subsequently. The adhered solvent molecules were dried by a high-pressure gas gun, and then, the substrates were enclosed in argon gas-filled ultraviolet ozone cleaner for 15 min to remove the organic residues attached to the surface of the substrate and to improve the hydrophilicity of the surface with more of dangling polar chemical bonds. This protocol is very important for the spreading of the precursor solutions to the homogeneous films. (2) Spin coating of WO_3 films. The precursor solutions should be prepared in advance. Firstly, 2 g W (tungsten powder, 99.8%) was added to 10 mL H_2O_2 (hydrogen peroxide, 30%) several times, and their reaction was accelerated at 70 °C; the solution was at a speed of 300 rpm mixed with 2 h at 70 °C and filter with water-based filter head to prepare a saturated solution of metatuntungstic acid. Next, we used a platinum filament to decompose the uncreated hydrogen peroxide and added absolute ethyl alcohol to it. It was allowed to evaporate for 0.5 h at 80 °C and then we obtained the orange transparent colloidal sol. The substrates were mounted on the spinner by vacuum, and then 100 μL of the precursor solutions were dripped onto the substrates. (3) Thermal treatment. The as-prepared WO_3 film precursor solutions were spin-coated on an FTO substrate at a speed of 3000 rpm for 30 s, followed by annealing at 500 °C for 2 h. Then, the films were cooled to room temperature naturally.

Based on the above experiments, we prepared Li_xWO_3 films with different lithium-ion insertion amounts. We used a three-electrode method to color the films. The counter electrode using Pt is the most stable [72]; the Pt sheet is parallel to the WO_3 film, the working electrode is WO_3 film, and the reference electrode is Ag/AgCl. The amount of lithium embedded is controlled by the voltage and time applied to the film. We applied "1.5 V, 1 s", "2 V, 1 s", "2.5 V, 1 s", "3 V, 1 s", "3.5 V, 1 s", "4 V, 1 s" and "4 V, 10 s" to the obtained WO_3 films in Multi-potential step method of the electrochemical workstation, respectively. After removal, the sample surface was rinsed with alcohol and blown dry. With the data from the electrochemical workstation, we calculated the insertion levels of Li ions and obtained the films with different insertion levels. the "$Li_{0.10}WO_3$", "$Li_{0.15}WO_3$", "$Li_{0.20}WO_3$", "$Li_{0.25}WO_3$", "$Li_{0.30}WO_3$"and ""$Li_{0.35}WO_3$" corresponding to ca. 13, ca. 28, ca. 47, ca. 64, ca. 79, and ca. 93 mC/cm^2 of charge, respectively [73].

4.3. Characterization

The characterization details were described as follows: (1) Structure. The phases of the films were determined by x-ray diffraction (XRD) technique with all the data collected from a Rigaku D/max−2500 (Tokyo, Japan) diffractometer with CuKα radiation (λ = 0.15406 nm) operating at a beam voltage of 40 kV and current of 200 mA, and a diffraction angle step of 0.02° in the angle range of 5–50°. (2) Raman spectroscopy. The Raman spectra were recorded using a HORIBA LabRAM HR Confocal Raman microscopy (Kyoto, Japan). The spectra were collected in the range of 50–1000 cm^{-1} at a spectral resolution of 1 cm^{-1} laser excitation with Spectra Physics 473 M ion laser (Kyoto, Japan). (3) Electrochromic performance. The visible transmittance/absorption spectra were measured using a StellarNet Black-Comet-SR Series fiber spectrometer (Tampa, FL, USA). The standardization of the lithiation process of thin films is achieved with the CHI 760E electrochemical workstation from CH Instruments Inc. (Shanghai, China). This is achieved by controlling the excitation voltage and time. The measurements were preceded by five activation cycles in the same electrolyte. The test range in $LiPF_6$ electrolyte is −0.3 V to 0.8 V; the scan rate is 100 mV/s.

4.4. Optical Properties

Spectroscopic ellipsometry is a simple, non-invasive, surface-sensitive technique based on measuring the relative phase change in reflected and polarized light from a sample to characterize the thin-film optical constants and other properties (e.g., thickness, interface conditions, and roughness) [31–33]. To measure the optical constants and dispersion of WO_3 film during Li^+ ion migration, the ellipsometry data were recorded as a function of the photon energy in the range of 1.5–5 eV by a HORIBA Jobin yvon UVISEL (Kyoto, Japan) spectroscopic ellipsometry system equipped at an incident angle of 70° in air, and the $I_s = \sin2\Psi\sin\Delta$ and $I_c = \sin2\Psi\cos\Delta$. The ellipsometry data were acquired with different degrees of Li^+ ion insertion. The optical parameters were obtained by simulating the measured wavelength/energy dependent Delta/Psi curves by the modeling in the DeltaPsi 2 software. The inversion of the ellipsometric data was performed using a five-phase representative model of the samples that added an EC active layer and roughness layer on the model of FTO. In this model, the thickness of the roughness layer, the thickness of the active layer, and the dielectric function of the active layer were unknown, while the roughness layer was modeled by a mixture of 50% air and 50% active material according to the Bruggeman effective-medium approximation (EMA) [74–77]. The unknown parameters were determined by minimizing the mean squared difference between the generated Is and I_c data and the experimental data. All solutions were found to be Kramers−Kronig consistent [78].

Author Contributions: B.Z.: Conceptualization, methodology, data curation, writing—original draft, investigation. S.J.: investigation, data curation. J.W.: conceptualization, supervision. M.Y.: conceptualization, review and editing, supervision. X.C.: supervision, project administration, funding acquisition. All authors have read and agreed to the published version of the manuscript.

Funding: This research was supported by the National Natural Science Foundation of China (Grant No. 11904031).

Data Availability Statement: The data that support the findings of this study are available from the authors upon reasonable request.

Conflicts of Interest: The authors declare that they have no known competing financial interests or personal relationships that could have appeared to influence the work reported in this paper.

erences

Martínez-Espert, A.; Montagud-Martínez, D.; Ferrando, V.; Furlan, W.D.; Monsoriu, J.A. Assessment of a new trifocal diffractive corneal inlay for presbyopia correction using an adaptive optics Visual Simulator. *Photonics* **2022**, *9*, 135. [CrossRef]
Fuh, Y.-K.; Lin, M.-X. Adaptive optics correction of a tunable fluidic lens for ophthalmic applications. *Opt. Commun.* **2013**, *308*, 100–104. [CrossRef]

3. Li, J.; Zhang, K.; Du, J.; Li, F.; Yang, F.; Yan, W. Double-sided telecentric zoom optical system using adaptive liquid lenses. *Express* **2023**, *31*, 2508. [CrossRef]
4. Zhang, D.-Y.; Justis, N.; Lo, Y.-H. Fluidic adaptive zoom lens with high zoom ratio and widely tunable field of view. *Opt. Com* **2005**, *249*, 175–182. [CrossRef]
5. Kim, M.K. Adaptive optics by incoherent digital holography. *Opt. Lett.* **2012**, *37*, 2694. [CrossRef]
6. Lenk, L.; Sinzinger, S. Diffractive Alvarez-Lohmann lenses for correcting aberrations of tunable membrane lenses. *Opt. Eng.* *62*, 035103. [CrossRef]
7. Huang, H.; Zhao, Y. Optofluidic lenses for 2D and 3D imaging. *J. Micromech. Microeng.* **2019**, *29*, 073001. [CrossRef]
8. Schmidt, K.; Koukourakis, N.; Czarske, J. Assignment of focus position with convolutional neural networks in adaptive based axial scanning for confocal microscopy. *Appl. Sci.* **2022**, *12*, 661. [CrossRef]
9. Koukourakis, N.; Finkeldey, M.; Stürmer, M.; Leithold, C.; Gerhardt, N.C.; Hofmann, M.R.; Wallrabe, U.; Czarske, J.W.; Fisch Axial scanning in confocal microscopy employing adaptive lenses (CAL). *Opt. Express* **2014**, *22*, 6025. [CrossRef]
10. Hands, P.J.W.; Tatarkova, S.A.; Kirby, A.K.; Love, G.D. Modal liquid crystal devices in optical tweezing: 3D control and oscilla potential wells. *Opt. Express* **2006**, *14*, 4525. [CrossRef]
11. Theofanidou, E.; Wilson, L.; Hossack, W.J.; Arlt, J. Spherical aberration correction for optical tweezers. *Opt. Commun.* **2004**, 145–150. [CrossRef]
12. Liu, Z.; Fulda, P.; Arain, M.A.; Williams, L.; Mueller, G.; Tanner, D.B.; Reitze, D.H. Feedback control of optical beam sp profiles using thermal lensing. *Appl. Opt.* **2013**, *52*, 6452. [CrossRef]
13. Mai, V.; Kim, H. Non-Mechanical Beam Steering and adaptive beam control using variable focus lenses for free-space op communications. *J. Light. Technol.* **2021**, *39*, 7600–7608. [CrossRef]
14. Gökçe, M.C.; Ata, Y.; Baykal, Y. Fiber coupling efficiency in ocean with adaptive optics corrections. *J. Opt. Soc. Am. B.* **202** 949. [CrossRef]
15. Courteville, A. Optimization of single-mode fiber coupling efficiency with an adaptive membrane mirror. *Opt. Eng.* **2002** 1073. [CrossRef]
16. Guoqiang, L. Adaptive Lens. In *Progress in Optics*; Elsevier: Amsterdam, The Netherlands, 2010; pp. 199–283.
17. Liu, C.; Chu, F.; Zheng, Y.-W.; Wang, Q.-H. Adaptive micro-projection system based on liquid lens. *Opt. Laser Technol.* **2023**, 109734. [CrossRef]
18. Algorri, J.F.; Zografopoulos, D.C.; Urruchi, V.; Sánchez-Pena, J.M. Recent advances in adaptive liquid crystal lenses. *Crystals* *9*, 272. [CrossRef]
19. Wu, Q.; Zhang, H.; Jia, D.; Liu, T. Recent development of tunable optical devices based on liquid. *Molecules* **2022**, *27*, [CrossRef]
20. Cheng, Y.; Cao, J.; Tang, X.; Hao, Q. Optical zoom imaging systems using adaptive liquid lenses. *Bioinspir. Biomim.* **2021** 041002. [CrossRef]
21. Xue, S.; Gao, G.; Zhang, Z.; Jiang, X.; Shen, J.; Wu, G.; Dai, H.; Xu, Y.; Xiao, Y. Nanoporous WO_3 gasochromic films for gas sen *ACS Appl. Nano Mater.* **2021**, *4*, 8368–8375. [CrossRef]
22. He, Y.; Li, W.; Han, N.; Wang, J.; Zhang, X. Facile flexible reversible thermochromic membranes based on micro/nanoencapsul phase change materials for wearable temperature sensor. *Appl. Energy* **2019**, *247*, 615–629. [CrossRef]
23. Guan, L.; Yang, Y.; Jia, F.; Gao, G. Highly transparent and stretchable hydrogels with rapidly responsive photochromic perform for UV-irradiated optical display devices. *React. Funct. Polym.* **2019**, *138*, 88–95. [CrossRef]
24. Ma, D.; Lee-Sie Eh, A.; Cao, S.; Lee, P.S.; Wang, J. Wide-spectrum modulated electrochromic smart windows based on MnO_2 films. *ACS Appl. Mater. Interfaces.* **2022**, *14*, 1443–1451. [CrossRef]
25. Guo, J.; Jia, H.; Shao, Z.; Jin, P.; Cao, X. Fast-switching WO_3-based electrochromic devices: Design, fabrication, and applicat *Acc. Mater. Res.* **2023**, *4*, 438–447. [CrossRef]
26. In, Y.R.; Kim, Y.M.; Lee, Y.; Choi, W.Y.; Kim, S.H.; Lee, S.W.; Moon, H.C. Ultra-low power electrochromic heat shutters thro tailoring diffusion-controlled behaviors. *ACS Appl. Mater. Interfaces* **2020**, *12*, 30635–30642. [CrossRef]
27. Wang, Z.; Wang, X.; Cong, S.; Geng, F.; Zhao, Z. Fusing electrochromic technology with other advanced technologies: A roadmap for future development. *Mater. Sci. Eng. R Rep.* **2020**, *140*, 100524. [CrossRef]
28. Yuan, G.; Hua, C.; Huang, L.; Defranoux, C.; Basa, P.; Liu, Y.; Song, C.; Han, G. Optical characterization of the coloration proce electrochromic amorphous and crystalline WO_3 films by spectroscopic ellipsometry. *Appl. Surf. Sci.* **2017**, *421*, 630–635. [Cross
29. Zhong, Q.; Dahn, J.R.; Colbow, K. lithium intercalation into WO_3 and the phase diagram of Li_xWo_3. *Phys. Rev. B* **1992**, *46*, [CrossRef]
30. Zheng, H.; Ou, J.Z.; Strano, M.S.; Kaner, R.B.; Mitchell, A.; Kalantar-zadeh, K. Nanostructured tungsten oxide–proper synthesis, and applications. *Adv. Funct. Mater.* **2011**, *21*, 2175–2196. [CrossRef]
31. Jauncey, G.E.M. The scattering of X-rays and Bragg's Law. *Proc. Natl. Acad. Sci. USA* **1924**, *10*, 57–60. [CrossRef] [PubMed]
32. Zhou, K.; Wang, H.; Liu, J.; Zhao, B.; Zhang, Q.; Yan, H. Ions Transport control in electrochromic WO_3 film for the cyclic stab study. *J. Electrochem. Soc.* **2020**, *167*, 106502. [CrossRef]
33. Locherer, K.R.; Swainson, I.P.; Salje, E.K.H. Transition to a new tetragonal phase of WO_3: Crystal structure and distor parameters. *J. Phys. Condens. Matter* **1999**, *11*, 4143–4156. [CrossRef]

Denis, S.; Gautier, E.; Simon, A.; Beck, G. Stress–phase-transformation interactions–basic principles, modelling, and calculation of internal stresses. *Mater. Sci. Technol.* **1985**, *1*, 805–814. [CrossRef]
Cazzanelli, E.; Vinegoni, C.; Mariotto, G.; Kuzmin, A.; Purans, J. Raman study of the phase transitions sequence in pure WO_3 at high temperature and in H_xWO_3 with variable hydrogen content. *Solid State Ion.* **1999**, *123*, 67–74. [CrossRef]
Scott, J.F. Raman Spectrum of SnO_2. *J. Chem. Phys.* **1970**, *53*, 852–853. [CrossRef]
He, Y.; Li, Y.; Yu, J.; Qian, Y. Chemical control synthesis of nanocrystalline SnO_2 by hydrothermal reaction. *Mater. Lett.* **1999**, *40*, 23–26. [CrossRef]
Lu, Z.; Kanan, S.M.; Tripp, C.P. Synthesis of high surface area monoclinic WO_3 particles using organic ligands and emulsion-based methods. *J. Mater. Chem.* **2002**, *12*, 983–989. [CrossRef]
Daniel, M.; Desbat, B.; Lassegues, J.; Gerand, B.; Figlarz, M. Infrared and Raman study of WO_3 tungsten trioxides and $WO_3•xH_2O$ tungsten trioxide tydrates. *J. Solid State Chem.* **1987**, *67*, 235–247. [CrossRef]
Mukherjee, R.; Prajapati, C.S.; Sahay, P.P. Tailoring the microstructural, optical, and electrical properties of nanocrystalline WO_3 thin films using Al doping. *J. Mater. Eng. Perform.* **2014**, *23*, 3141–3151. [CrossRef]
Shigesato, Y.; Murayama, A.; Kamimori, T.; Matsuhiro, K. Characterization of evaporated amorphous WO_3 films by raman and ytir spectroscopies. *Appl. Surf. Sci.* **1988**, *33*, 804–811. [CrossRef]
Lee, S.-H.; Cheong, H.M.; Zhang, J.-G.; Mascarenhas, A.; Benson, D.K.; Deb, S.K. Electrochromic mechanism in a-WO_{3-y} thin films. *Appl. Phys. Lett.* **1999**, *74*, 242–244. [CrossRef]
Lee, S.-H.; Cheong, H.M.; Tracy, C.; Mascarenhas, A.; Benson, D.K.; Deb, S.K. Raman spectroscopic studies of electrochromic a-WO_3. *Electrochim. Acta* **1999**, *44*, 3111–3115. [CrossRef]
Lee, S.H.; Cheong, H.M.; Tracy, C.E.; Mascarenhas, A.; Czanderna, A.W.; Deb, S.K. Electrochromic coloration efficiency of a-WO_{3-y} thin films as a function of oxygen deficiency. *Appl. Phys. Lett.* **1999**, *75*, 1541–1543. [CrossRef]
Lee, S.-H.; Cheong, H.M.; Liu, P.; Smith, D.; Tracy, C.E.; Mascanrenhas, A.; Pitts, J.R.; Deb, S.K. Gasochromic mechanism in a-WO_3 thin films based on Raman spectroscopic studies. *J. Appl. Phys.* **2000**, *88*, 3076–3078. [CrossRef]
Lu, D.Y.; Chen, J.; Zhou, J.; Deng, S.Z.; Xu, N.S.; Xu, J.B. Raman spectroscopic study of oxidation and phase transition in $W_{18}O_{49}$ nanowires. *J. Raman Spectrosc.* **2007**, *38*, 176–180. [CrossRef]
Lu, D.Y.; Chen, J.; Chen, H.J.; Gong, L.; Deng, S.Z.; Xu, N.S.; Liu, Y.L. Raman study of thermochromic phase transition in tungsten trioxide nanowires. *Appl. Phys. Lett.* **2007**, *90*, 041919. [CrossRef]
Boulova, M.; Lucazeau, G. Crystallite Nanosize Effect on the structural transitions of WO_3 studied by raman spectroscopy. *J. Solid State Chem.* **2002**, *167*, 425–434. [CrossRef]
Lee, S.H.; Cheong, H.M.; Tracy, C.E.; Mascarenhas, A.; Pitts, J.R.; Jorgensen, G.; Deb, S.K. Alternating current impedance and Raman spectroscopic study on electrochromic a-WO_3 films. *Appl. Phys. Lett.* **2000**, *76*, 3908–3910. [CrossRef]
Wen, R.-T.; Granqvist, C.G.; Niklasson, G.A. Eliminating degradation and uncovering ion-trapping dynamics in electrochromic WO_3 thin films. *Nat. Mater.* **2015**, *14*, 996–1001. [CrossRef]
O'brien, N.A.; Gordon, J.; Mathew, H.; Hichwa, B.P. Electrochromic coatings–applications and manufacturing issues. *Thin Solid Film.* **1999**, *345*, 312–318. [CrossRef]
Song, B.; Gu, H.; Fang, M.; Chen, X.; Jiang, H.; Wang, R.; Zhai, T.; Ho, Y.T.; Liu, S. Layer-dependent dielectric function of wafer-scale 2D MoS_2. *Adv. Opt. Mater.* **2019**, *7*, 1801250. [CrossRef]
Perez, I.; Faudoa, J.C.M.; Acuña, J.R.A.; Galindo, J.T.E. Influence of lithium interstitial doping on the optoelectronic properties of NiO and WO_3. *Comp. Mater. Sci.* **2021**, *190*, 110248. [CrossRef]
Xu, X.; Randorn, C.; Efstathiou, P.; Irvine, J.T.S. A red metallic oxide photocatalyst. *Nat. Mater.* **2012**, *11*, 595–598. [CrossRef]
Bullett, D.W. Bulk and surface electron states in WO_3 and tungsten bronzes. *J. Phys. C Solid State Phys.* **1983**, *16*, 2197–2207. [CrossRef]
Yoshimatsu, K.; Soma, T.; Ohtomo, A. Insulator-to-metal transition of WO_3 epitaxial films induced by electrochemical Li-ion intercalation. *Appl. Phys. Express.* **2016**, *9*, 075802. [CrossRef]
Walkingshaw, A.D.; Spaldin, N.A.; Artacho, E. Density-functional study of charge doping in WO_3. *Phys. Rev. B.* **2004**, *70*, 165110. [CrossRef]
Fayad, A.M.; Ouis, M.A.; ElBatal, F.H.; ElBatal, H.A. Shielding behavior of gamma-irradiated MoO_3 or WO_3-doped lead phosphate glasses assessed by optical and FT infrared absorption spectral measurements. *Silicon* **2018**, *10*, 1873–1879. [CrossRef]
Madhavi, V.; Kondaiah, P.; Hussain, O.; Uthanna, S. Structural, optical and electrochromic properties of RF magnetron sputtered WO_3 thin films. *Phys. B-Condens. Matter* **2014**, *454*, 141–147. [CrossRef]
Hjelm, A.; Granqvist, C.G.; Wills, J.M. Electronic structure and optical properties of WO_3, Li_xWO_3, Na_xWO_3, and H_xWO_3. *Phys. Rev. B* **1996**, *54*, 2436–2445. [CrossRef]
Gu, C.; Jia, A.-B.; Zhang, Y.-M.; Zhang, S.X.-A. Emerging electrochromic materials and devices for future displays. *Chem. Rev.* **2022**, *122*, 14679–14721. [CrossRef]
Mardare, C.C.; Hassel, A.W. Review on the versatility of tungsten oxide coatings. *Phys. Status Solidi A* **2019**, *216*, 1900047. [CrossRef]
Xu, B.; Chen, J.; Ding, Z.; Hu, J.; Zhang, Y.; Li, H.; Wang, H. The progress and outlook of multivalent-ion-based electrochromism. *Small Sci.* **2023**, *3*, 2300025. [CrossRef]

64. Hu, Y.Q.; Mao, X.; Yang, H.; Wang, M.; Qin, G.Q.; Long, G.L. Demonstration of Yb^{3+}-doped and Er^{3+}/Yb^{3+}-codoped on-c microsphere lasers. *Opt. Express.* **2021**, *29*, 25663. [CrossRef]
65. Kahle, H.; Bek, R.; Heldmaier, M.; Schwarzbäck, T.; Jetter, M.; Michler, P. High optical output power in the UVA range frequency-doubled, strain-compensated AlGaInP-VECSEL. *Appl. Phys. Express.* **2014**, *7*, 092705. [CrossRef]
66. Pretel, H.; Oliveira, J.; Lizarelli, R.; Ramalho, L. Evaluation of dental pulp repair using low level laser therapy (688 nm 785 nm) morphologic study in capuchin monkeys. *Laser Phys. Lett.* **2009**, *6*, 149–158. [CrossRef]
67. Imamura, T.; Tatehara, S.; Takebe, Y.; Tokuyama, R.; Ohshima, T.; Maeda, N.; Satomura, K. Antibacterial and antifungal effec 405 nm monochromatic laser on endodontopathogenic microorganisms. *Int. J. Photoenergy* **2014**, *2014*, 387215. [CrossRef]
68. Hilfiker, J.N.; Singh, N.; Tiwald, T.; Convey, D.; Smith, S.M.; Baker, J.H.; Tompkins, H.G. Survey of methods to characterize absorbing films with Spectroscopic Ellipsometry. *Thin Solid Film.* **2008**, *516*, 7979–7989. [CrossRef]
69. Yuan, M.; Yuan, L.; Hu, Z.; Yu, Z.; Li, H.; Barea, E.M.; Bisquert, J.; Meng, X. In Situ Spectroscopic ellipsometry for thermochro $CsPbI_3$ phase evolution portfolio. *J. Phys. Chem. C* **2020**, *124*, 8008–8014. [CrossRef]
70. Yuan, L.; Yuan, M.; Xu, H.; Hou, C.; Meng, X. Moisture-stimulated reversible thermochromic $CsPbI_{3-x}Br_x$ films: In- spectroscopic-resolved structure and optical properties. *Appl. Surf. Sci.* **2022**, *573*, 151484. [CrossRef]
71. Chen, C.; Wu, D.; Yuan, M.; Yu, C.; Zhang, J.; Li, C.; Duan, Y. Spectroscopic ellipsometry study of $CsPbBr_3$ perovskite thin fi prepared by vacuum evaporation. *J. Phys. D Appl. Phys.* **2021**, *54*, 224002. [CrossRef]
72. Zeller, H.R.; Beyeler, H.U. Electrochromism and local order in amorphous WO_3. *Appl. Phys.* **1977**, *13*, 231–237. [CrossRef]
73. Zimmer, A.; Gilliot, M.; Tresse, M.; Broch, L.; Tillous, K.E.; Boulanger, C.; Stein, N.; Horwat, D. Coloration mechanism electrochromic Na_xWO_3 thin films. *Opt. Lett.* **2019**, *44*, 1104. [CrossRef] [PubMed]
74. Logothetidis, S. Surface-roughness and grain-boundary effects on the optical properties of low-pressure chemical-vapor-deposi silicon thin films by spectroscopic ellipsometry. *J. Appl. Phys.* **1989**, *65*, 2416–2426. [CrossRef]
75. Valyukh, I.; Green, S.; Arwin, H.; Niklasson, G.A.; Wäckelgård, E.; Granqvist, C.G. Spectroscopic ellipsometry characterizatio electrochromic tungsten oxide and nickel oxide thin films made by sputter deposition. *Sol. Energ. Mat. Sol. C* **2010**, *94*, 724– [CrossRef]
76. Fares, H.; Jlassi, I.; Elhouichet, H.; Férid, M. Investigations of thermal, structural, and optical properties of tellurite glass w WO_3 adding. *J. Non-Cryst. Solids* **2014**, *396–397*, 1–7. [CrossRef]
77. Egbo, K.O.; Shil, S.K.; Kwok, C.G.; Wang, Y.; Liu, C.P.; Yu, K.M. Band alignment of wide bandgap NiO/MoO_3 and NiO/WO_3 heterojunctions studied by high-resolution X-ray photoelectron spectroscopy. *J. Alloy Compd.* **2021**, *876*, 160136. [CrossRef]
78. Tompkins, H.G.; Irene, E.A. (Eds.) *Handbook of Ellipsometry*; William Andrew Pub.: Norwich, NY, USA; Springer: Heidelb Germany, 2005.

micromachines

MDPI

cle

·oadband Near-Infrared Emission from Bi/Cr Co-Doped luminosilicate Glasses

wen Song and Min Zhang *

School of Physics and Astronomy, China West Normal University, Nanchong 637002, China; songshiwen97@163.com
* Correspondence: zmzmi1987@cwnu.edu.cn

Abstract: Bismuth-doped aluminosilicate glass has garnered significant attention due to its unique ultra-wide luminescence properties in the near-infrared (NIR) band. Enhancing the NIR luminescence of Bi-doped glass remains challenging. To achieve Bi-doped glass with more excellent luminescent properties, a series of Bi/Cr co-doped glasses were prepared, and the optical and structural properties of the samples were observed. The results indicate that low-concentration Cr doping broadens the luminescence range of Bi/Cr co-doped glass samples. The luminescence peak of Bi in the samples is at 1230 nm, while the peak of Cr is around 804 nm. The addition of an appropriate amount of Bi_2O_3 can enhance the NIR luminescence of Bi and Cr in the sample, realizing the energy conversion between Bi and Cr. Bi/Cr co-doped is a novel approach for achieving broadband NIR luminescence in glass materials.

Keywords: bismuth; aluminosilicate glasses; chromium; broadband NIR emission

1. Introduction

With the rapid arrival of the fifth-generation communication technology (5G), the development of communications has placed higher demands on the information transmission capabilities of optical communication systems. The near-infrared (NIR) band, spanning from 1000 to 1600 nm, represents a crucial window for optical communications [1–3]. However, traditional optical amplifiers struggle to cover this entire spectrum. Traditional optical fiber amplifiers typically use rare-earth-doped materials, which generally exhibit narrowband emission [4,5]. For instance, the widely utilized erbium-doped fiber amplifier in the traditional sense can only cover the 1530 to 1600 nm range, owing to its special 4f-4f forbidden transition [6]. Consequently, new optical fiber materials that can cover the 1000 to 1600 nm range are urgently needed.

In recent years, bismuth-doped glasses have been widely applied as gain materials in tunable fiber lasers operating in the NIR spectral region due to their long fluorescence lifetime and ultra-broadband NIR emission characteristics [7–9]. Additionally, Bi-doped glass fibers have achieved laser output in the range of 1000–1600 nm [10–12]. Consequently, Bi-doped glasses are considered promising candidates for next-generation broadband fiber amplifiers and lasers to address the capacity demands of modern communication systems. Despite significant advancements, few Bi-doped glass materials achieve both high luminescence intensity and broad spectral range in the NIR region. Recently, various methods have been tried in the search for efficient luminescence of Bi-doped materials. The topochemical reduction of bismuth using reducing agents such as AlN, Si_3N_4, and SiC in Bi-doped germanate glass has successfully expanded Bi luminescence across the entire NIR region [13–15]. However, achieving a delicate balance between the reducing and glass melting processes is challenging. The reduction of Bi^{3+} ions to low-valent Bi luminescent centers can also be achieved by high-energy irradiation through light-activated redox, although this often results in the creation of defects that can complicate experimental procedures [13]. Therefore, it is necessary to find new methods to enhance the NIR luminescence of Bi ions and broaden the luminescence range.

The transition metal ion Cr^{3+} is a highly effective NIR emission activator in luminescent materials [16–18]. The $3d^3$ transitions of Cr^{3+} ions are highly sensitive to changes in the crystal field, enabling wavelength adjustment [19]. Moreover, in a weak crystal field environment, Cr^{3+} ions undergo the $^4T_2 \rightarrow {}^4A_2$ transition, covering the NIR region of 650~1400 nm [20]. To date, Bi NIR luminescence properties have been extensively studied in phosphate [21,22], silicate [23,24], borate [12] and germanate [25,26] glass systems. Silicate glass exhibits more stable chemical and mechanical properties compared to other glass matrices. In addition, silicate glass provides a weak crystalline environment, which can make Cr^{3+} ions produce NIR emission. A large number of experiments have demonstrated that the NIR luminescence of Bi can be optimized by adjusting the Al content, and that CaO effectively lowers the glass melting point. Therefore, this study combines these advantages to prepare optimized samples.

In this study, Bi/Cr co-doped aluminosilicate glass was successfully prepared by the high-temperature melting method. At low Cr_2O_3 doping concentrations, Bi/Cr luminescence centers coexist, and the NIR luminescence intensity can be tuned by varying the Bi doping concentration. The NIR luminescence properties and energy transfer mechanism of Bi/Cr co-doped glass were systematically investigated.

2. Experimental Procedures

2.1. Materials Preparation

Glass samples were synthesized by conventional melt-quenching technique with compositions of $(60-x)SiO_2$-$10Al_2O_3$-30CaO-$0.5Bi_2O_{3-x}Cr_2O_3$ (in mol%, x = 0, 0.02, 0.04, 0.06, 0.08), $(60-y)SiO_2$-$10Al_2O_3$-30CaO-yBi_2O_3-$0.1Cr_2O_3$ (in mol%, y = 0, 0.5, 1) and $(60-z)SiO_2$-$10Al_2O_3$-30CaO-zBi_2O_3-$0.005Cr_2O_3$ (in mol%, z = 0, 0.5, 1, 1.5, 2). Commercial SiO_2, Al_2O_3, CaO, Bi_2O_3, and Cr_2O_3 were employed as starting materials of high chemical purity (99.99%). The powders were weighted accurately and ground thoroughly in an agate mortar. The well-ground stoichiometric chemicals were put into an alumina crucible and melted at 1600 °C in air for 1.5 h by a high temperature lifting furnace. The melt was poured onto a preheated (~200 °C) stainless-steel plate and pressed by another plate to improve the cooling rate and form solid glasses. All glasses were annealed at 750 °C for 5 h to remove thermal strains. Samples were polished for further optical characterization, with a size of $10 \times 10 \times 1$ mm^3.

2.2. Characterization

The absorption spectra for all samples were measured by an ultraviolet-visible-near-infrared spectrophotometer (UV–VIS–NIR, UV–3600i Plus, Shimadzu, Tokyo, Japan) in the range of 300–1200 nm. The fluorescence spectra were recorded by a steady-state fluorescence spectroscopy spectrometer (FLS1000, Edinburgh Instruments, Edinburgh, UK). The densities of glass were tested by the Archimedes principle with distilled water as the medium (MDJ–300T, Beijing, China). The Fourier transform infrared (FTIR) spectra were performed on a FTIR spectrometer (Nicolet 6700, Thermo Fisher Scientific, Waltham, MA, USA) by dispersing the sample powders uniformly on KBr particles within the range 400–2000 cm^{-1}. X-ray diffraction spectroscopy (XRD) of the samples were performed by an X-ray diffractometer (TD–3500, Dandong Tongda, Dandong, China).

3. Results and Discussion

Figure 1a shows the transmittance spectra for $0.5Bi_2O_3-xCr_2O_3$ (x = 0, 0.02, 0.04, 0.06, 0.08) glass samples. Three absorption peaks at ~362, ~440 and 640 nm are noticed in the visible range. According to previous studies, the absorption bands at 362 nm are assigned to the $^4A_2 \rightarrow {}^2A_1$ transition of Cr^{6+}, while the peaks at 440 and 640 nm are contributed by the $^4A_2 \rightarrow {}^4T_1$ and $^4A_2 \rightarrow {}^4T_2$ transitions of Cr^{3+} [27–29]. With the increasing concentration of Cr_2O_3, the absorbance at 362 nm, 440 nm, and 640 nm gradually increases. Notably, no significant Bi-related characteristic absorption peaks are detected in Figure 1a, suggesting a limited presence of Bi NIR emission centers in the glass samples.

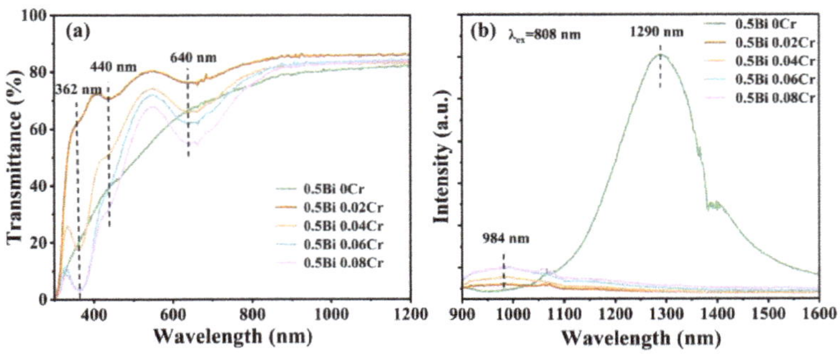

Figure 1. (**a**) Transmittance spectra of $0.5Bi_2O_{3-x}Cr_2O_3$ (x = 0, 0.02, 0.04, 0.06, 0.08) glass samples. (**b**) The NIR emission spectra of $0.5Bi_2O_{3-x}Cr_2O_3$ glasses under the excitation of 808 nm.

The NIR emission spectra for $0.5Bi_2O_3{-}xCr_2O_3$ (x = 0, 0.02, 0.04, 0.06, 0.08) glass samples under 808 nm excitation at room temperature are portrayed in Figure 1b. For the Bi-doped glass sample, the emission peak located at 1290 nm might be assigned to the $^2D_{3/2} \rightarrow {}^4S_{3/2}$ transition of Bi^0 [30,31]. However, with the addition of chromium, the Bi-related NIR luminescence peak vanishes, and a new peak appears at 984 nm, which increases in intensity with higher chromium content. According to previous reports, the luminescence of 984 nm is from the $^4T_2 \rightarrow {}^4A_2$ transition of Cr^{3+} [32]. Under the 808 nm, part of the energy is absorbed by a number of Cr^{3+} ions and consumed through non-radiative transition processes, significantly reducing the energy available for Bi ions, leading to the quenching of the 1290 nm emission peak. Therefore, the emission peak at 1290 nm of Bi^0 that was quenched may be due to the high Cr_2O_3 concentration.

To further confirm the above result, the concentration of Cr_2O_3 was increased to 0.1 mol, and a series of xBi_2O_3-$0.1Cr_2O_3$ (x = 0, 0.5, 1) samples were prepared. The NIR luminescence spectra of glass samples under 466 nm excitation are shown in Figure 2. The results confirm that Bi luminescence is quenched at a Cr_2O_3 concentration of 0.1 mol, consistent with the observations in Figure 1b. This behavior aligns with previous studies, suggesting a strong dependence on Cr_2O_3 doping concentration.

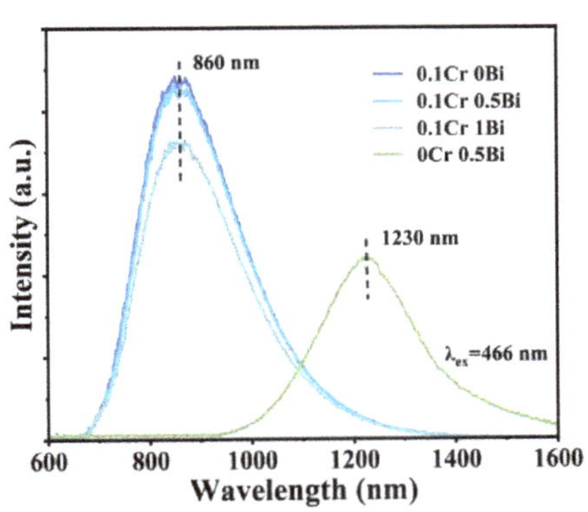

Figure 2. The NIR emission spectra of xBi_2O_3-$0.1Cr_2O_3$ (x = 0, 0.5, 1) and $0.5Bi_2O_3$-$0Cr_2O_3$ glasses under the excitation of 466 nm.

Considering that too high a concentration of Cr ions can quench the luminescence of Bi, a low concentration of Cr was selected for doping. In structurally stable glass samples, the Bi ion concentration is also one of the important factors affecting the luminescence performance of the glass samples. In order to achieve co-doping of Bi and Cr for enhanced joint luminescence and to further explore the optimal glass component, the doping concentration of Cr_2O_3 was maintained at 0.005 mol, while varying the doping concentrations of Bi_2O_3 were used to study the structural properties and luminescence behavior of the glasses.

The variation of density and molar volume with the content of is shown in Figure 3. The measured density of samples increased monotonically from 2.67 g/cm^3 to 2.93 g/cm^3 as Bi_2O_3 replaced SiO_2. The increasing density of samples may be due to the higher molecular weight of Bi_2O_3 than of SiO_2 [33]. Molar volume (V_m), as an important parameter to describe the compactness of glass, can be obtained by the following formula:

$$V_m = \sum x_i \cdot M_i / \rho \tag{1}$$

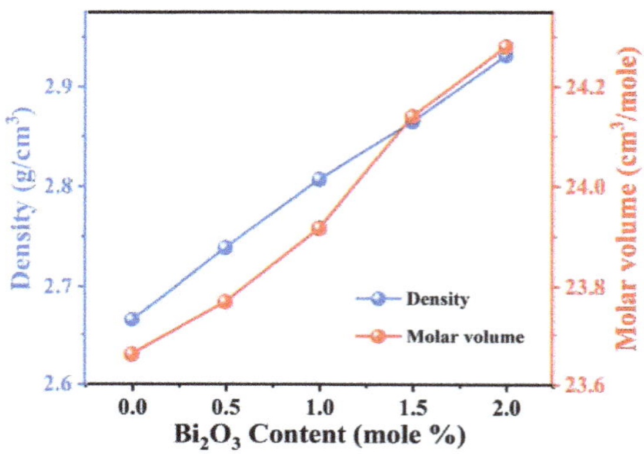

Figure 3. Variation of density and molar volume with the content of Bi_2O_3.

Here, x_i is the molar fraction of oxide composition, M_i is the molecular mass of oxide of nominal compositions of sample, ρ is the density of glass.

The molar volume of the samples can be calculated by Equation (1) to show a monotonically decreasing trend, which is exactly the same as the density trend in the samples. Normally, density and molar volume show opposite trends, but the result of the present work is anomalous. The increase in molar volume with increasing Bi_2O_3 content is most likely attributed to the replacement of a smaller ionic radius Si by a larger ionic radius Bi, which points to the change in the structure of glass [34,35].

FTIR spectroscopy can be applied to observe chemical bonds and unit structure information in glass. Figure 4a is the FTIR spectra of the xBi_2O_3-$0.005Cr_2O_3$ glasses. Three obvious absorption peaks can be noticed in the range of 400–2000 cm^{-1} in Figure 4a. The Si-O asymmetrical stretching vibration peak at 966 cm^{-1} is dominant, and the stability of the relative intensity of the absorption peak indicates that the number of non-bridging oxygens (NBOs) remains basically unchanged, which also implies that the silicon network has not depolymerized. The stretching vibrations of Al-O (~728 cm^{-1}) in the isolated AlO_4 structural unit increase as the bismuth concentration increases. The absorption peak at 450 cm^{-1} gradually shifted from 450 cm^{-1} to 440 cm^{-1}, which may be the formation of new Si-O-Al links in the silicon network structure, as shown in Figure 4b.

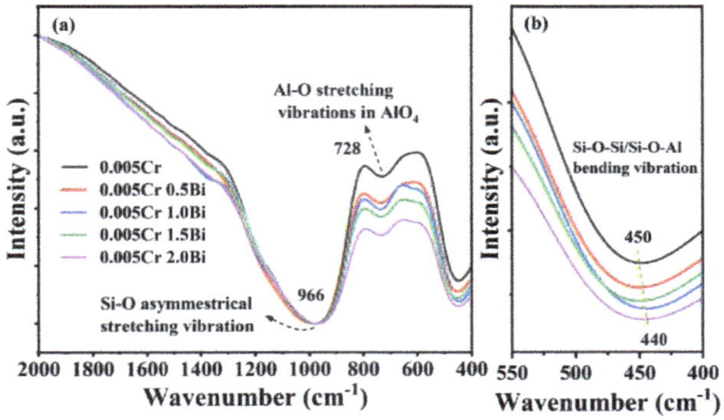

Figure 4. (**a**) FTIR spectra of xBi_2O_3-$0.005Cr_2O_3$ glasses (x = 0, 0.5, 1, 1.5, 2). (**b**) Amplified FTIR in the range of 400–550 cm^{-1} showing the variation of the Si-O-Al vibration peaks with the concentration of Bi_2O_3.

Figure 5a is the transmittance spectrum of the sample of 0.005 Cr_2O_3. In the Cr-doped sample, the absorption peak intensity of Cr ions decreases but does not disappear, as the Cr_2O_3 doping concentration is reduced. In Figure 5b, with the increase in the Bi_2O_3 content in the samples, the Bi-related absorption peak shows an enhancing trend, and there is also a weak absorption peak related to Cr ions. According to previous reports, once Bi is introduced into aluminosilicate glass at high temperature, Bi ions with different valences are generated and reach valence equilibrium in the glass structure [36]. This means that different Bi related absorption peaks may be observed in the glass samples. The weak peak at 725 nm may be attributed to the $^3P_0 \rightarrow {}^3P_2$ transition of Bi^+. The absorption peak at 470 nm is mainly caused by the $^4S_{3/2} \rightarrow {}^2P_{1/2}$ transition of Bi^0 [37]. With increasing Bi_2O_3 content, the absorbance at 470 nm increases gradually, which is also a symbol of the increase in the number of Bi luminescence centers. A careful observation of the peak at 470 nm reveals that the two sides of the absorption peak are asymmetric, which may be related to the $^3P_0 \rightarrow {}^3P_2$ transition of Bi^+. Meanwhile, the weak absorption peak of the $^4A_2 \rightarrow {}^4T_2$ transition of Cr^{3+} at 640 nm can also be observed, which implies that the luminescent centers of both Bi and Cr may exist simultaneously in the Bi/Cr co-doped glasses. The inset is the photo of xBi_2O_3-$0.005Cr_2O_3$ (x = 0, 0.5, 1, 1.5, 2) glass samples. The color of samples changes from light green to reddish brown with the addition of Bi_2O_3, which indicates an increase in the number of Bi luminescence centers.

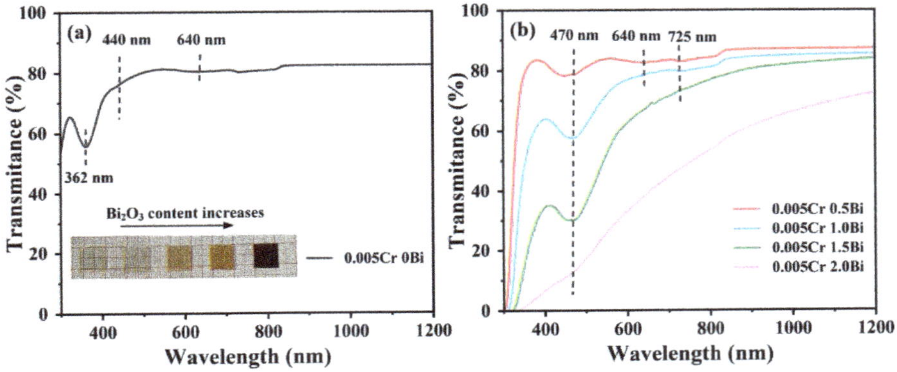

Figure 5. (**a**) Transmittance spectrum of $0.005Cr_2O_3$ glass and inset are images of xBi_2O_3-$0.005Cr_2O_3$ glasses (x = 0, 0.5, 1, 1.5, 2). (**b**) Transmittance spectra of xBi_2O_3-$0.005Cr_2O_3$ glasses (x = 0.5, 1, 1.5, 2).

Figure 6a shows the emission spectra of the glass samples under 466 nm excitation. Two emission peaks in the NIR range are observed at 804 nm and 1230 nm. The peak at 1230 nm is composed of two luminescent centers at 1100 nm and 1300 nm. The emissions at 1100 nm and 1300 nm correspond to the $^3P_1 \rightarrow {}^3P_0$ transition of Bi^+ and the $^2D_{3/2} \rightarrow {}^4S_{3/2}$ transition of Bi^0, respectively [31,38]. Under the excitation at 466 nm, Bi^+ and Bi^0 can be excited simultaneously, and the NIR emission leaps of Bi^+ and Bi^0 overlap. The emission peak at 1230 nm may be the result of the overlapping transitions of Bi^+ and Bi^0, which also indicates that there are two Bi-related NIR luminescence centers in the glass sample at the same time. The asymmetric luminescence peak at 1230 nm also confirms the result. By comparing Figure 2, it can be found that the emission peak around 800 nm is not observed in the sample doped with 0.5 mol Bi alone. Therefore, in the Bi/Cr co-doped samples, the emission peak near 804 nm may be the $^4T_2 \rightarrow {}^4A_2$ transition of Cr^{3+}, and the peak position blue-shifts with the change in Bi doping concentration. In other words, Bi^0, Bi^+, Cr^{3+} are the main luminescence centers in the glass samples, and energy exchange exists between them. The relationship between Bi doping concentration and emission peak intensity is shown in Figure 6b. With the increase in Bi doping concentration, the emission peak intensity at 1230 nm and 804 nm shows a trend of increasing first and then decreasing. The luminescence spectra of Bi-doped, Cr-doped, and Bi/Cr co-doped glass samples under 466 nm laser excitation are shown in Figure 7a. The addition of a small amount of Cr results in an increasing trend in Bi luminescence intensity. Meanwhile, compared with Bi-doped samples, the luminescence range of samples co-doped with appropriate amounts of Bi_2O_3 and a small amount of Cr_2O_3 is expanded, which involves the energy transfer between the ions mentioned above. Figure 7b is a simplified diagram of Bi^+, Bi^0, and Cr^{3+} derivatives. Bi^+, Bi^0, and Cr^{3+} are excited from the ground state to $^1D_2$, $^2P_{1/2}$, and $^4T_1$ excitation levels by 466 nm light, and then Bi^+ relaxes non-radiatively to the $^3P_1$ level. During the Bi^+ ions relaxation, Bi^+ transfers energy to Cr^{3+}. When Cr^{3+} ions transition from $^4A_2$ level, part of the energy is consumed by non-radiative relaxation, and the rest of the energy can be emitted in the form of transition to form an emission peak at 804 nm, as shown in Figure 6a. The process of Bi^+ transferring energy to Bi^0 further forms the luminescence peak of 1230 nm [39]. The mechanism of energy transfer from Bi^+ to Bi^0 and Cr^{3+} can be described as $^3P_1$ (Bi^+) + $^4S_{3/2}$ (Bi^0) $\rightarrow$ $^3P_0$ (Bi^+) + $^2D_{3/2}$ (Bi^0) and $^1D_2$ (Bi^+) + $^4T_2$ (Cr^{3+}) $\rightarrow$ $^3P_1$ (Bi+) + $^4T_1$ (Cr^{3+}), which can help the luminescence at 804 nm and 1230 nm. The luminescence peaks at these two positions can be attributed to the energy transfer induced by the co-doping of Bi and Cr, which broadens the luminescence bandwidth through this doping mechanism.

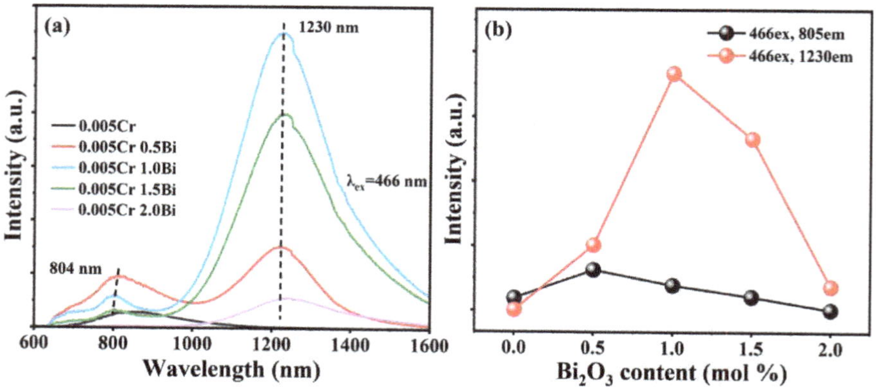

Figure 6. (**a**) Luminescence spectra of xBi_2O_3-$0.005Cr_2O_3$ glass samples excited by 466 nm (x = 0, 0.5, 1, 1.5, 2). (**b**) Dependence of luminescence intensity on the content of Bi_2O_3.

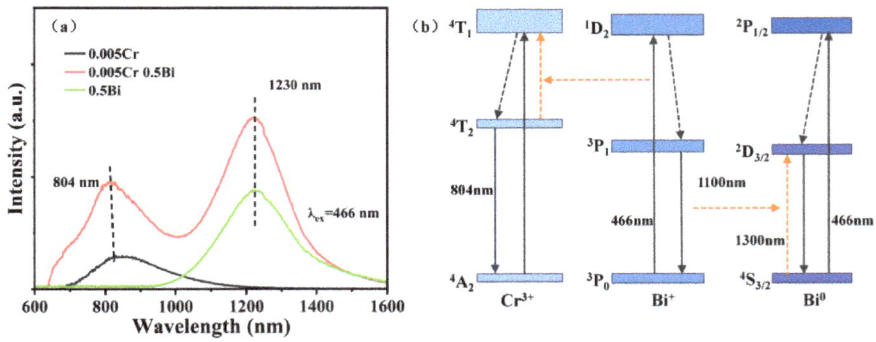

Figure 7. (**a**) Luminescence spectra of 0.005Cr, 0.005Cr-0.5Bi and 0.5Bi glass samples excited by 466 nm. (**b**) Simplified energy-level diagram of Cr^{3+}, Bi^{+}, and Bi^{0} and the possible transfer process between Cr^{3+}, Bi^{+}, and Bi^{0}.

4. Conclusions

Bi/Cr co-doped aluminosilicate glasses were successfully prepared using a high-temperature melting process. FTIR results show that the glass structure remains stable as the Bi_2O_3 content increases. Emission spectra results show that the addition of low concentration Cr_2O_3 to Bi-doped glasses can cause the luminescence centers of Bi and Cr to exist simultaneously in the samples, which means that the luminescence range of samples is broadened by the addition of a small amount of Cr_2O_3. With the increase in Bi_2O_3 doping content, the luminescence intensity of bismuth ions can increase first. When the Bi_2O_3 doping concentration is higher than 1 mol%, the luminescence of Bi at 1230 nm is quenched. The luminescence peak of Cr at 804 nm has a similar trend to the luminescence peak of Bi at 1230 nm. The experimental results indicate that an appropriate increase in Bi concentration can enhance the Bi NIR emission, and there is energy exchange between Bi and Cr.

Author Contributions: Conceptualization, M.Z.; methodology, M.Z.; software, S.S.; validation, M.Z.; formal analysis, S.S.; investigation, S.S.; resources, M.Z.; data curation, S.S.; writing—original draft preparation, S.S.; writing—review and editing, M.Z.; visualization, S.S.; supervision, M.Z.; project administration, M.Z.; funding acquisition, M.Z. All authors have read and agreed to the published version of the manuscript.

Funding: This research was funded by Sichuan Science and Technology Program grant number 2023YFG0086, Nanchong Science and Technology Project grant number 21GXJS0014.

Data Availability Statement: The original contributions presented in the study are included in the article, further inquiries can be directed to the corresponding author.

Conflicts of Interest: There are no conflicts to declare.

ferences

Hilbert, M.; Lopez, P. The world's technological capacity to store, communicate, and compute information. *Science* **2011**, *332*, 60–65. [CrossRef] [PubMed]

Diddams, S.A.; Hollberg, L.; Mbele, V. Molecular fingerprinting with the resolved modes of a femtosecond laser frequency comb. *Nature* **2007**, *445*, 627–630. [CrossRef] [PubMed]

Jetschke, S.; Unger, S.; Schwuchow, A.; Leich, M.; Kirchhof, J. Efficient Yb laser fibers with low photodarkening by optimization of the core composition. *Opt. Express* **2008**, *16*, 15540–15545. [CrossRef]

Yan, D.; Zhu, J.; Wu, H.; Yang, Z.; Qiu, J.; Song, Z.; Yu, X.; Yang, Y.; Zhou, D.; Yin, Z.; et al. Energy transfer and photoluminescence modification in Yb–Er–Tm triply doped $Y_2Ti_2O_7$ upconversion inverse opal. *J. Mater. Chem.* **2012**, *22*, 18558–18563. [CrossRef]

Tanabe, S. Rare-earth-doped glasses for fiber amplifiers. *C. R. Chim.* **2002**, *5*, 815–824. [CrossRef]

Xu, Y.; Qi, J.; Ren, J.; Chen, G.; Huang, F.; Li, Y.; Lu, S.; Dai, S. Luminescence and energy transfer in Er^{3+}/Nd^{3+} ion-codoped Ge–In–S–CsBr chalcohalide glasses. *Mater. Res. Bull.* **2013**, *48*, 4733–4737. [CrossRef]

7. Yasushi, F.; Masahiro, N. Infrared Luminescence from Bismuth-doped Silica Glass. *Jpn. Soc. Appl. Phys.* **2001**, *40*, 279– [CrossRef]
8. Qiu, J. Bi-Doped Glass for Photonic Devices. *Int. J. Appl. Glass Sci.* **2015**, *6*, 275–286. [CrossRef]
9. Bufetov, I.A.; Melkumov, M.A.; Firstov, S.V.; Riumkin, K.E.; Shubin, A.V.; Khopin, V.F.; Guryanov, A.N.; Dianov, E.M. Bi-Do Optical Fibers and Fiber Lasers. *IEEE J. Sel. Top. Quantum Electron.* **2014**, *20*, 111–125. [CrossRef]
10. Dianov, E.M.; Dvoyrin, V.V.; Mashinsky, V.M.; Umnikov, A.A.; Yashkov, M.V.; Gur'yanov, A.N. CW bismuth fibre laser. *Quan Electron.* **2005**, *35*, 1083–1084. [CrossRef]
11. Li, X.; Cao, J.; Peng, M. The origin of the heterogeneous distribution of bismuth in aluminosilicate laser glasses. *J. Am. Ceram.* **2018**, *101*, 2921–2929. [CrossRef]
12. Meng, X.G.; Qiu, J.R.; Peng, M.Y.; Chen, D.P.; Zhao, Q.Z.; Jiang, X.W.; Zhu, C.S. Infrared broadband emission of bismuth-do barium-aluminum-borate glasses. *Opt. Express* **2005**, *13*, 1635–1642. [CrossRef]
13. Cao, J.; Peng, J.; Wang, L.; Luo, H.; Wang, X.; Xiong, P.; Wang, Y.; Peng, M. Broadband NIR emission from multiple Bi center nitridated borogermanate glassesviatailoring local glass structure. *J. Mater. Chem. C* **2019**, *7*, 2076–2084. [CrossRef]
14. Cao, J.; Xu, S.; Zhang, Q.; Yang, Z.; Peng, M. Ultrabroad Photoemission from an Amorphous Solid by Topochemical Reduct *Adv. Opt. Mater.* **2018**, *6*, 1801059. [CrossRef]
15. Cao, J.; Li, L.; Wang, L.; Li, X.; Zhang, Z.; Xu, S.; Peng, M. Creating and stabilizing Bi NIR-emitting centers in low Bi con materials by topo-chemical reduction and tailoring of the local glass structure. *J. Mater. Chem. C* **2018**, *6*, 5384–5390. [CrossR
16. Wei, Y.; Dang, P.; Dai, Z.; Li, G.; Lin, J. Advances in Near-Infrared Luminescent Materials without Cr^{3+}: Crystal Structure Des Luminescence Properties, and Applications. *Chem. Mater.* **2021**, *33*, 5496–5526. [CrossRef]
17. Zhong, J.; Zhuo, Y.; Du, F.; Zhang, H.; Zhao, W.; Brgoch, J. Efficient and Tunable Luminescence in $Ga_{2-x}In_xO_3$: Cr^{3+} Near-Infrared Imaging. *ACS Appl. Mater. Interfaces* **2021**, *13*, 31835–31842. [CrossRef]
18. Zhang, L.; Dong, L.; Xu, Y.; Yin, S.; You, H. Effect of R^+ (R = Li, Na and K) codoping on the luminescence enhancemen broadband NIR $BaZrGe_3O_9$: Cr^{3+} phosphors for NIR LED. *J. Lumin.* **2021**, *236*, 118084. [CrossRef]
19. Mao, N.; Liu, S.; Song, Z.; Yu, Y.; Liu, Q. A broadband near-infrared phosphor $Ca_3Y_2Ge_3O_{12}$: Cr^{3+} with garnet structure. *J. Al Compd.* **2021**, *863*, 158699. [CrossRef]
20. Liu, C.; Xia, Z.; Chen, M.; Molokeev, M.S.; Liu, Q. Near-infrared luminescence and color tunable chromophores based Cr^{3+}-doped mullite-type $Bi_2(Ga,Al)_4O_9$ solid solutions. *Inorg. Chem.* **2015**, *54*, 1876–1882. [CrossRef]
21. Denker, B.; Galagan, B.; Osiko, V.; Sverchkov, S.; Dianov, E. Luminescent properties of Bi-doped boro-alumino-phosphate glas *Appl. Phys. B* **2006**, *87*, 135–137. [CrossRef]
22. Qiu, Y.Q.; Kang, J.; Li, C.X.; Dong, X.Y.; Zhao, C.L. Broadband near-infrared luminescence in bismuth borate glasses. *Laser P* **2009**, *20*, 487–492. [CrossRef]
23. Jiang, Z.; Dai, N.; Yang, L.; Peng, J.; Li, H.; Li, J.; Liu, W. Effects of Al_2O_3 composition on the near-infrared emission in Bi-do and Yb–Bi-codoped silicate glasses for broadband optical amplification. *J. Non-Cryst. Solids* **2014**, *383*, 196–199. [CrossRef]
24. Wan, R.; Song, Z.; Li, Y.; Liu, Q.; Zhou, Y.; Qiu, J.; Yang, Z.; Yin, Z.; Wang, Q.; Zhou, D. Influence of alkali metal ions on ther stability of Bi-activated NIR-emitting alkali-aluminoborosilicate glasses. *Chin. Opt. Lett.* **2014**, *12*, 111601.
25. Zhang, N.; Sharafudeen, K.N.; Dong, G.; Peng, M.; Qiu, J. Mixed Network Effect of Broadband Near-Infrared Emissior Bi-Doped B_2O_3-GeO_2 Glasses. *J. Am. Ceram. Soc.* **2012**, *95*, 3842–3846. [CrossRef]
26. Xu, Z.-S.; Xu, C.; Guo, Q.-B.; Liu, X.-F.; Qiu, J.-R. Effect of TeO_2 on Near-Infrared Emission from Bismuth Centers in Borogerman Glasses. *J. Am. Ceram. Soc.* **2016**, *99*, 760–764. [CrossRef]
27. Xiao, Z.; Zhang, B.; Li, J.; Liang, H.; Luo, Z.; Lu, A.; Zhang, L.; You, W.; Han, L.; Xia, L. Ultra-broadband NIR luminescence chromium doped magnesium aluminosilicate glass as future light sources for spectroscopy applications. *J. Non-Cryst. Solids* **2** *578*, 121362. [CrossRef]
28. Yusub, S.; Rao, D.K. The role of chromium ions on dielectric and spectroscopic properties of LiO-PbO-BO-PO glasses. *J. Non-Cr Solids* **2014**, *398*, 1–9. [CrossRef]
29. Othman, A.M.; Abd El-Fattah, Z.M.; Farouk, M.; Moneep, A.M.; Hassan, M.A. Optical spectroscopy of chromium dop bismuth-lithium borate glasses. *J. Non-Cryst. Solids* **2021**, *558*, 120665. [CrossRef]
30. Minh Hau, T.; Wang, R.; Yu, X.; Zhou, D.; Song, Z.; Yang, Z.; He, X.; Qiu, J. Near-infrared broadband luminescence and ene transfer in Bi–Tm–Er co-doped lanthanum aluminosilicate glasses. *J. Phys. Chem. Solids* **2012**, *73*, 1182–1186. [CrossRef]
31. E, F.; Zhang, D.; Ye, R.; Hua, Y.; Xu, S.; Huang, F. Structural engineering of germanosilicate glass network for enhanced Bi: N luminescence. *Opt. Mater.* **2019**, *95*, 109222. [CrossRef]
32. Chen, J.; Yu, Y.; Luo, H.; Lin, L.; Zhou, S. Simultaneous control of the chemical state and local environment of Cr dopant in gl for extended emission. *J. Am. Ceram. Soc.* **2017**, *101*, 159–166. [CrossRef]
33. Mohan, S.; Thind, K.S.; Sharma, G.; Gerward, L. Spectroscopic investigations of Nd^{3+} doped flouro- and chloro-borate glas *Spectrochim. Acta. A Mol. Biomol. Spectrosc.* **2008**, *70*, 1173–1179. [CrossRef]
34. Saad, N.; Haouari, M.; Bulou, A.; Hadi Kassiba, A.; Ben Ouada, H. Structural and optical properties of Cr^{3+} embedded i P_2O_5–B_2O_3–ZnO–BaF_2–AlF_3 fluoroborophosphate glasses. *Mater. Chem. Phys.* **2018**, *212*, 461–470. [CrossRef]
35. Ahmad, F. Study the effect of alkali/alkaline earth addition on the environment of borochromate glasses by means of spectrosco analysis. *J. Alloys Compd.* **2014**, *586*, 605–610. [CrossRef]

Li, X.; Cao, J.; Wang, L.; Peng, M. Predictable tendency of Bi NIR emission in Bi-doped magnesium aluminosilicate laser glasses. *J. Am. Ceram. Soc.* **2017**, *101*, 1159–1168. [CrossRef]
Cao, J.; Li, X.; Wang, L.; Zhang, Z.; Xu, S.; Peng, M. New strategy to enhance the broadband near-infrared emission of bismuth-doped laser glasses. *J. Am. Ceram. Soc.* **2018**, *101*, 2297–2304. [CrossRef]
Li, X.; Peng, M.; Cao, J.; Yang, Z.; Xu, S. Distribution and stabilization of bismuth NIR centers in Bi-doped aluminosilicate laser glasses by managing glass network structure. *J. Mater. Chem. C* **2018**, *6*, 7814–7821. [CrossRef]
Chen, F.; Wang, Y.; Chen, W.; Xiong, P.; Jiang, B.; Zhou, S.; Ma, Z.; Peng, M. Regulating the Bi NIR luminescence behaviours in fluorine and nitrogen co-doped germanate glasses. *Mater. Adv.* **2021**, *2*, 4743–4751. [CrossRef]

micromachines

MDPI

Article

Advancing Lithium-Ion Batteries' Electrochemical Performanc Ultrathin Alumina Coating on $Li(Ni_{0.8}Co_{0.1}Mn_{0.1})O_2$ Cathode Materials

Mehdi Ahangari, Fan Xia, Benedek Szalai, Meng Zhou * and Hongmei Luo *

Department of Chemical and Materials Engineering, New Mexico State University, Las Cruces, NM 88003, USA; mahani92@nmsu.edu (M.A.); fxia@nmsu.edu (F.X.); szalaib@nmsu.edu (B.S.)
* Correspondence: mzhou@nmsu.edu (M.Z.); hluo@nmsu.edu (H.L.)

Abstract: Ni-rich $Li(Ni_xCo_yMn_z)O_2$ ($x \geq 0.8$)-layered oxide materials are highly promising as cathode materials for high-energy-density lithium-ion batteries in electric and hybrid vehicles. However, their tendency to undergo side reactions with electrolytes and their structural instability during cyclic lithiation/delithiation impairs their electrochemical cycling performance, posing challenges for large-scale applications. This paper explores the application of an Al_2O_3 coating using an atomic layer deposition (ALD) system on Ni-enriched $Li(Ni_{0.8}Co_{0.1}Mn_{0.1})O_2$ (NCM811) cathode material. Characterization techniques, including X-ray diffraction, scanning electron microscopy, and transmission electron microscopy, were used to assess the impact of alumina coating on the morphology and crystal structure of NCM811. The results confirmed that an ultrathin Al_2O_3 coating was achieved without altering the microstructure and lattice structure of NCM811. The alumina-coated NCM811 exhibited improved cycling stability and capacity retention in the voltage range of 2.8–4.5 V at a 1 C rate. Specifically, the capacity retention of the modified NCM811 was 5%, 9.11%, and 11.28% higher than the pristine material at operating voltages of 4.3, 4.4, and 4.5 V, respectively. This enhanced performance is attributed to reduced electrode–electrolyte interaction, leading to fewer side reactions and improved structural stability. Thus, NCM811@Al_2O_3 with this coating process emerges as a highly attractive candidate for high-capacity lithium-ion battery cathode materials.

Keywords: Ni-rich cathode; surface coating; lithium-ion batteries; atomic layer deposition

1. Introduction

Rechargeable lithium-ion batteries (LIBs) are pivotal in the evolution of microelectronics and serve as primary power sources for portable electronic devices. Their superior energy density, both in terms of weight and volume compared to other rechargeable battery technologies, has made them ubiquitous, integral, and essential components of modern life [1–3]. The burgeoning demand for electric vehicles (EVs) and hybrid electric vehicles (HEVs) has positioned LIBs as a promising solution to meet the requirements for high energy and power density [4,5]. Among all types of cathode materials, $LiNi_xCo_yMn_zO_2$ (NCM, where x + y + z = 1)-layered oxide materials have attracted attention due to their low cost, high capacity, and long stability [6]. In particular, $Li(Ni_{0.8}Co_{0.1}Mn_{0.1})O_2$, as a crucial component of lithium-ion batteries, has garnered considerable attention due to its high reversible specific capacity (200 mAh g^{-1}) resulting from the double-redox reaction of Ni^{2+}/Ni^{4+} and its relatively favorable cost profile [7]. Following the commercial success of NCM with moderate nickel content, it has been demonstrated that producing Ni-rich NCM-layered oxides ($\geq$0.8) with higher nickel and lower cobalt content is an effective strategy to enhance the cathode's specific capacity and operating voltage while reducing costs [8,9]. This approach is particularly well-suited for practical applications in the EV sector. Nevertheless, the practical adoption of Ni-enriched NCM materials faces challenges such as rapid capacity decay and increased impedance upon cycling due to pronounced structural instability and an unstable electrode–electrolyte interface [10–12]. The main mechanisms for electrochemical degradation are as follows: (i) irreversible migration of Ni ions to Li sites due to the similar ionic radii of Ni^{2+} (0.69 Å) and Li^{+} (0.76 Å), leading to Ni/Li mixing [13]; (ii) the reaction of residual lithium on the surface with moisture in the air, forming passivating layers such as Li_2CO_3 and LiOH on the cathode surface [14]; (iii) the dissolution of transition metals, resulting in interfacial resistance and capacity deterioration; and (iv) irreversible detrimental phase transitions among three hexagonal structures (H1, H2, and H3) that manifest when Ni-rich cathode materials undergo repeated Li insertion/extraction, particularly during highly charged/discharged states, leading to microcrack propagation and pulverization [15,16]. The mentioned drawbacks hinder the commercialization of high-Ni-content cathode materials in LIBs.

Surface modification has emerged as a great strategy for enhancing the electrochemical performance of Ni-rich cathode materials and resolving the mentioned technical issues, as a continuous coating can effectively decrease the occurrence of parasitic surface reactions and stabilize both surface and bulk structure of the materials [17,18]. A suitable coating is characterized by properties such as conformality, low thickness, integrity, and continuity [19]. Metal oxides such as TiO_2 [20], ZrO_2 [21], Al_2O_3 [22], and MgO [23] are the most common materials used for their low cost and high protection properties. Among these, Al_2O_3 has been widely utilized as a coating for cathode active materials in LIBs, which can be applied using wet chemistry or atomic layer deposition (ALD) techniques [24]. The ALD method has earned a good reputation for producing ultrathin films with atomic-level control using sequential, self-limiting surface reactions [25]. It has been reported that ZrO_2 [26], TiO_2 [27], and MgF_2 [28] coatings applied to cathode active materials using ALD have enhanced the stability of the surface-modified samples.

Herein, an ultrathin Al_2O_3 coating (2 nm) using an ALD machine was adopted for the surface of NCM811 cathode active material for LIBs using trimethylaluminum (TMA) and H_2O in a process consisting of two half-reaction steps. The resultant Al_2O_3-coated NCM811 was assembled as cathode electrodes, demonstrating significantly improved stability compared to pristine electrodes upon prolonged cycling.

2. Materials and Methods

2.1. Materials Synthesis

NCM811 powder was produced by the solid-state method. All the reagents used in this research were purchased from Sigma-Aldrich company (St. Louis, MO, USA). A mixture of high-purity $NiSO_4.6H_2O$, $CoSO_4{\cdot}7H_2O$, and $MnSO_4{\cdot}H_2O$ was dissolved in distilled water (1 mol/L), maintaining a molar ratio of Ni:Co:Mn = 8:1:1. Then, a 2 mol/L NaOH solution was added to the transition metals solution under a N_2 atmosphere while mixing the two solutions. The pH was adjusted to 10.0–11.0 using $NH_3{\cdot}H_2O$, and the temperature was maintained at 50 °C for 24 h. The resultant powder was then filtered and washed using distilled water. Following this, the obtained powder ($Ni_{0.8}Co_{0.1}Mn_{0.1}(OH)_2$) was kept in a vacuum oven at 120 °C for 24 h. Finally, $Ni_{0.8}Co_{0.1}Mn_{0.1}(OH)_2$ and $LiOH.H_2O$ were mixed at a 1.00:1.05 molar ratio and ball-milled, followed by calcination of the mixture at 750 °C for 12 h under a pure oxygen atmosphere in a tube furnace.

Al_2O_3 on NCM811 was performed in an Angstrom-Dep ALD machine (Albuquerque, NM, USA), which was equipped with a rotary reactor. NCM811 powder served as the active material to be coated with Al_2O_3. Nine grams of NCM811 powder were loaded into the rotary reactor. N_2 gas was employed as the carrier and purging gas, with the temperature of TMA and water set at 90 °C, while the batch reactor was maintained at 120 °C. The two half-reaction steps consisted of injecting TMA and H_2O with 0.2 and 0.5 s purging, respectively, between each pulse, and purging of 30 s after the last pulse. Three ALD growth cycles were used to prepare the Al_2O_3-coated powders. However, the resulting film thickness was 2 nm, which is approximately 6–7 times thicker than expected from ALD. This suggests that chemical vapor deposition (CVD) might contribute significantly to the growth process. The deviation in thickness is likely due to the short pulse and purge times used during the process.

2.2. Material Characterizations

Powder X-ray diffraction (XRD) analysis was conducted using a Bruker AXS D8 (Billerica, MA, USA) Focus diffractometer equipped with a LynxEye position-sensitive detector (PSD), utilizing Cu Kα radiation (λ = 0.15406 nm) and a 0.2 mm slit (Empyrean XRD PANalytical, UK) on pristine and alumina-coated NCM811. The diffraction pattern was recorded in the 2θ range of 10–80° with a step size of 0.02° and a count time of 3 s per step. Morphological characteristics and particle size were examined using a field emission scanning electron microscope (Hitachi FESEM Model SU7000, Tokyo, Japan) equipped with an elemental energy dispersive spectroscopy (EDS) detector. The chemical composition of samples was determined by EDS. A transmission electron microscope (TEM) operated at 200 kV was employed to identify the thickness and continuity of the coating on the powder's surface.

2.3. Electrochemical Measurements

The as-prepared materials underwent electrochemical measurements in CR2032 coin cells at room temperature. Cathode electrodes were prepared using a slurry coating procedure. The slurry comprised 80 wt.% NCM811 (active material), 10 wt.% Super-P carbon black (conductive agent), and 10 wt.% polyvinylidene difluoride (PVDF, binder) dissolved in N-methyl-2-pyrrolidene (NMP). The solution was stirred overnight with a magnetic stirrer, and the cathode electrodes were fabricated by tape-casting the mixed slurry onto battery-grade aluminum foil using the doctor blade method. After tape casting, the cathodes were dried overnight at 120 °C in a vacuum oven. The CR2032 coin cells were assembled in an argon-filled glove box under a dry argon atmosphere and <0.1 ppm water and oxygen. The electrolyte used was 1 M $LiPF_6$ dissolved in a 1:1 mixture of ethylene carbonate (EC) and diethyl carbonate (DEC). Lithium foil served as the anode, and a Celgard 2400 membrane acted as the separator. The cathode electrode mass loading was approximately 2.3–2.5 mg cm^{-2}. Following assembly, the cells rested for 10 h before electrochemical characterization.

For the rate capability test, cells were charged in the galvanostatic mode to cutoff voltages of 4.3, 4.4, and 4.5 V with varying current densities (0.1, 0.2, 0.5, 1.0, 2.0, and 5.0 C, 1 C = 200 mAh g^{-1}, where g refers to the mass loading of the active material) followed by discharging to 2.8 V at the same rate as charging using a NEWARE battery testing machine. A 10 min resting period was applied prior to each step. Long-term cycling was conducted at 1 C for 150 cycles. The half-cells were cycled at 0.1 C for 3 cycles before cycling at 1 C.

Three-electrode cells were employed for cyclic voltammetry (CV) and electrochemical impedance spectroscopy (EIS), with lithium foil as the reference and counter electrode in a three-electrode cell system. Cyclic voltammograms of the electrodes were recorded within the potential range of 2.8–4.6 V at a scanning rate of 0.1 mV s^{-1}. The samples were cycled at 0.05 C for 3 cycles before CV measurements. EIS measurements were performed in a frequency range of 100 kHz to 10 mHz with a perturbation amplitude of ±10 mV when the samples were charged to the upper cutoff voltage. The measurements were conducted using a CHI 660E electrochemical workstation.

3. Results and Discussion

3.1. X-ray Diffraction Analysis

Pristine and coated samples were analyzed by XRD to evaluate crystal ordering and determine any impurities and phase changes during powder synthesis and the Al_2O_3 deposition process. As shown in Figure 1, the XRD patterns of both samples indicate that the powder structures are well indexed to the layered hexagonal α-$NaFeO_2$ type belonging to the R-3m space group. There are no significant shifts in peak positions or any new phases related to impurities. The I(003)/I(104) ratio of both samples is approximately the same and exceeds 1.2, 1.78, and 1.71 for uncoated and coated samples, respectively. Any change in the I(003)/I(104) ratio is attributed to changes in lithium content in the unit cell, which typically occurs during additional heat treatment in the coating process [29]. It is generally accepted that a peak intensity ratio greater than 1.2 signifies a good layered structure and a lower degree of Li^+/Ni^{2+} cation mixing in the lattice [30]. Moreover, the obvious splitting of the (006), (102) and (108), (110) peaks indicates the maintenance of a well-ordered layered structure as well as a high degree of crystallinity [31]. Additionally, there is no shift in the (003) reflection in the coated sample, which significantly implies two points: first, powder synthesis is highly accurate, and second, the Al_2O_3 coating process has no effect on the host crystal structure of NCM811 powder and does not introduce a new phase into the particles. This might be related to the fact that the coating deposition process was performed at a low temperature (120 °C), which inhibited Al diffusion into the lattice structure, ensuring no changes occurred in the crystal structure during the coating synthesis. The distinct diffraction peak of the Al_2O_3 phase is not detected in the NCM811@Al_2O_3 XRD pattern. This suggests that an ultrathin alumina coating covered the NCM811 particle surfaces and that the coating has a very low content and is an amorphous structure, which would be expected for a growth temperature of 120 °C.

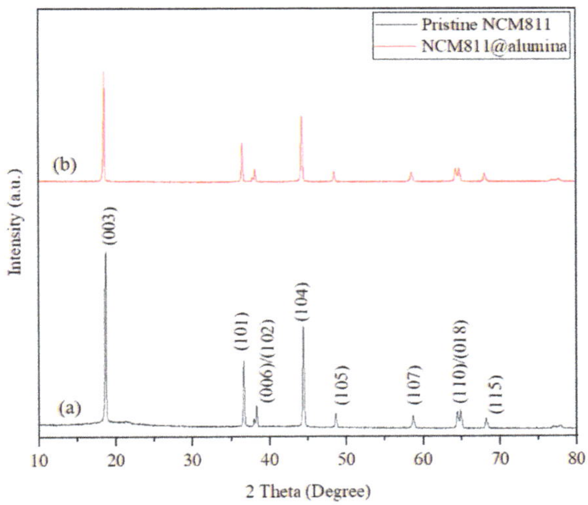

Figure 1. XRD patterns comparing the (a) pristine and (b) NCM811 surface modified with Al_2O_3.

3.2. Morphology

SEM images (Figure 2a,b) were captured to analyze the surface morphology and microstructure of the samples. Both samples exhibited secondary microspherical-like shape particles due to the aggregation of numerous nanometer-sized primary particles. Notably, the surface morphology of NCM811 particles remained unchanged, indicating a thin Al_2O_3 layer. EDS mapping of transition metals and Al (Figure 2c,d) was conducted to determine the element distribution on the particle surface. The images confirm the homogeneous distribution of transition metals and Al in the coated sample, indicating the desirable conformality and uniformity of the applied coating. TEM observation (Figure 2e) reveals a smooth and uniform Al_2O_3 coating achieved on the NCM811 particles, with an ultrathin thickness. The coating thickness is critical; too thick a coating inhibits Li^+ ion intercalation and extraction, while too thin a coating is ineffective in protecting the cathode material from reactions with the electrolyte [32]. More measurements for the coating thickness are provided in Figure S1.

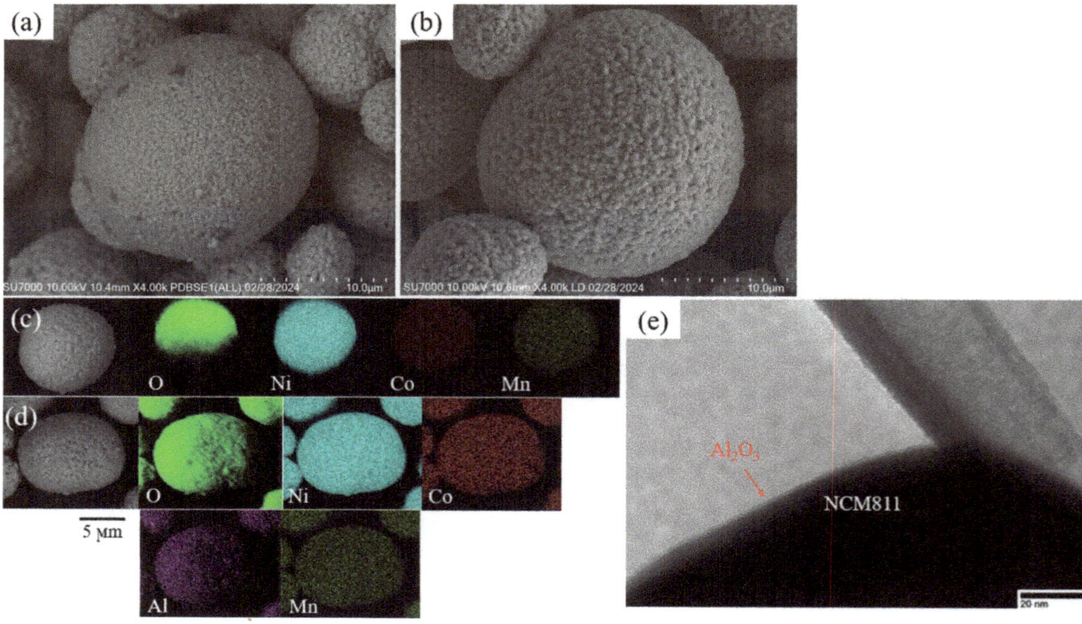

Figure 2. SEM images showing (**a**) pristine and (**b**) NCM811 coated with Al_2O_3, along with EDS-mapping of (**c**) pristine NCM811 and (**d**) Al_2O_3-coated NCM811 particles. Additionally, TEM images depicting the Al_2O_3-coated NCM811 particle are shown in (**e**).

3.3. Electrochemical Properties

The impact of Al_2O_3 using the ALD coating system on the enhancement of the electrochemical performance of the NCM811 cathode material was assessed through cycling at 1 C across the voltage ranges of 2.8–4.3, 4.4, and 4.5 V vs. Li/Li^+ at 25 °C, as depicted in Figure 3. The results reveal that the coated sample demonstrated superior electrochemical performance across all upper cutoff voltages. Particularly noteworthy is the significant difference in capacity retention as the upper cutoff voltage increased, with values of 82.00%, 75.68%, and 63.64% for pristine NCM811 compared to 87.00%, 84.79%, and 74.92% for the modified sample at 4.3, 4.4, and 4.5 V, respectively. Furthermore, both samples exhibited Coulombic Efficiency (CE) exceeding 95%. At the critical 4.5 V cutoff voltage, the Al_2O_3-coated sample delivered a discharge capacity of 198.93 mAh g^{-1} at 1 C discharge current, slightly lower than pristine NCM811, which yielded 200.66 mAh g^{-1}. After 150 cycles, the former delivered 149.04 mAh g^{-1}, while the latter achieved 127.69 mAh g^{-1}. The diminished capacity retention of the uncoated sample compared to the coated counterpart can be attributed to the high polarization and deterioration of the interface structure of NCM811, particularly under high upper cutoff voltage operation [33]. Indeed, the uniform nano-Al_2O_3 coating reduced the electrode–electrolyte reaction at the interface and delayed the formation of a thick passive layer, thereby postponing polarization at the material interface. Additionally, the formation of HF in the electrolyte leads to damage to cathode materials, while an ultrathin layer can significantly prolong the onset of severe degradation of the active material [34]. The relevant electrochemical stability results corresponding to this experiment are provided in Table 1. The data in Table 1 suggest that although the discharge capacity of the samples in the first few cycles is higher at 4.5 V cutoff potential, the capacity loss is comparable. Therefore, an operating voltage of 4.4 V can be identified as the optimized working condition for NCM811@Al_2O_3, striking a balance between capacity retention and high-capacity delivery. One plausible explanation for the poor cycling stability of NCM811 when charging to a 4.5 V potential is electrolyte oxidation and the side reactions between the cathode and electrolyte [35].

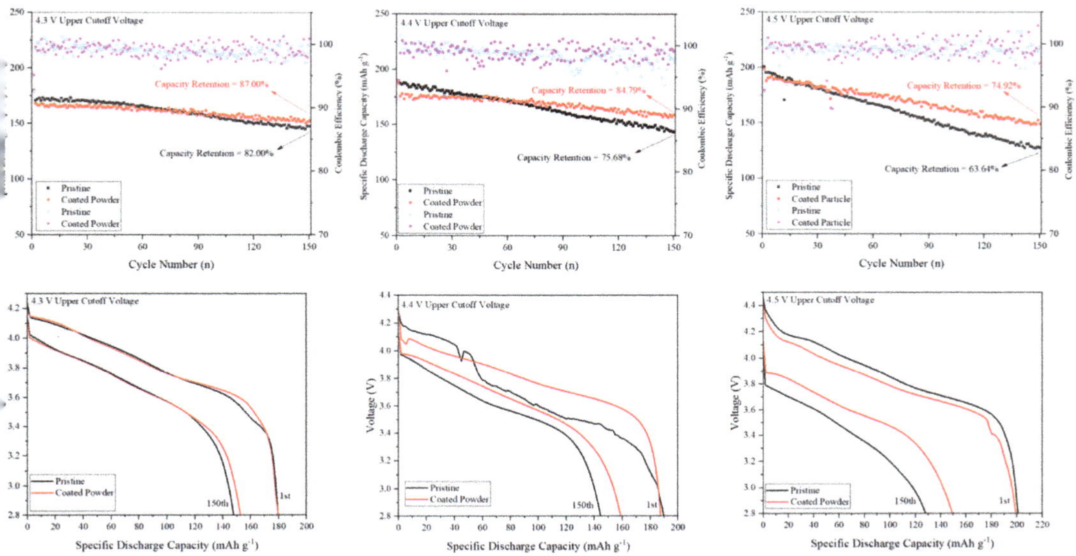

Figure 3. Electrochemical properties of bare and Al_2O_3-coated NCM811 over 150 cycles (1 C) at 4.3, 4.4, and 4.5 upper cutoff voltages. The upper figures compare the cycling stability of the samples, while the lower ones illustrate the voltage fading after 150 cycles.

Table 1. Electrochemical data for stability of bare and Al_2O_3-coated NCM811 over 150 cycles between 2.8 and 4.5 V.

Sample	1st Discharge Capacity (mAh g^{-1})	150th Discharge Capacity (mAh g^{-1})	Capacity Retention (%)	Voltage Fading (V)	Upper Cutoff Voltage (V)
Pristine NCM811	179.46	147.51	82.00	0.0986	4.3
	189.62	144.52	75.68	0.1559	4.4
	200.66	127.69	63.64	0.3522	4.5
Al_2O_3-Coated NCM811	179.61	152.25	87.00	0.0710	4.3
	187.23	158.74	84.79	0.1413	4.4
	198.93	149.04	74.92	0.2930	4.5

Voltage fading in Ni-rich NCM cathode materials occurs during cycling, primarily due to the destruction of the crystal structure and irreversible phase transitions. Galvanostatic voltage profiles recorded with a discharge current density of 1 C for both pristine and coated samples for the 1st and 150th cycles are illustrated in Figure 4 to compare the voltage decay over cycling. It is evident that, for all samples, the discharge profiles gradually shifted to lower voltage plateaus during cycling, accompanied by a decrease in discharge capacity. This phenomenon is attributed to the dissolution of transition metals from the active material due to structural and interfacial instabilities [36]. Moreover, the voltage fading increased as the upper cutoff voltage increased. Specifically, the pristine NCM811 experienced voltage fading of 0.0986 V, 0.1559 V, and 0.3522 V during cycling at cutoff voltages of 4.3 V, 4.4 V, and 4.5 V, respectively. In contrast, the Al_2O_3-coated sample exhibited lower voltage decay in this experiment, with values of 0.0710 V, 0.1413 V, and 0.2930 V for the corresponding cutoff voltages. The reduced voltage decay during cycling can be attributed to the successful suppression of polarization in the NCM811 electrode by the Al_2O_3 coating, which improved the structural stability and phase reversibility of the active material while protecting the bulk from direct contact with the electrolyte [37].

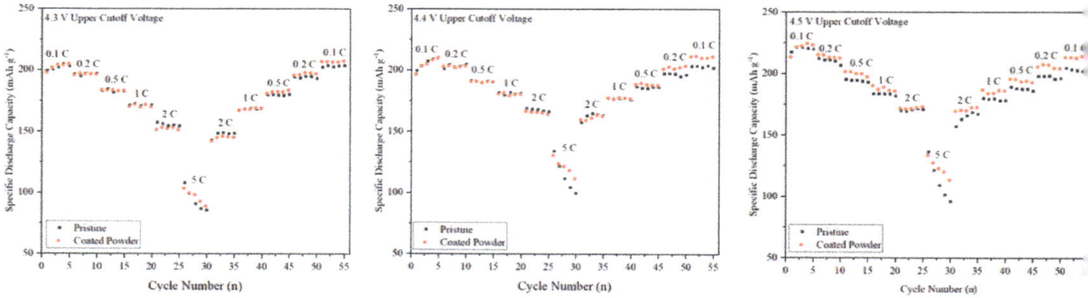

Figure 4. Rate capability for pristine and NCM811@Al_2O_3 at different cutoff voltages.

The results of rate capability performance for the samples at various charging cutoff voltages are shown in Figure 4. The samples were charged and discharged at current densities ranging from 0.1 to 5 C. Due to the low diffusion efficiency of Li^+ at high current densities, the discharge capacity of both cathode materials decreased as the discharge current density increased. However, the coated sample exhibited higher discharge capacity, especially at higher rates (2 and 5 C) and upper cutoff voltages (4.4 and 4.5 V).

At low current densities and in the initial cycles (1–25), both samples demonstrated similar reversible discharge capacity. In contrast, at higher rates (5 C) and in later cycles (26–55), the coated sample delivered more capacity, attributed to the improved stability of the surface structure. For instance, at the 4.5 V charging cutoff, pristine and coated NCM811 exhibited a reversible capacity of 217.28 and 213.54 mAh g^{-1} in the first cycle, respectively. The difference in initial discharge capacity might be due to the electrochemical inactivity of Al_2O_3. Over multiple charging and discharging cycles at different rates, the Al_2O_3-coated NCM811 demonstrated an impressive reversible capacity of 210.90 mAh g^{-1}, compared to 202.16 mAh g^{-1} for the pristine NCM811.

It is worth noting that while the Al_2O_3 coating did not significantly inhibit Li^+ ion diffusion at different rates, it enhanced the stability of the sample, enabling it to deliver more capacity than the pristine sample, irrespective of the charging voltage. The cycling and rate performances of the samples confirm that the structural stability of NCM811 cathode materials has been effectively improved by the Al_2O_3 coating.

3.4. EIS Measurements

An EIS test was conducted to further investigate the positive influence of the Al_2O_3 coating on the NCM811 cathode material. The Nyquist plots of the samples and the equivalent electrical circuit at the charged state at various upper cutoff voltages before cycling and after 150 cycles are illustrated in Figure 5. Each EIS plot consists of two semicircles and a straight line in different frequency regions. In the circuit, R_s, R_{SEI}, and R_{ct} represent the electrolyte resistance, the film resistance due to the solid electrolyte interface, and the charge transfer resistance related to the interface between the electrolyte and the electrode, respectively. In addition, the straight line at a low frequency corresponds to Warburg impedance (W), which is related to Li^+ diffusion in the particle [38].

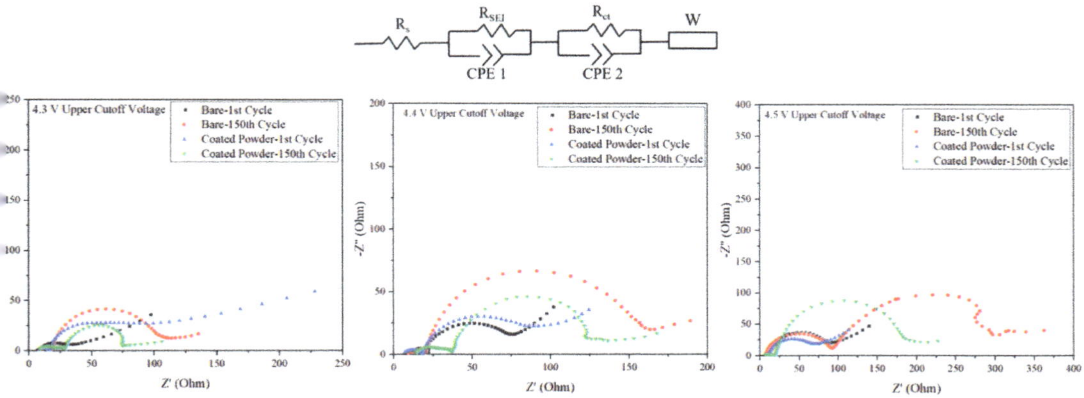

Figure 5. EIS curves of charged bare and Al_2O_3-coated NCM811 at 4.3–4.5 voltage range before cycling and after 150 cycles.

The R_s value for all samples is nearly the same, as identical electrolytes were used for this experiment. The R_{ct} value is listed in Table 2. Although the R_{ct} values for both electrodes were nearly the same before cycling, the value increased after cycling. However, the increase in charge transfer resistance was significantly higher for the uncoated sample compared to the coated one. This sharp increase in R_{ct} correlates with the surface distortion induced by side reactions with the electrolyte. In contrast, the alumina-coated NCM experienced a lower increase in R_{ct} indicating a superior charge-transfer rate during long-term cycling. The EIS results are consistent with the cycling and rate capability results. It is worth noting that cycling at a 4.5 V upper cutoff potential induced a significant increase in R_{ct}, which can justify the drastic capacity retention deterioration over cycling.

Table 2. R_{ct} values for pristine and Al_2O_3-coated NCM811 before cycling and after 150 cycles.

Sample	R_{ct} before Cycling (Ω)	R_{ct} after 150 Cycles (Ω)	Upper Cutoff Voltage (V)
Pristine NCM811	34.18	114.8	4.3
	75.88	164.00	4.4
	91.34	307.10	4.5
Al_2O_3-Coated NCM811	62.59	75.58	4.3
	92.97	126.53	4.4
	72.42	205.8	4.5

3.5. Cyclic Voltammetry Measurements

To further understand the electrochemical behavior of NCM811 and NCM811@Al_2O_3, cyclic voltammograms of the cathode electrodes were recorded over 15 cycles within a 2.8–4.6 V voltage range at a scan rate of 0.1 mV s^{-1}, as shown in Figure 6. The curves for both samples exhibited similar profiles, indicating that the Al_2O_3 coating did not participate in the electrochemical reactions. The main paired peaks correspond to the typical phase transitions for layered oxide cathode materials.

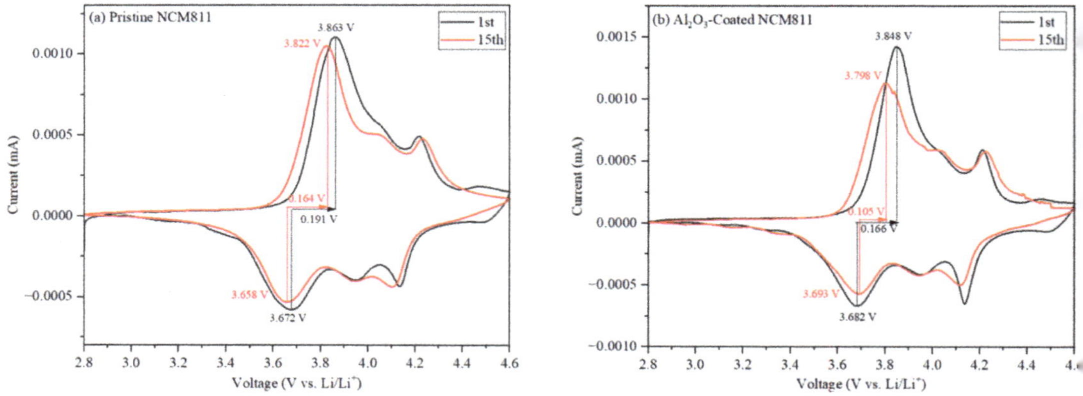

Figure 6. CV curves of (**a**) pristine and (**b**) Al_2O_3-coated NCM811 at a scan rate of 0.1 mV s^{-1} within a 2.8–4.6 V voltage range.

The potential difference between the oxidation and reduction peaks is a key kinetic factor influenced by the formation of the cathode solid electrolyte interface or by side reactions on the electrode surface. This potential difference reflects the reversibility of the electrochemical reaction [30]. For the coated sample, the oxidation and reduction peaks are located at 3.848 V and 3.682 V, respectively, with a corresponding potential difference of 0.166 V for the first cycle. In contrast, the uncoated sample has a higher potential difference of 0.191 V.

After 15 cycles, the coated sample exhibited a smaller voltage gap, indicating higher structural stability and a greater degree of reversibility. These results reveal that the ultrathin Al_2O_3 coating effectively reduced electrochemical polarization and enhanced electrochemical performance. This improvement is attributed to the coating's ability to suppress side reactions at the electrode/electrolyte interface.

4. Conclusions

In this manuscript, NCM811 cathode active material was synthesized, and an ultrathin Al_2O_3 coating was established on the NCM811 surface using the ALD technique. The alumina coating was demonstrated to exist on the surface of the cathode particles without altering their morphology or crystal structure, attributed to the low-temperature coating process. This coating significantly improved the cycling stability and rate capability of the cathode electrode, which is related to the enhanced structural stability of NCM811 powder. Additionally, the coating suppressed irreversible phase transitions and side reactions at the electrode–electrolyte interface, confirmed by EIS and CV measurements. Electrochemical measurements showed that cycling at higher upper cutoff voltages drastically reduced capacity retention for pristine NCM811. However, Al_2O_3 surface treatment stabilized the cathode material's structure, with a 2.8–4.4 V cutoff potential identified as the optimized cycling condition for NCM811@Al_2O_3 in this research.

Supplementary Materials: The following supporting information can be downloaded at: https://www.mdpi.com/article/10.3390/mi15070894/s1, Figure S1. Al_2O_3 ALD Coating thickness from two different locations of NCM811@Al_2O_3 particles.

Author Contributions: Conceptualization, M.A.; methodology, F.X.; writing—original draft M.A.; writing—review and editing, B.S.; analysis and supervision, H.L. and M.Z. All authors have read and agreed to the published version of the manuscript.

Funding: The research was funded by the National Science Foundations: NSF-OIA-2119688, NSF-CMMI-2215982 and NSF-CHE-2216473.

Data Availability Statement: The original contributions presented in the study are included in the article, further inquiries can be directed to the corresponding author.

Conflicts of Interest: The authors declare no conflicts of interest.

erences

Levartovsky, Y.; Wu, X.; Erk, C.; Maiti, S.; Grinblat, J.; Talianker, M.; Aurbach, D. Enhancement of structural, electrochemical, and thermal properties of Ni-rich $LiNi_{0.85}Co_{0.1}Mn_{0.05}O_2$ cathode materials for Li-ion batteries by Al and Ti doping. *Batter. Supercaps* **2021**, *4*, 221–231. [CrossRef]

Yang, J.; Xia, Y. Suppressing the phase transition of the layered Ni-rich oxide cathode during high-voltage cycling by introducing low-content Li_2MnO_3. *ACS Appl. Mater. Interfaces* **2016**, *8*, 1297–1308. [CrossRef]

Xiao, W.; Nie, Y.; Miao, C.; Wang, J.; Tan, Y.; Wen, M. Structural design of high-performance Ni-rich $LiNi_{0.83}Co_{0.11}Mn_{0.06}O_2$ cathode materials enhanced by Mg^{2+} doping and Li_3PO_4 coating for lithium ion battery. *J. Colloid Interface Sci.* **2022**, *607*, 1071–1082. [CrossRef] [PubMed]

Liu, Y.; Fan, X.; Luo, B.; Zhao, Z.; Shen, J.; Liu, Z.; Xiao, Z.; Zhang, B.; Zhang, J.; Ming, L.; et al. Understanding the enhancement effect of boron doping on the electrochemical performance of single-crystalline Ni-rich cathode materials. *J. Colloid Interface Sci.* **2021**, *604*, 776–784. [CrossRef] [PubMed]

Heenan, T.M.M.; Wade, A.; Tan, C.; Parker, J.E.; Matras, D.; Leach, A.S.; Robinson, J.B.; Llewellyn, A.; Dimitrijevic, A.; Jervis, R.; et al. Identifying the origins of microstructural defects such as cracking within Ni-rich NMC811 cathode particles for lithium-ion batteries. *Adv. Energy Mater.* **2020**, *10*, 2002655. [CrossRef]

Negi, R.S.; Culver, S.P.; Wiche, M.; Ahmed, S.; Volz, K.; Elm, M.T. Optimized atomic layer deposition of homogeneous, conductive Al_2O_3 coatings for high-nickel NCM containing ready-to-use electrodes. *Phys. Chem. Chem. Phys.* **2021**, *23*, 6725–6737. [CrossRef] [PubMed]

Lee, H.B.; Hoang, T.D.; Byeon, Y.S.; Jung, H.; Moon, J.; Park, M.-S. Surface stabilization of Ni-rich layered cathode materials via surface engineering with $LiTaO_3$ for lithium-ion batteries. *ACS Appl. Mater. Interfaces* **2022**, *14*, 2731–2741. [CrossRef]

Wu, F.; Tian, J.; Su, Y.; Wang, J.; Zhang, C.; Bao, L.; He, T.; Li, J.; Chen, S. Effect of Ni^{2+} content on lithium/nickel disorder for Ni-rich cathode materials. *ACS Appl. Mater. Interfaces* **2015**, *7*, 7702–7708. [CrossRef]

She, S.; Zhou, Y.; Hong, Z.; Huang, Y.; Wu, Y. Surface coating of NCM-811 cathode materials with g-C_3N_4 for enhanced electrochemical performance. *ACS Omega* **2022**, *7*, 24851–24857. [CrossRef] [PubMed]

Li, Y.; Tan, C.; Wei, S.; Cui, L.; Fan, X.; Pan, Q.; Lai, F.; Zheng, F.; Wang, H.; Li, Q. Stable surface construction of the Ni-rich $LiNi_{0.8}Mn_{0.1}Co_{0.1}O_2$ cathode material for high-performance lithium-ion batteries. *J. Mater. Chem. A* **2020**, *8*, 21649–21660. [CrossRef]

Dose, W.M.; Temprano, I.; Allen, J.P.; Björklund, E.; O'Keefe, C.A.; Li, W.; Mehdi, B.L.; Weatherup, R.S.; De Volder, M.F.L.; Grey, C.P. Electrolyte reactivity at the charged Ni-rich cathode interface and degradation in Li-ion batteries. *ACS Appl. Mater. Interfaces* **2022**, *14*, 13206–13222. [CrossRef] [PubMed]

Yang, H.; Gao, R.-M.; Zhang, X.-D.; Liang, J.-Y.; Meng, X.-H.; Lu, Z.-Y.; Cao, F.-F.; Ye, H. Building a Self-Adaptive Protective Layer on Ni-Rich Layered Cathodes to Enhance the Cycle Stability of Lithium-Ion Batteries. *Adv. Mater.* **2022**, *34*, 2204835. [CrossRef]

Gu, M.; Belharouak, I.; Zheng, J.; Wu, H.; Xiao, J.; Genc, A.; Amine, K.; Thevuthasan, S.; Baer, D.R.; Zhang, J.-G.; et al. Formation of the spinel phase in the layered composite cathode used in Li-ion batteries. *ACS Nano* **2013**, *7*, 760–767. [CrossRef]

Xiong, X.; Wang, Z.; Yue, P.; Guo, H.; Wu, F.; Wang, J.; Li, X. Washing effects on electrochemical performance and storage characteristics of $LiNi_{0.8}Co_{0.1}Mn_{0.1}O_2$ as cathode material for lithium-ion batteries. *J. Power Sources* **2013**, *222*, 318–325. [CrossRef]

Ryu, H.-H.; Park, K.-J.; Yoon, C.S.; Sun, Y.-K. Capacity fading of Ni-rich $Li[Ni_xCo_yMn_{1-x-y}]O_2$ ($0.6 \leq x \leq 0.95$) cathodes for high-energy-density lithium-ion batteries: Bulk or surface degradation? *Chem. Mater.* **2018**, *30*, 1155–1163. [CrossRef]

Li, T.; Yuan, X.-Z.; Zhang, L.; Song, D.; Shi, K.; Bock, C. Degradation mechanisms and mitigation strategies of nickel-rich NMC-based lithium-ion batteries. *Electrochem. Energy Rev.* **2020**, *3*, 43–80. [CrossRef]

Chen, T.; Wang, F.; Li, X.; Yan, X.; Wang, H.; Deng, B.; Xie, Z.; Qu, M. Dual functional $MgHPO_4$ surface modifier used to repair deteriorated Ni-rich $LiNi_{0.8}Co_{0.15}Al_{0.05}O_2$ cathode material. *Appl. Surf. Sci.* **2019**, *465*, 863–870. [CrossRef]

Feng, Z.; Rajagopalan, R.; Sun, D.; Tang, Y.; Wang, H. In-situ formation of hybrid Li_3PO_4-$AlPO_4$-$Al(PO_3)_3$ coating layer on $LiNi_{0.8}Co_{0.1}Mn_{0.1}O_2$ cathode with enhanced electrochemical properties for lithium-ion battery. *Chem. Eng. J.* **2020**, *382*, 122959. [CrossRef]

Xiong, L.; Sun, M.; Xu, Y.; Du, X.; Xiao, X. Synthesis of carbon-coated Li_2MnO_3 cathode material with enhanced rate capability for lithium-ion batteries. *Solid State Ion.* **2018**, *325*, 170–175. [CrossRef]

Chen, Y.; Zhang, Y.; Chen, B.; Wang, Z.; Lu, C. An approach to application for $LiNi_{0.6}Co_{0.2}Mn_{0.2}O_2$ cathode material at high cutoff voltage by TiO_2 coating. *J. Power Sources* **2014**, *256*, 20–27. [CrossRef]

Tao, T.; Chen, C.; Yao, Y.; Liang, B.; Lu, S.; Chen, Y. Enhanced electrochemical performance of ZrO_2 modified $LiNi_{0.6}Co_{0.2}Mn_{0.2}O_2$ cathode material for lithium-ion batteries. *Ceram. Int.* **2017**, *43*, 15173–15178. [CrossRef]

Cheng, R.; Zhou, M.; Zhang, K.; Xu, F.; Sun, L.; Lin, Q.; Li, H. Enhanced electrochemical performance of $LiNi_{1/3}Co_{1/3}Mn_{1/3}O_2$ cathode material by Al_2O_3 surface coating derived via NH_2-MIL-53(Al) MOF. *J. Alloys Compd.* **2021**, *875*, 159956. [CrossRef]

23. Ma, F.; Wu, Y.; Wei, G.; Qiu, S.; Qu, J. Enhanced electrochemical performance of $LiNi_{0.8}Co_{0.1}Mn_{0.1}O_2$ cathode via wet-chem coating of MgO. *J. Solid State Electrochem.* **2019**, *23*, 2213–2224. [CrossRef]
24. Neudeck, S.; Mazilkin, A.; Reitz, C.; Hartmann, P.; Janek, J.; Brezesinski, T. Effect of low-temperature Al_2O_3 ALD coating Ni-rich layered oxide composite cathode on the long-term cycling performance of lithium-ion batteries. *Sci. Rep.* **2019**, *9*, 5 [CrossRef] [PubMed]
25. Lee, Y.; Sun, H.; Young, M.J.; George, S.M. Atomic layer deposition of metal fluorides using HF–pyridine as the fluorine precu *Chem. Mater.* **2016**, *28*, 2022–2032. [CrossRef]
26. Kong, J.-Z.; Wang, S.-S.; Tai, G.-A.; Zhu, L.; Wang, L.-G.; Zhai, H.-F.; Wu, D.; Li, A.-D.; Li, H. Enhanced electrochem performance of $LiNi_{0.5}Co_{0.2}Mn_{0.3}O_2$ cathode material by ultrathin ZrO_2 coating. *J. Alloys Compd.* **2016**, *657*, 593–600. [Cross]
27. Li, X.; Liu, J.; Meng, X.; Tang, Y.; Banis, M.N.; Yang, J.; Hu, Y.; Li, R.; Cai, M.; Sun, X. Significant impact on cathode performanc lithium-ion batteries by precisely controlled metal oxide nanocoatings via atomic layer deposition. *J. Power Sources* **2014**, 57–69. [CrossRef]
28. Kraytsberg, A.; Drezner, H.; Auinat, M.; Shapira, A.; Solomatin, N.; Axmann, P.; Wohlfahrt-Mehrens, M.; Ein-Eli, Y. Atomic La Deposition of a Particularized Protective MgF_2 Film on a Li-ion Battery $LiMn_{1.5}Ni_{0.5}O_4$ Cathode Powder Material. *ChemNano* **2015**, *1*, 577–585. [CrossRef]
29. Lee, S.M.; Oh, S.H.; Ahn, J.P.; Cho, W.I.; Jang, H. Electrochemical properties of ZrO_2-coated $LiNi_{0.8}Co_{0.2}O_2$ cathode material *Power Sources* **2006**, *159*, 1334–1339. [CrossRef]
30. Yang, G.; Pan, K.; Yan, Z.; Yang, S.; Peng, F.; Liang, J.; Lai, F.; Wang, H.; Zhang, X.; Li, Q. Fully coating of $Mg_3B_2O_6$ in nonaque solution on Ni-rich $LiNi_{0.8}Co_{0.1}Mn_{0.1}O_2$ secondary particles to improve cycling stability of lithium-ion batteries. *Chem. En* **2023**, *452*, 139405. [CrossRef]
31. Zhang, B.; Dong, P.; Tong, H.; Yao, Y.; Zheng, J.; Yu, W.; Zhang, J.; Chu, D. Enhanced electrochemical performance $LiNi_{0.8}Co_{0.1}Mn_{0.1}O_2$ with lithium-reactive Li_3VO_4 coating. *J. Alloys Compd.* **2017**, *706*, 198–204. [CrossRef]
32. Huang, X.; Zhu, W.; Yao, J.; Bu, L.; Li, X.; Tian, K.; Lu, H.; Quan, C.; Xu, S.; Xu, K.; et al. Suppressing structural degradat of Ni-rich cathode materials towards improved cycling stability enabled by a Li_2MnO_3 coating. *J. Mater. Chem. A* **202** 17429–17441. [CrossRef]
33. Li, L.; Chen, Z.; Zhang, Q.; Xu, M.; Zhou, X.; Zhu, H.; Zhang, K. A hydrolysis-hydrothermal route for the synthesis of ultrat $LiAlO_2$-inlaid $LiNi_{0.5}Co_{0.2}Mn_{0.3}O_2$ as a high-performance cathode material for lithium-ion batteries. *J. Mater. Chem. A* **201** 894–904. [CrossRef]
34. Fan, Q.; Lin, K.; Yang, S.; Guan, S.; Chen, J.; Feng, S.; Liu, J.; Liu, L.; Li, J.; Shi, Z. Constructing effective TiO_2 nano-coating high-voltage Ni-rich cathode materials for lithium-ion batteries by precise kinetic control. *J. Power Sources* **2020**, *477*, 228 [CrossRef]
35. Du, M.; Yang, P.; He, W.; Bie, S.; Zhao, H.; Yin, J.; Zou, Z.G.; Liu, J. Enhanced high-voltage cycling stability of Ni-r $LiNi_{0.8}Co_{0.1}Mn_{0.1}O_2$ cathode coated with Li_2O–$2B_2O_3$. *J. Alloys Compd.* **2019**, *805*, 991–998. [CrossRef]
36. Kong, J.-Z.; Chen, Y.; Cao, Y.-Q.; Wang, Q.-Z.; Li, A.-D.; Li, H.; Zhou, F. Enhanced electrochemical performance of Ni-r $LiNi_{0.6}Co_{0.2}Mn_{0.2}O_2$ coated by molecular layer deposition derived dual-functional C-Al_2O_3 composite coating. *J. Alloys Com* **2019**, *799*, 89–98. [CrossRef]
37. Yao, W.; Liu, Y.; Li, D.; Zhang, Q.; Zhong, S.; Cheng, H.; Yan, Z. Synergistically enhanced electrochemical performance of Ni-r cathode materials for lithium-ion batteries by K and Ti Co-modification. *J. Phys. Chem. C* **2020**, *124*, 2346–2356. [CrossRef]
38. Sim, S.-J.; Lee, S.-H.; Jin, B.-S.; Kim, H.-S. Effects of lithium tungsten oxide coating on $LiNi_{0.90}Co_{0.05}Mn_{0.05}O_2$ cathode mate for lithium-ion batteries. *J. Power Sources* **2021**, *481*, 229037. [CrossRef]

micromachines

MDPI

cle

ıMn$_2$O$_4$/V$_2$CT$_x$ Composites Prepared as an Anode Material via igh-Temperature Calcination Method for Optimized -Ion Batteries

[1,2], Yu Wang [3], Xinyuan Pei [1,2], Chunhe Zhou [1,2], Qing Zhao [1,2], Ming Lu [1,2,*], Wenjuan Han [1,2] and Li Wang [1,2,*]

1 Key Laboratory of Functional Materials Physics and Chemistry of the Ministry of Education, Jilin Normal University, Changchun 130103, China; liji@jlnu.edu.cn (J.L.); peixyyy@163.com (X.P.); zhouch342@163.com (C.Z.); zhaoqing19991229@126.com (Q.Z.); hanwj@jlnu.edu.cn (W.H.)

2 Joint Laboratory of MXene Materials, Jilin Normal University, Changchun 130103, China

3 State Key Laboratory of Superhard Materials, College of Physics, Jilin University, Changchun 130012, China; yu_w20@mails.jlu.edu.cn

* Correspondence: luming@jlnu.edu.cn (M.L.); jlsdwangli@126.com (L.W.)

Abstract: The ZnMn$_2$O$_4$/V$_2$CT$_x$ composites with a lamellar rod-like bond structure were successfully synthesized through high-temperature calcination at 300 °C, aiming to enhance the Li storage properties of spinel-type ZnMn$_2$O$_4$ anode materials for lithium-ion batteries. Moreover, even though the electrode of the composites obtained at 300 °C had a nominal specific capacity of 100 mAh g^{-1}, it exhibited an impressive specific discharge capacity of 163 mAh g^{-1} after undergoing 100 cycles. This represents an approximate increase of 64% compared to that observed in the pure ZnMn$_2$O$_4$ electrode (99.5 mAh g^{-1}). The remarkable performance of the composite can be credited to the collaborative impact between ZnMn$_2$O$_4$ and V$_2$CT$_x$, leading to a substantial improvement in its lithium ion storage capacity. Therefore, this study offers valuable insights into developing cost-effective, safe, and easily prepared anode materials.

Keywords: ZnMn$_2$O$_4$; V$_2$CT$_x$ MXene; composite; lithium-ion batteries

1. Introduction

Recently, human society has witnessed a remarkable surge in progress, which has been accompanied by two significant challenges in the energy and environmental sectors that present major obstacles to the formulation of a national economic strategy [1,2]. Lithium-ion batteries (LIBs) have become widely used in various industries due to their exceptional characteristics, including a high capacity for storing energy, a minimal loss of charge over time, a lightweight construction, and an environmentally friendly nature. They play an indispensable role as vital energy storage devices that drive technological advancements and foster sustainable development [3–5]. Although LIBs exhibit exceptional performance in various aspects, they also possess certain drawbacks such as safety concerns, limited cycle life, temperature sensitivity, and capacity decay. Nevertheless, through continuous advancements in science and technology, along with technological enhancements, diligent efforts are being made to overcome these challenges. The ultimate goal is to further enhance the performance and safety of LIBs while promoting their extensive utilization across diverse fields [6–8].

Mixed transition metal oxides (MTMOs), exhibiting a spinel structure AB$_2$O$_4$, where A denotes divalent metal ions, such as Cu^{2+}, Zn^{2+}, Mn^{2+}, and Mg^{2+}, and B represents trivalent metal ions, including Fe^{3+}, Al^{3+}, Co^{3+}, etc., have garnered significant attention owing to their remarkable electrochemical properties [9], being conversion anode materials exhibiting a theoretical capacity ranging from 700 to 1000 mAh g^{-1}, involving the transfer of at least eight moles of electrons during the initial discharge process. Among the various MTMOs, ZnMn$_2$O$_4$ exhibits two distinct lithium intercalation mechanisms (conversion and alloying), allowing it to achieve an initial discharge treatment with a theoretical capacity of 1008 mAh g^{-1}, surpassing other MTMOs without zinc [10,11]. In comparison to most Co- or Fe-based oxides, $ZnMn_2O_4$ operates within a voltage range of only 1.2–1.5 V (vs. Li^+/Li), thereby expanding its potential window to battery applications [12]. Moreover, its abundant raw material sources, cost-effectiveness, and near-non-toxicity contribute to its suitability for large-scale utilization [13]. However, the inevitable volume changes during charge and discharge pose challenges in practical applications by causing structural collapse and comminution of the anode material, as well as reducing its capacity [14]. Additionally, similar to other MTMOs, $ZnMn_2O_4$ exhibits low electronic conductivity, which further hampers its performance. To address these limitations effectively, various strategies involve combining $ZnMn_2O_4$ with auxiliary materials, such as carbon nanotubes (CNTs) or graphene, that facilitate electron and ion transport while mitigating adverse effects associated with volumetric changes occurring throughout battery cycling [15,16].

The MXene material is a two-dimensional (2D) layered material characterized by its highly conductive core and exceptional dispersibility in aqueous solutions owing to its hydrophilic surface [17]. Its unique nature makes fleck-like 2D MXene commonly used for constructing high-level structures for LiBs [18,19]. In this study, V_2CT_x MXene was utilized to enhance the low conductivity, slow ion diffusion, and volume expansion of $ZnMn_2O_4$. An effective method for synthesizing a $ZnMn_2O_4/V_2CT_x$ composite electrode is reported, aiming to (1) develop a simple and applicable approach for preparing composite materials and (2) characterize the correlation between the morphology, structure, and electrochemical characteristics of $ZnMn_2O_4/V_2CT_x$. Additionally, rational energy storage mechanisms have been proposed based on cyclic voltammetry tests. This work provides valuable data supporting the enhancement of the electrochemical properties of $ZnMn_2O_4$.

2. Experimental Methods

2.1. Materials

The V_2AlC powder was supplied by 11 technology Co., Ltd. (China). Zinc sulfate heptahydrate ($ZnSO_4 \cdot 7H_2O$), anhydrous oxalic acid ($C_2H_2O_4$), and hydrochloric acid (HCl) were supplied by Shanghai Aladdin Biochemical Technology Co., Ltd. (China). Ammonium fluoride (NH_4F) and polyethylene glycol 400 ($H(OCH_2CH_2)_n$ OH) were supplied by Sinopharm Chemical Reagent Co., Ltd. (China). Manganese chloride ($MnCl_2 \cdot 4H_2O$) was supplied by Tianjin Damao Chemical Reagent Factory. All chemicals and reagents utilized in the research are analyzed-grade or higher purity, obviating the need for addition purification.

2.2. Preparation Methods

2.2.1. Preparation of $ZnMn_2O_4$ Micron Rods

The mixture of 20 mL anhydrous ethanol, 10 mL polyethylene glycol 400, and 10 mL deionized water was supplemented with 0.18 g of anhydrous oxalic acid ($C_2H_2O_4$). Subsequently, the solution was further enriched with 0.144 g of zinc sulfate heptahydrate ($ZnSO_4 \cdot 7H_2O$) and 0.198 g of manganese chloride ($MnCl_2 \cdot 4H_2O$), followed by stirring at a temperature of 300 K for a period of 6 h. After being subjected to centrifugation at a speed of 5000 rpm for 5 min, the resulting precipitates underwent three washes with pure water and two washes with ethanol. Subsequently, they were dried in a preheated oven maintained at a temperature of 60 °C for up to 12 h to obtain precursor materials in the form of rods. The obtained precursor underwent calcination in a muffle furnace starting from room temperature until reaching a final temperature of 400 °C at an increment rate of 5 °C/min, followed by holding it steady for two hours prior to a natural cooling process, resulting in the formation of $ZnMn_2O_4$ micron rods labeled as the ZMO sample.

2.2.2. Preparation of V_2CT_x MXene by Hydrochloric Acid and Ammonium Fluoride Etching Method

The plastic beaker with a capacity of 50 mL was filled by combining 20 mL of hydrochloric acid and an equal volume of deionized water, and NH_4F powder particles weighing 2.0 g were introduced into the mixture. The solution underwent magnetic stirring for a duration of 0.5 h at a temperature of 300 K. Subsequently, 2.0 g of V_2AlC powder was gently added to the beaker and stirred for an additional 5 min. The mixture obtained was subsequently transferred into a 100 mL autoclave lined with Teflon and underwent hydrothermal treatment at a temperature of 120 °C for a period of 24 h. Afterwards, the remaining black sediment underwent several rinses with deionized water until the liquid above it had a pH value of $\geq$5. This was followed by an extra wash using alcohol. The damp black sediment produced from this procedure underwent filtration at a speed of 4500 rpm for 5 min, followed by dehydration in a vacuum oven set to 60 °C for a period of 12 h, resulting in the acquisition of the V_2CT_x MXene specimen.

2.2.3. Preparation of $ZnMn_2O_4/V_2CT_x$ Composite by High Temperature Calcination Method

In this method, 1.0 g $ZnMn_2O_4$ and 1.0 g V_2CT_x MXene were ground together in a grinder until achieving homogeneity. The resulting mixture was divided into thirds: one portion remained uncalcined and labeled as the ZMO/V_2CT_x sample; while the other two portions were calcined by air in the muffle furnace at heating rates of 5 °C/min, and the temperature reached 300 °C and 400 °C, respectively, and was kept for 2 h before natural cooling. The corresponding samples were designated as ZMO/V_2CT_x-300 and ZMO/V_2CT_x-400.

2.3. *Characterization*

The XRD patterns were acquired utilizing a Rigaku D/Max-2500 X-ray diffractometer (Japan) within the range of 5–65°. Raman spectroscopic investigations were conducted employing a Renishaw in Via RFS100/S micro-Raman spectrometer (Britain) with an excitation wavelength of 514 nm. Fourier transform infrared (FTIR) spectroscopy analysis was performed using a Thermo Scientific Nicolet iS50 spectrophotometer (USA). The surface morphology and composition analysis were carried out utilizing a JEOL JSM-7800F scanning electron microscope (SEM) (Japan). The surface composition was determined through X-ray photoelectron spectroscopy (XPS) using an Escalab 250Xi spectrometer (USA).

2.4. *Electrochemical Tests*

The electrochemical performance of the prepared samples was investigated employing the 2032 button battery technique, wherein a copper sheet was utilized as the substrate for coating the sample to create an anode, while lithium ions were employed as the reference electrode.

The active ingredient (sample obtained), conductive material (acetylene black), and binder (PVDF) were combined in a ratio of 8:1:1 by weight and dissolved in N-methyl pyrrolidone (NMP) solvent to create a slurry. This slurry was then applied onto circular copper sheets with a radius of 6 mm, followed by vacuum drying at 60 °C for 12 h. The electrochemical performance of the prepared electrodes was examined using cyclic voltammetry (CV) and constant current charge–discharge (GCD) measurements, conducted on electrochemical workstations (Chenhua CHI660E and Metrohm Autolab PGSTAT302N) (Switzerland), as well as a battery test system NEWARE CT-4008T (China). All electrochemical experiments utilized approximately 1 mg of active ingredients, and the mass loading of the active materials in the electrode was 1 mg/cm^2.

3. Results and Discussion

3.1. Structural Characterization

In Figure 1a, an anode composed of $ZnMn_2O_4/V_2CT_x$ and a reference electrode consisting of a lithium sheet were utilized to fabricate LIBs. The two parallel electrodes were separated by a filter paper.

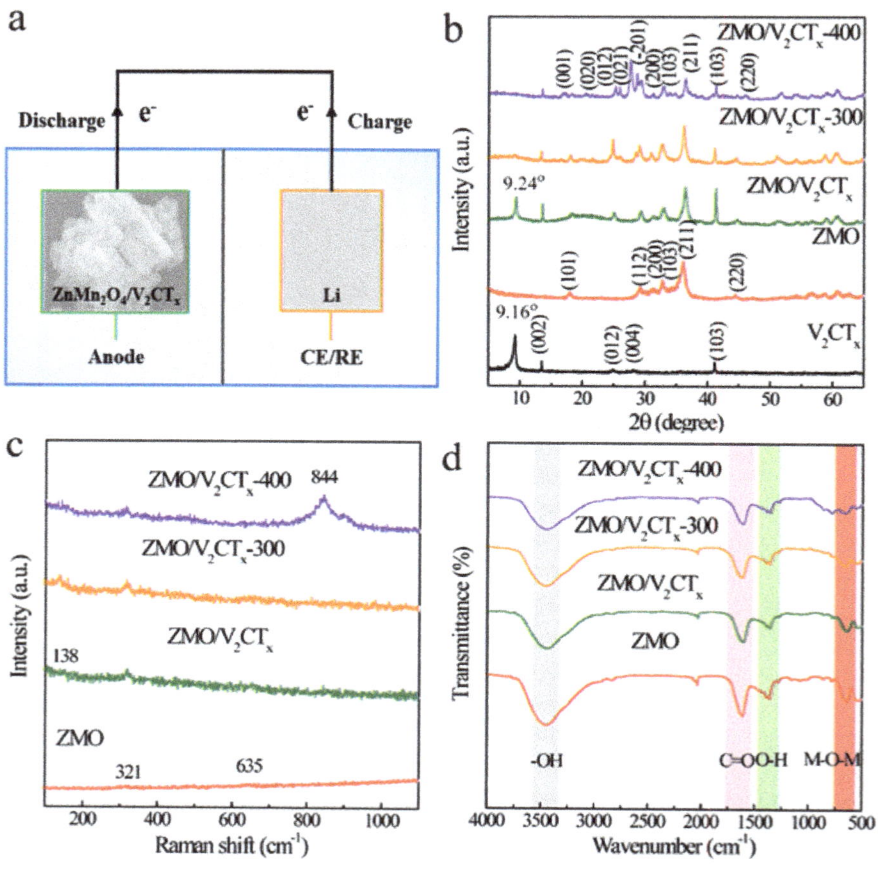

Figure 1. Battery configuration diagram of $ZnMn_2O_4/V_2CT_x$ (**a**); XRD pattern of V_2CT_x MXene, ZMO, and $ZnMn_2O_4/V_2CT_x$ (**b**); Raman spectra of ZMO and $ZnMn_2O_4/V_2CT_x$ (**c**); FTIR spectra of ZMO and $ZnMn_2O_4/V_2CT_x$ (**d**).

Figure 1b clearly illustrates the diffraction peaks of ZMO, which match well with standard PDF#71-2499. After etching V_2AlC (PDF#29-0101) with NH_4F+HCl, the characteristic diffraction peak at approximately 9.16° can be attributed to the (002) plane of V_2CTx MXene. The results demonstrate successful fabrication of pure $ZnMn_2O4$ and V_2CT_x MXene materials. For the ZMO/V_2CT_x sample, observed peaks correspond to crystal planes for both $ZnMn_2O4$ and V_2CT_x MXene without any significant impurity peak detected. It is noteworthy that compared to V_2CT_x alone, in the ZMO/V_2CT_x composite material, there is a slight shift in position from 9.16° to 9.24° for peak (002), indicating a more compact and stacked structure resulting from the grinding process on V_2CT_x MXene. Upon calcination at 300 °C, the XRD pattern shows no apparent changes except for a gradual decrease in diffraction peak intensity of V_2CT_x MXene, suggesting structural breakdown at high temperatures. However, when the temperature reaches 400 °C, the characteristic peaks strength of $ZnMn_2O_4$ gradually increases, along with the appearance of MnV_7O_{14} phase (PDF#89-0484). As shown in Figure 1c, the Raman spectroscopy tests demonstrate that the 321 cm^{-1} Raman peak observed in the ZMO electrode is a result of the vibration of the Zn-O bond within the tetrahedral structure of $ZnFe_2O_4$. Additionally, it can be inferred from the presence of a 635 cm^{-1} Raman peak that there are symmetric stretching vibrations occurring in MnO_6 groups, indicating the existence of a spinel phase. The peak at around 138 cm^{-1} represents a characteristic feature of V_2CT_x MXene, confirming successful preparation of the $ZnMn_2O_4/V_2CT_x$ composite sample. Additionally, a characteristic peak near 844 cm^{-1} confirms the synthesis of $Mn_2V_2O_7$ for ZMO/V_2CT_x-400 and is consistent with the XRD results. The FTIR spectra for synthesized ZMO and $ZnMn_2O_4/V_2CT_x$ are plotted over a range spanning from 400 to 4000 cm^{-1} and are shown in Figure 1d. In terms of pure ZMO (as seen in figure), an observed peak appears between regions ranging from 500 to 700 cm^{-1}, which may be due to M-O-M (where M = Zn, Mn) bonding; weak bands at approximately 1001.84 cm^{-1} correspond to -OH vibrations; strong absorption around 1613 cm^{-1} arises from C-O stretching vibrations; and finally, the peak located at 3436 cm^{-1} corresponds to -OH stretching vibrations. The FTIR spectra for ZMO/V_2CT_x, ZMO/V_2CT_x-300, and ZMO/V_2CT_x-400 are generally consistent with those obtained for pure ZMO, except for the overall weaker intensity.

The morphology of ZMO and $ZnMn_2O_4/V_2CT_x$ composites is depicted in Figure 2, revealing the binding between the micrometer rods of ZMO and the lamellae of V_2CT_x MXene. As the temperature increases, the bonding becomes more pronounced, leading to a spherical structure at 400 °C, with severe aggregation phenomena observed. The presence of Zn, Mn, O, V, and C elements on the sample surface is confirmed through elemental mapping (Figure 3).

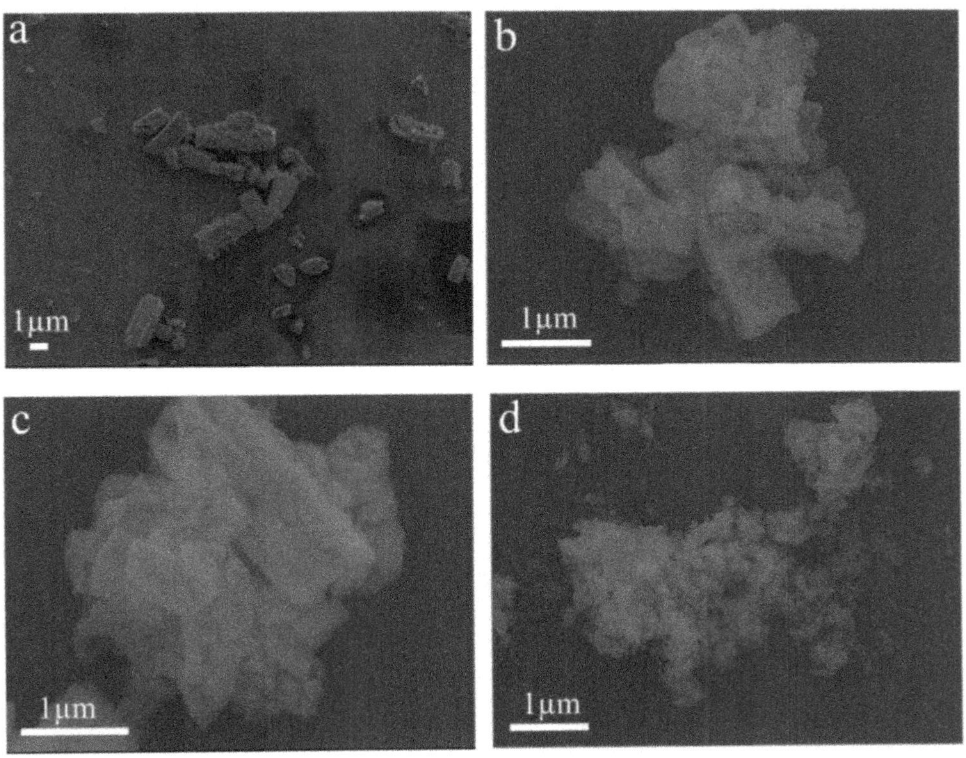

Figure 2. SEM images of ZMO (**a**); ZMO/V_2CT_x (**b**); ZMO/V_2CT_x-300 (**c**); and ZMO/V_2CT_x-400 (**d**).

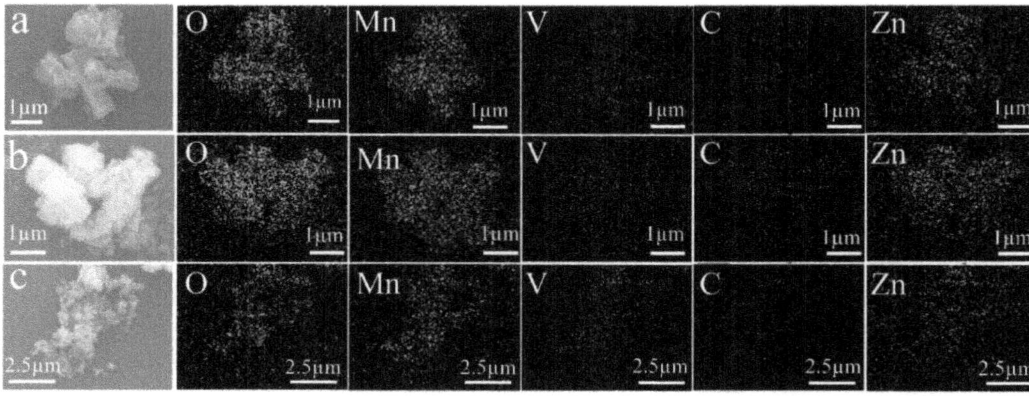

Figure 3. Elemental mappings of ZMO/V_2CT_x (**a**); ZMO/V_2CT_x-300 (**b**); and ZMO/V_2CT_x-400 (**c**).

The change in the valence state of ZMO/V_2CT_x-300 was investigated using XPS. The obtained spectra from the investigation (Figure 4a) indicate the presence of elements, including zinc (Zn), manganese (Mn), oxygen (O), vanadium (V), and carbon (C).

In Figure 4b, the observed peaks corresponding to energy levels of 1022.1 and 1045.05 eV can be assigned to the Zn 2*p* 3/2 and Zn 2*p* 1/2 orbitals, respectively. Similarly, in Figure 4c, the Mn 2*p* spectra exhibit two separate peaks at energies around 641.65 and 653.95 eV, suggesting the presence of distinct states corresponding to Mn 2*p* 3/2 and Mn 2*p* 1/2. In particular, it is worth noting that the peak corresponding to Mn^{3+} can be further resolved into four distinct peaks at approximate binding energies of around 641.6 eV (653.4 eV) and around binding energies of approximately 642.95 eV (654.8 5eV), which are attributed to different oxidation states: Mn^{3+} and Mn^{4+}. The analysis of the O1s spectrum (Figure 4d) exhibited three separate peaks at binding energies of 530.15, 530.65, and 531.95 eV, respectively, suggesting the existence of diverse oxygen components on the surface. These peaks correspond to Mn-bonded lattice oxygen (OL), oxygen vacancy sites (Ov), and surface-absorbed hydroxyl groups (-OH). The oxygen vacancies act as adsorption sites for various oxygen species like O^-, O^{2-}, and Mn-OH. The presence of adsorbed surface oxygen components is responsible for the observed peak at 530.65 eV, which affects the oxygen vacancy state of the active substance. The V 2*p* spectrum depicted in Figure 4e displays two distinct peaks at approximately 517.45 and 525.25 eV, representing the binding energies of V 2*p* 3/2 and V 2*p* 1/2, respectively. Specifically, the V 2*p* spectrum can be accurately modeled with four peaks: V^{4+} at around 525.4 and 517.85 eV, as well as the coexisting V^{5+} at approximately 524.5 and 517.25 eV. Furthermore, the C1s spectra (Figure 4f) were fitted using a three-peak model with assignments made for C=O, C-O, and C-C at energies of about 286.55, 285, and 284.6 eV, respectively. The XPS results confirm the successful synthesis of the ZMO/V_2CT_x-300 composite.

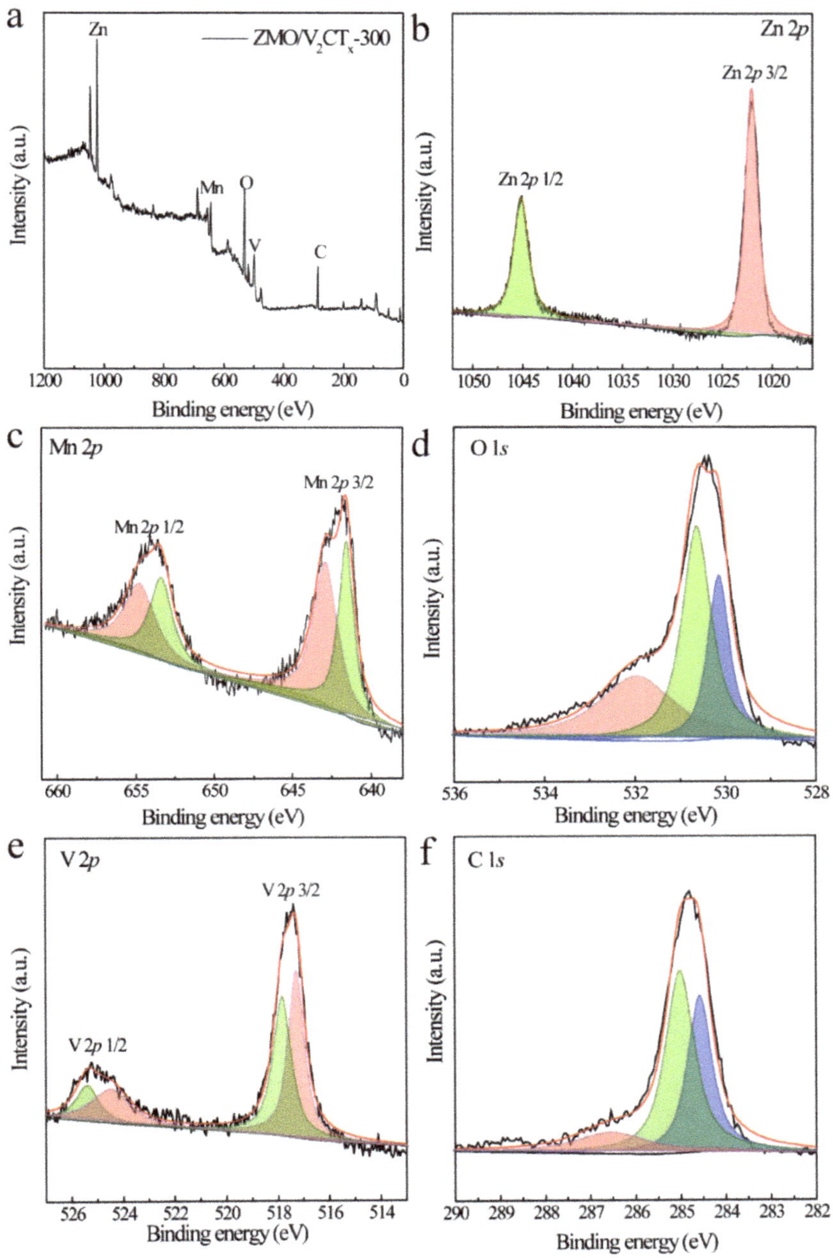

Figure 4. XPS spectra of ZMO/V_2CT_x−300: XPS profiles (**a**); Zn 2*p* spectra (**b**); Mn 2*p* spectra (**c**); O 1*s* spectra (**d**); V 2*p* spectra (**e**); C 1*s* spectra (**f**). Black is the experimental curve, and red is the fitting line.

3.2. Electrochemical Performance

The cyclic behavior of the ZMO and $ZnMn_2O_4/V_2CT_x$ electrodes is depicted in Figure 5a. The ZMO/V_2CT_x-300 electrode demonstrates superior cycling performance compared to ZMO, exhibiting a remarkable reversible capacity of 163 mAh g^{-1} after undergoing 100 cycles. In contrast, the capacity achieved by the ZMO electrode is lower at 99.5 mAh g^{-1}. It is important to mention that the Coulombic efficiency of the ZMO/V_2CT_x-300 electrode remains consistently close to 100% throughout continuous cycling, providing additional evidence supporting the electrochemical stability of this composite. The distinct advantages can be ascribed to the composite nature of the material and its calcination process, which enhance ion diffusion pathways and afford a greater abundance of active sites. Figure 5b illustrates the average discharge capacities at various current rates, where the values of 212.9, 169.4, 129.1, 72.4, and 29.7 mAh g^{-1} correspond to current rates of 0.5, 1, 2, 5, and 10 C, respectively. Furthermore, after undergoing cycling at different current densities, electrodes demonstrated discharge capacities around approximately 167.99 and 216.3 mAh g^{-1} when operated at rates of 1 and 0.5 C, respectively. The electrochemical charge–discharge curve shown in Figure 5c depicts the performance of ZMO/V_2CT_x-300 electrodes when discharged and charged at a rate of 1 C. In the second cycle, these electrodes exhibited a discharge capacity of 396.7 mAh g^{-1}, which remained consistent at a specific capacity of 163 mAh g^{-1}, even after undergoing 100 cycles. Additionally, it is worth noting that the charge–discharge curves have maintained their shape since the second cycle, indicating remarkable periodic electrochemical properties of the material.

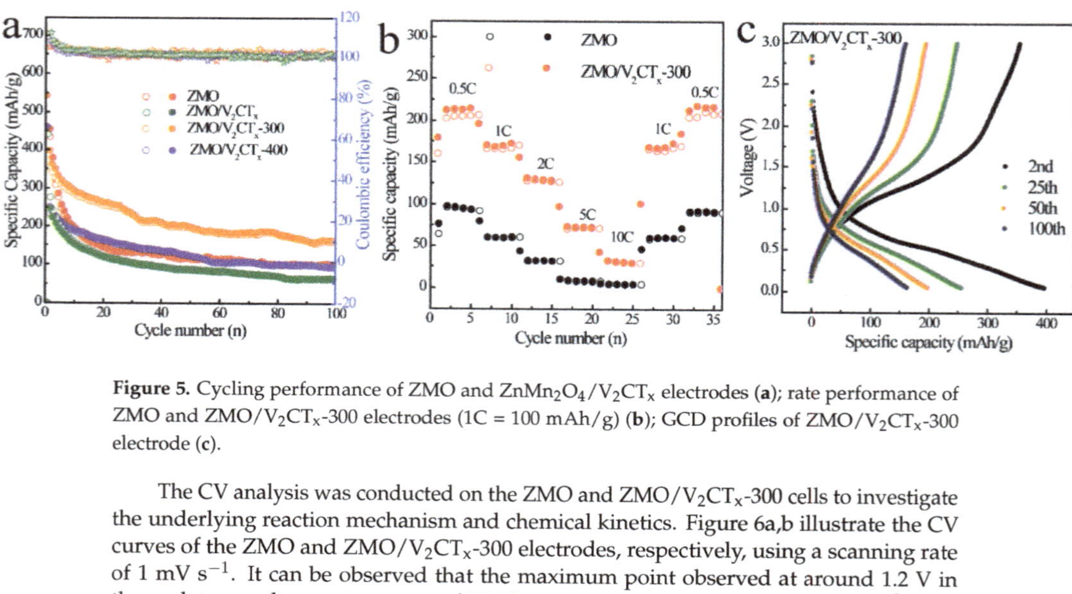

Figure 5. Cycling performance of ZMO and $ZnMn_2O_4/V_2CT_x$ electrodes (**a**); rate performance of ZMO and ZMO/V_2CT_x-300 electrodes (1C = 100 mAh/g) (**b**); GCD profiles of ZMO/V_2CT_x-300 electrode (**c**).

The CV analysis was conducted on the ZMO and ZMO/V_2CT_x-300 cells to investigate the underlying reaction mechanism and chemical kinetics. Figure 6a,b illustrate the CV curves of the ZMO and ZMO/V_2CT_x-300 electrodes, respectively, using a scanning rate of 1 mV s^{-1}. It can be observed that the maximum point observed at around 1.2 V in the cyclotron voltammetry curve of ZMO corresponds to the conversion of Mn^{3+} into Mn^{2+}. Similarly, the peak at approximately 0.3 V corresponds to the transformation of both Mn^{2+} and Zn^{2+} ions into elemental forms of Mn and Zn, respectively, resulting in the subsequent creation of the Li-Zn alloy at this specific potential. This can be represented by Equation [20]:

$$ZnMn_2O_4 + 8Li^+ + 8e^- \leftrightarrow Zn + 2Mn + 4Li_2O, \quad (1)$$

$$Zn + xLi^+ + xe^- \leftrightarrow Li_xZn\ (0 < x < 1). \quad (2)$$

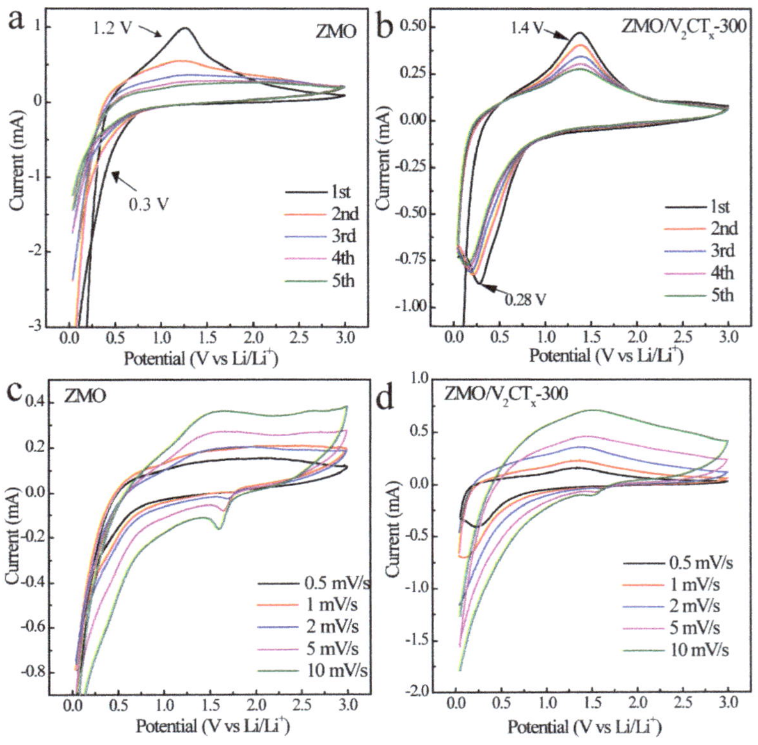

Figure 6. CV curves of ZMO and ZMO/V_2CT_x-300 electrodes at 1 mV s^{-1} (**a**,**b**); CV curves of ZMO and ZMO/V_2CT_x-300 electrodes at different scan rates (**c**,**d**).

The subsequent similarity of the curves demonstrates the robustness of the ZMO electrochemical properties. In comparison to Figure 6a,b exhibits certain alterations, such as a shift in its oxidation peak from 1.2 to 1.4 V, which aligns more closely with the oxidation peak position observed for V_2CT_x. The corresponding reaction equations for this transformation are provided below [21]:

$$V_2C/V_2CY_2 + xLi^+ + xe^- \leftrightarrow V_2CLi_x/V_2CY_2Li_x\ (Y = F/OH). \quad (3)$$

The chemical kinetics of the cell were investigated using cyclovoltammetry at various scan rates (0.5, 1, 2, 5, and 10 mV s^{-1}), as depicted in Figure 6c,d. It was noted in the CV spectrum that a higher scanning rate led to a displacement of the oxidation peak towards higher potential, a movement of the reduction peak towards lower potential, and an enlargement of the current area due to Li^+ polarization during continuous insertion and extraction processes.

4. Conclusions

In summary, $ZnMn_2O_4/V_2CT_x$ composites were synthesized through high-temperature calcination. The results demonstrate that the structure and morphology of the $ZnMn_2O_4/V_2CT_x$ material undergo alterations upon different temperature treatments. At 400 °C, the $Mn_2V_2O_7$ phase emerges, leading to a transformation in the morphology of the $ZnMn_2O_4/V_2CT$ material from rod-like bonds to spherical structures and subsequently scaffold-like structures. Despite undergoing 100 cycles at a nominal specific capacity of 100 mAh g^{-1}, the ZMO/V_2CT_x-300 electrode exhibits an enduring specific discharge capacity of 163 mAh g^{-1}. This signifies an estimated surge of around 64% in contrast to the discharge capability (99.5 mAh g^{-1})

witnessed in the electrode composed solely of $ZnMn_2O_4$. These findings suggest that recombination and calcination processes effectively enhance both ionic transport capacity and conductivity within $ZnMn_2O_4$ materials. This research offers valuable perspectives on the effective utilization of AB_2O_4-based materials and MXene materials as electrodes with exceptional performance.

Author Contributions: Conceptualization, M.L. and L.W.; methodology, J.L., Y.W. and X.P.; software, C.Z.; validation, Q.Z. and W.H.; formal analysis, M.L.; investigation, J.L.; resources, L.W.; data curation, J.L.; writing—original draft preparation, J.L.; writing—review and editing, L.W.; visualization, M.L.; supervision, M.L.; project administration, L.W.; funding acquisition, L.W. All authors have read and agreed to the published version of the manuscript.

Funding: The authors gratefully acknowledge the financial support provided by the Development Plan of Science and Technology of Jilin Province (20230101281JC) and the Science and Technology Research Project of Education Department of Jilin Province (JJKH20230513KJ, JJKH20230492KJ).

Data Availability Statement: The raw/processed data required to reproduce these findings cannot be shared at this time as the data also forms part of an ongoing study.

Conflicts of Interest: The authors declare that there are no conflict of interest.

References

1. Zhu, L.; Li, F.; Yao, T.; Liu, T.; Wang, J.; Li, Y.; Lu, H.; Qian, R.; Liu, Y.; Wang, H. Electrospun $MnCo_2O_4$ Nanotubes High-Performance Anode Materials for Lithium-Ion Batteries. *Energy Fuels* **2020**, *34*, 11574–11580. [CrossRef]
2. Deng, Z.; He, M.; Feng, Y.; Xiong, D. A fluorine-doped $MnFe_2O_4$ nanorod/carbon composite as an anode material for h performance lithium-ion batteries. *Int. J. Electrochem. Sci.* **2020**, *15*, 4203–4217. [CrossRef]
3. Liang, C.; Chen, J.; Yu, K.; Jin, W. $ZnMn_2O_4$ spheres anchored on jute porous carbon for use as a high-performance anode mate in lithium-ion batteries. *J. Alloys Compd.* **2021**, *878*, 160445. [CrossRef]
4. Rajoba, S.J.; Kale, R.D.; Kulkarni, S.B.; Parale, V.G.; Patil, R.; Olin, H.; Park, H.-H.; Dhavale, R.P.; Phadatare, M. Synthesis Electrochemical Performance of Mesoporous $NiMn_2O_4$ Nanoparticles as an Anode for Lithium-Ion Battery. *J. Compos. Sci.* 2 5, 69. [CrossRef]
5. Lu, H.; Qian, R.; Zhu, L.; Yao, T.; Li, C.; Li, L.; Wang, H. Phase structure engineering of $MnCo_2O_x$ within electrospun car nanofibers towards high-performance lithium-ion batteries. *J. Colloid Interface Sci.* **2022**, *607*, 171–180. [CrossRef]
6. Zhang, F.; Jia, Z.; Wang, C.; Feng, A.; Wang, K.; Hou, T.; Liu, J.; Zhang, Y.; Wu, G. Sandwich-like silicon/$Ti_3C_2T_x$ MXene compo by electrostatic self-assembly for high performance lithium ion battery. *Energy* **2020**, *195*, 117047. [CrossRef]
7. Fan, P.; Liu, H.; Marosz, V.; Samuels, N.T.; Suib, S.L.; Sun, L.; Liao, L. High Performance Composite Polymer Electrolytes Lithium-Ion Batteries. *Adv. Funct. Mater.* **2021**, *31*, 2101380. [CrossRef]
8. Li, P.; Kim, H.; Myung, S.-T.; Sun, Y.-K. Diverting Exploration of Silicon Anode into Practical Way: A Review Focused Silicon-Graphite Composite for Lithium Ion Batteries. *Energy Storage Mater.* **2021**, *35*, 550–576. [CrossRef]
9. Dong, L.; Hao, J.; Liu, H.; Shi, W.; Yang, J.; Lian, J. Three-Dimensional $ZnMn_2O_4$ Nanoparticles/Carbon Cloth Anodes High-Performance Flexible Lithium-Ion Batteries. *ChemistrySelect* **2020**, *5*, 2372–2378. [CrossRef]
10. Chen, Y.; Xu, Y.; Li, Z.; Zhang, W.; Zheng, M.; Zhang, H. Biomass-mediated synthesis of carbon-supported $ZnMn_2O_4$ nanoparti as high-performance anode materials for lithium-ion batteries. *Colloids Surf. A Physicochem. Eng. Asp.* **2020**, *600*, 2738–2 [CrossRef]
11. Min, X.; Zhang, Y.; Yu, M.; Wang, Y.; Yuan, A.; Xu, J. A hierarchical dual-carbon supported $ZnMn_2O_4$/C composite as an an material for Li-ion batteries. *J. Alloys Compd.* **2021**, *877*, 160242. [CrossRef]
12. Chen, S.; Yao, M.; Wang, F.; Wang, J.; Zhang, Y.; Wang, Y. Facile microemulsion synthesis of mesoporous $ZnMn_2O_4$ submicrocu as high-rate and long-life anodes for lithium ion batteries. *Ceram. Int.* **2019**, *45*, 5594–5600. [CrossRef]
13. Pang, F.; Hou, S.; Wang, P.; Liu, M.; Luo, Y.; Zhao, L. β-MnO_2/Metal-Organic Framework Derived Nanoporous ZnMn Nanorods as Lithium-Ion Battery Anodes with Superior Lithium-Storage Performance. *Chem. Eur. J.* **2019**, *25*, 5043–5 [CrossRef] [PubMed]
14. Sun, M.; Li, S.; Zou, J.; Cui, Z.; Zhang, Q.; Schiettekatte, F.; Barba, D.; Liu, B.; Wang, Y. Graphene-Wrapped $ZnMn_2O_4$ Nanoparti with Enhanced Performance as Lithium-Ion Battery Anode Materials. *Nano* **2020**, *15*, 2050117. [CrossRef]
15. Tang, Q.; Shi, Y.; Ding, Z.; Wu, T.; Wu, J.; Mattick, V.; Yuan, Q.; Yu, H.; Huang, K. Three-dimensional hierarchical graphene CNT-coated spinel $ZnMn_2O_4$ as a high-stability anode for lithium-ion batteries. *Electrochim. Acta* **2020**, *338*, 135853. [CrossR
16. Kommu, P.; Singh, G.; Chakra, C.S.; Jana, S.; Kumar, V.; Bhattacharyya, A. Preparation of $ZnMn_2O_4$ and $ZnMn_2O_4$/graph nano composites by combustion synthesis for their electrochemical properties. *Mater. Sci. Eng. B* **2020**, *261*, 114647. [CrossRe
17. Lu, M.; Zhang, Y.; Chen, J.; Han, W.; Zhang, W.; Li, H.; Zhang, X.; Zhang, B. K^+ alkalization promoted Ca^{2+} intercalation in V_2 MXene for enhanced Li storage. *J. Energy Chem.* **2020**, *49*, 358–364. [CrossRef]

Liu, F.; Liu, Y.; Zhao, X.; Liu, K.; Yin, H.; Fan, L. Prelithiated V_2C MXene: A High-Performance Electrode for Hybrid Magnesium/Lithium-Ion Batteries by Ion Cointercalation. *Small* **2020**, *16*, 1906076. [CrossRef] [PubMed]
Zhang, R.; Xue, Z.; Qin, J.; Sawangphruk, M.; Zhang, X.; Liu, R. NiCo-LDH/Ti_3C_2 MXene hybrid materials for lithium ion battery with high-rate capability and long cycle life. *J. Energy Chem.* **2020**, *50*, 143–153. [CrossRef]
Feng, C.; Wang, W.; Chen, X.; Wang, S.; Guo, Z. Synthesis and electrochemical properties of $ZnMn_2O_4$ anode for lithium-ion batteries. *Electrochim. Acta* **2015**, *178*, 847–855. [CrossRef]
Hu, J.; Xu, B.; Ouyang, C.; Yang, S.A.; Yao, Y. Investigations on V_2C and V_2CX_2 (X = F, OH) Monolayer as a Promising Anode Material for Li Ion Batteries from First-Principles Calculations. *J. Phys. Chem. C* **2014**, *118*, 24274–24281. [CrossRef]

micromachines

MDPI

Review

Review of Layered Transition Metal Oxide Materials for Cathodes in Sodium-Ion Batteries

Mehdi Ahangari, Meng Zhou * and Hongmei Luo *

Department of Chemical and Materials Engineering, New Mexico State University, Las Cruces, NM 88003, USA
* Correspondence: mzhou@nmsu.edu (M.Z.); hluo@nmsu.edu (H.L.)

Abstract: The growing interest in sodium-ion batteries (SIBs) is driven by scarcity and the rising costs of lithium, coupled with the urgent need for scalable and sustainable energy storage solutions. Among various cathode materials, layered transition metal oxides have emerged as promising candidates due to their structural similarity to lithium-ion battery (LIB) counterparts and their potential to deliver high energy density at reduced costs. However, significant challenges remain, including limited capacity at high charge/discharge rates and structural instability during extended cycling. Addressing these issues is critical for advancing SIB technology toward industrial applications, particularly for large-scale energy storage systems. This review provides a comprehensive analysis of layered sodium transition metal oxides, focusing on their structural properties, electrochemical performance, and degradation mechanisms. Special attention is given to the intrinsic and extrinsic factors contributing to their instability, such as structural phase transitions, and cationic/anionic redox behavior. Additionally, recent advancements in material design strategies, including doping, surface modifications, and composite formation, are discussed to highlight the progress toward enhancing the stability and performance of these materials. This work aims to bridge the knowledge gaps and inspire further innovations in the development of high-performance cathodes for sodium-ion batteries.

Keywords: sodium-ion batteries; layered transition metal oxides; cathode materials

1. Introduction

The urgent global need for clean, affordable, and sustainable energy storage systems has become increasingly critical due to the escalating effects of climate change, the rapid depletion of fossil fuel resources, and rising environmental pollution. To address these challenges, the adoption of advanced technologies for energy storage and conversion has gained immense importance. Over the past few decades, renewable energy sources such as wind, solar, nuclear, and wave energy have emerged as promising alternatives. However, the intermittent nature of these energy sources, influenced by factors like time, weather, and location, has limited their widespread application, underscoring the need for efficient energy storage systems [1–4].

Electrochemical energy storage devices, including batteries, fuel cells, and supercapacitors, are among the most effective technologies for storing electricity and powering modern devices. Among these, rechargeable (secondary) batteries stand out for their superior energy and power density compared to fuel cells and supercapacitors. Since their commercialization by Sony in the 1990s, lithium-ion batteries (LIBs) have become a cornerstone of modern society, transforming how we interact with technology by enabling wireless and portable solutions. LIBs have dominated battery technology for decades, powering devices such as laptops, smartphones, and even electric and hybrid vehicles, significantly reducing dependence on fossil fuels [5–8].

Lithium's unique properties, including high voltage, energy density, and specific power compared to other elements (Figure 1a) make it an ideal material for battery applications, particularly in the automotive industry. However, the surging demand for LIBs has led to a steep rise in the cost of lithium-based precursors like lithium carbonate (Li_2CO_3) over the past decade (Figure 1b). Additionally, the geographically constrained lithium reserves, predominantly located in Latin America and Australia, have raised concerns about supply security and cost, especially for large-scale applications. These challenges highlight the need for alternative energy storage solutions that are cost-effective, widely available, and sustainable (Figure 1c) [9–11].

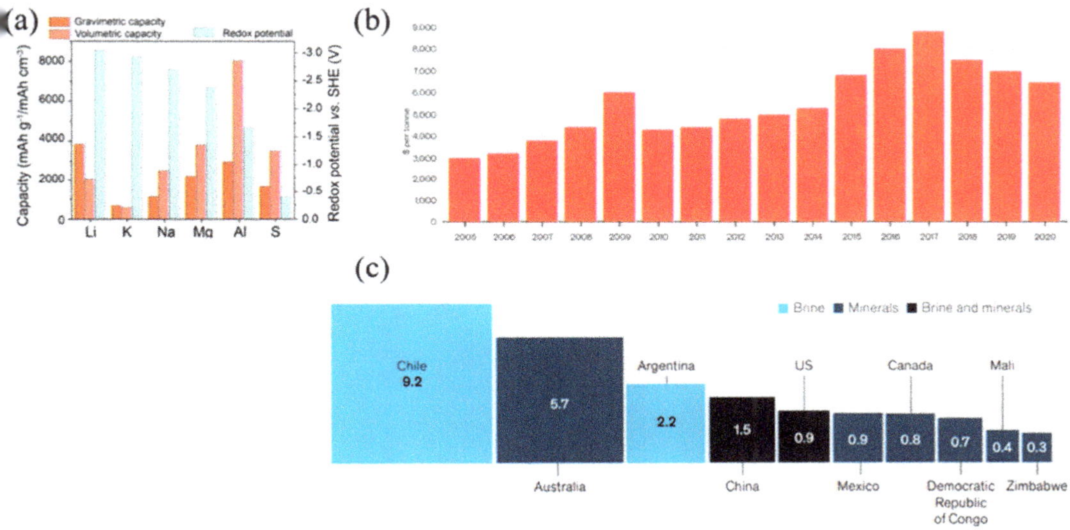

Figure 1. (**a**) Comparison of redox potentials and gravimetric/volumetric capacities for Li, K, Na, Mg, Al, and S. Reprinted from Ref. [9] with permission from Nature, (**b**) price trends of lithium carbonate from 2005 onward. Reprinted from Ref. [10], and (**c**) top 10 countries with the largest lithium reserves, measured in million metric tons. Reprinted from Ref. [11].

Sodium-ion batteries (SIBs) share a similar architecture and working principles with LIBs (Figure 2), making them a promising alternative. They are environmentally benign, cost-effective, and abundant, addressing several limitations associated with LIBs. Research into SIBs began in the early 1980s, but the successful commercialization of LIBs in the 1990s shifted focus away from sodium-based systems [12]. Table 1 provides a general comparison between sodium and lithium batteries, highlighting their key characteristics and differences.

In addition to SIBs, solid-state sodium batteries are emerging as a promising candidate for the future of large-scale energy storage applications. The organic electrolytes currently used in SIBs—such as propylene carbonate, ethylene carbonate, and dimethyl carbonate—struggle to meet the performance requirements of both cathodes and anodes, particularly in achieving high electrochemical performance. Moreover, the flammability of these liquid electrolytes raises significant safety concerns, as seen in lithium-ion batteries, particularly in large battery systems for vehicles and grid applications. Solid electrolytes offer a compelling solution by effectively mitigating the risks of short circuits caused by sodium dendrites and eliminating the hazards associated with flammable organic electrolytes. The use of solid-state batteries, which leverage the inherent non-flammability of solid electrolytes, is anticipated to address these safety issues comprehensively [13,14].

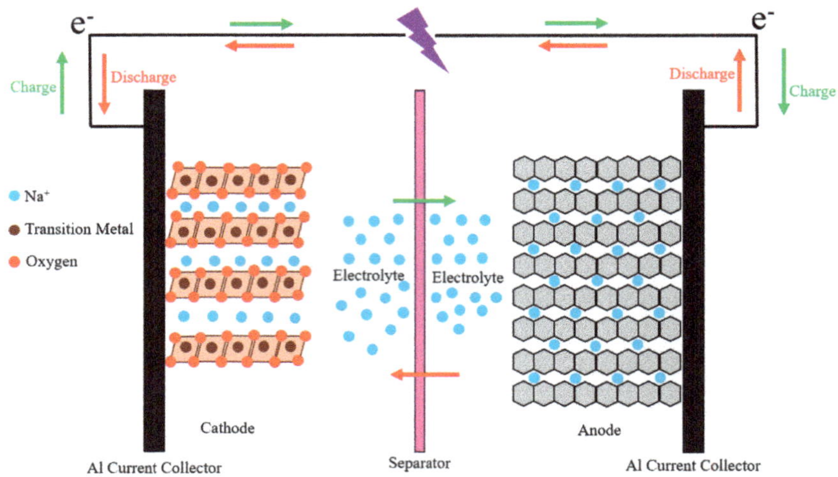

Figure 2. A schematic view of SIB.

Table 1. General comparison between sodium and lithium batteries.

Characteristic	Sodium Batteries	Lithium Batteries
Energy density	150–200 Wh kg^{-1}	200–300 Wh kg^{-1}
Nominal voltage	3.2–3.3 V	3.6–3.7 V
Environmentally friendly	Can be transported at zero volt	Should be always stored with a minimum charge
Availability	Mostly worldwide available at about 500 times higher the lithium	Availability is limited to some countries and not abundant
Raw material price	~330 U.S. dollar per metric ton for Na_2CO_3	~68,000 U.S. dollar per metric ton for Li_2CO_3
Lifespan	5–10 years	5–15 years

Several types of solid electrolytes have been explored for rechargeable sodium batteries, including Na-ion conductive ceramics, polymers, and ceramic–polymer composites. Among these, organic polymer electrolytes—created by dissolving sodium salts into a polymer matrix—have garnered significant attention. These materials offer high flexibility, excellent thermal stability, and the ability to form intimate contact with both electrodes and inorganic solid electrolytes. With moderate ionic conductivity (>10^{-4} S cm^{-1} at room temperature), polymer-based solid electrolytes remain one of the most widely investigated options for solid-state sodium batteries [15–18].

Sodium is one of the most abundant elements in the Earth's crust, with a Clarke number of 2.64 (Figure 3a). It exists in various forms such as sodium carbonate (Na_2CO_3), sodium chloride (NaCl), and sodium sulfate (Na_2SO_4), commonly found in minerals and brines. Sodium also exhibits an electrochemical redox potential close to that of lithium (−2.71 V for sodium versus −3.04 V for lithium, both measured against the standard hydrogen electrode), making it suitable for battery applications [19,20].

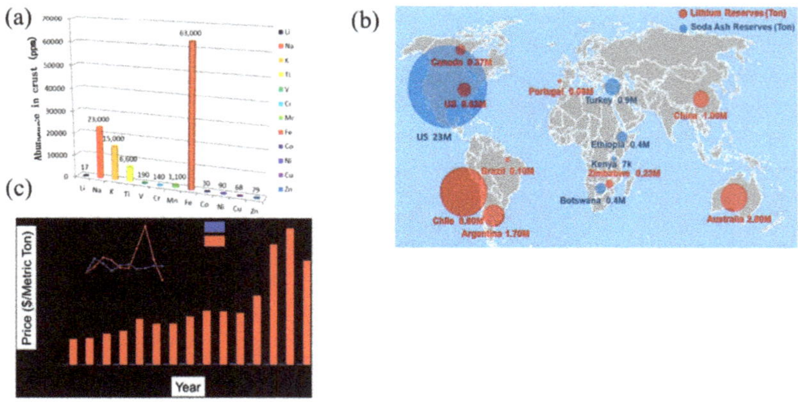

Figure 3. (**a**) Abundance of elements in the Earth's crust reprinted from Ref. [21] with permission from Elsevier, (**b**) global distribution of lithium and sodium reserves as of 2020, represented by red (lithium) and blue (soda ash) circles, with circle size proportional to reserve quantity in metric tons. Light blue shading indicates oceans as additional sodium sources from brine, and (**c**) price trends for sodium carbonate and lithium carbonate from 2005 to 2019, with an inset highlighting the percentage price changes of both materials over the past decade. Reprinted from Ref. [22] with permission from Wiley.

The abundance of sodium resources makes it a highly accessible material, with significantly lower costs compared to lithium. For example, the cost of trona, a key source of sodium carbonate, ranges between USD 135 and USD 165 per ton, whereas lithium carbonate, a precursor material for LIBs, cost approximately USD 5000 per ton in 2010. This stark contrast in resource availability and cost highlights sodium as a viable alternative for scalable and cost-effective energy storage systems (Figure 3b,c). Therefore, the cost per kilowatt-hour (USD/kWh) for SIBs can be significantly reduced due to the absence of expensive transition metals like cobalt and nickel in the cathode materials.

Additionally, aluminum can replace copper as the current collector for the anode because sodium does not react with aluminum. This allows the aluminum current collector to remain stable in the electrolyte at 0 V, unlike in LIBs, where lithium and copper form a binary alloy at low potentials (below 0.1 V vs. Li/Li^+).

However, sodium's larger ionic radius (1.02 A° compared to 0.76 A° for lithium) and heavier atomic mass (23 g mol^{-1} vs. 7 g mol^{-1} for lithium) contribute to slower ion kinetics during the (de)sodiation process and lower gravimetric capacity (1165 mAh g^{-1} compared to 3829 mAh g^{-1}). This leads to structural instability in the host materials, increased heat generation from exothermic reactions, and a lower theoretical energy density. While high energy density is not a primary requirement for stationary, in large-scale energy storage systems—such as those paired with solar panels where cost efficiency is prioritized—these issues still negatively affect the performance of SIBs [23–25].

Moreover, the solid electrolyte interphase (SEI) formed on the electrodes in SIBs is less stable due to sodium's lower Lewis acidity, leading to a high solubility of the SEI components. This results in incomplete coverage of the electrode surface, promoting side reactions and excessive heat generation. These factors highlight that the electrochemical behavior of SIBs cannot be directly extrapolated from LIBs. To overcome these challenges and improve performance, comprehensive studies are essential to understanding the unique electrochemical processes and underlying mechanisms of SIBs.

Cathode materials reported for SIBs include layered oxides, polyanions, organic compounds, and Prussian blue and its analogues. Despite their potential, these materials face persistent challenges such as low ionic conductivity and thermal instability, which limit their performance. Nonaqueous organic liquid electrolytes are predominantly used in SIBs due to their wide electrochemical stability window, high ionic conductivity, and efficient mass transfer at the electrolyte–electrode interface. However, they also present safety concerns, including flammability and potential for thermal runaway.

Identifying a cathode material with superior electrochemical properties—such as high capacity, long cycle life, and optimal operating voltage—remains a top priority for advancing SIB technology. Currently, cathode materials in SIBs typically exhibit specific capacities ranging from 120 to 200 mA h g^{-1}, with operating potentials between 2.6 and 3.2 V. These values are considerably lower than the oxidation potential of electrolytes, which exceeds 4.0 V. In contrast, key anode materials offer specific capacities of 300–600 mA h g^{-1}, with operating potentials between 0.1 and 0.6 V, which are close to the potential for sodium plating. Consequently, cathode materials not only represent a larger mass fraction in SIBs, but they also offer more potential for enhancing the overall voltage of the system. Therefore, cathodes play a critical role in determining the cost, safety, specific energy, cycle life, and specific power of SIBs. Unfortunately, the fundamental differences in intercalation chemistry between lithium and sodium often result in the underperformance of sodium-based analogues of lithium-based compounds. These differences underscore the need for tailored materials and innovative design strategies to optimize the performance of SIB cathodes [26–30].

The α-$NaFeO_2$ (R-3m) structured materials hold a pivotal role in the development of LIBs and serve as a foundation for similar materials in SIBs. These materials feature a layered structure composed of slabs of edge-sharing MO_6 octahedra, with alkali ions intercalated between the layers [31]. Similar to lithium-based transition metal oxides, such as $LiCoO_2$ and Li(Ni, Mn)O_2, sodium transition metal oxides (Na_xTMO_2, where TM = transition metals and $0.67 \leq x \leq 1$) are commonly used as positive electrodes in SIBs [32–34].

Typical sodium layered oxides can be synthesized in O3-, P2-, and P3-phases, which are determined by the surrounding Na environment and the number of unique oxide layer packings. The notation O3, P3, and P2 describe the coordination environment and stacking order: O indicates that sodium ions occupy octahedral (O) sites, while P signifies sodium ions in trigonal prismatic (P) sites. The numbers 3 and 2 represent the number of edge-sharing MO_6 octahedra in the oxygen stacking sequence, arranged in ABCABC or ABBA patterns, respectively [35]. The O3 phase typically occurs when $0.7 \leq x \leq 1$. The oxide layer stacking follows the ABCABC pattern, where all Na ions share one edge and one face with adjacent octahedra. When $0.6 < x < 0.7$, the P2 phase is known to occur, which features oxide layer stacking that follows the ABBA pattern, with Na ions sharing either entirely edge or entirely face positions with adjacent octahedra. P3 typically occurs when $x \sim 0.5$, the oxide layer stacking in this phase follows the ABBCCA pattern, where Na ions share one face with a transition metal octahedron (MO_6) and three edges with three adjacent MO_6 octahedra, as illustrated in Figure 4 [36]. The O3 phases belong to the R-3m space group, whereas P2 phases crystallize in the P63/mmc space group. Structural distortions in these phases, such as monoclinic (C2/m) or orthorhombic (Cmcm) modifications, are denoted by a prime symbol (for example, O3′ or O′3) [37,38].

The extraction of alkali ions during battery operation creates vacancies in the Li or Na layers, which can facilitate the migration of TMs into these sites. This migration leads to irreversible structural degradation and poor electrochemical performance in layered oxides. The phenomenon of TM migration has been extensively studied in LIBs cathode materials. For instance, Wang et al. [39] observed the migration of Ni and Co ions in

$LiNi_{1/3}Mn_{1/3}Co_{1/3}O_2$ and attributed it to structural distortions and the formation of irreversible phases.

Both experimental observations and theoretical calculations indicate that distortions in the MO_6 octahedron play a critical role in promoting TM migration. These distortions may arise from TM oxidation or oxygen oxidation during cycling [40,41]. A key characteristic of these oxidation processes is the alteration in TM-O bond lengths, which induces strain within the MO_6 octahedra or the O-TM-O slabs. This strain serves as a driving force for TM migration, ultimately impacting the structural integrity and electrochemical stability of the cathode material.

The cycling life of cathode materials in SIBs is primarily influenced by the structural stability of the materials, as well as the electrochemical stability of the compatible electrolytes. Electrochemical behavior is significantly influenced by the structure of the phase, not only because of the amount of Na in the pristine state but also due to the stability of each layer. The kinetics of sodium insertion and extraction are affected by the surrounding environment of Na within the layered structure, which impacts the overall performance of the material. The stability of each layer plays a crucial role in maintaining structural integrity during cycling, while the arrangement and interaction of Na ions within the structure influence the efficiency of ion transport and the electrochemical response. This review focuses on the phase transitions, structural stability, and electrochemical characteristics of transition metal layered oxide materials, as well as various strategies reported to enhance the stability of these materials for SIBs.

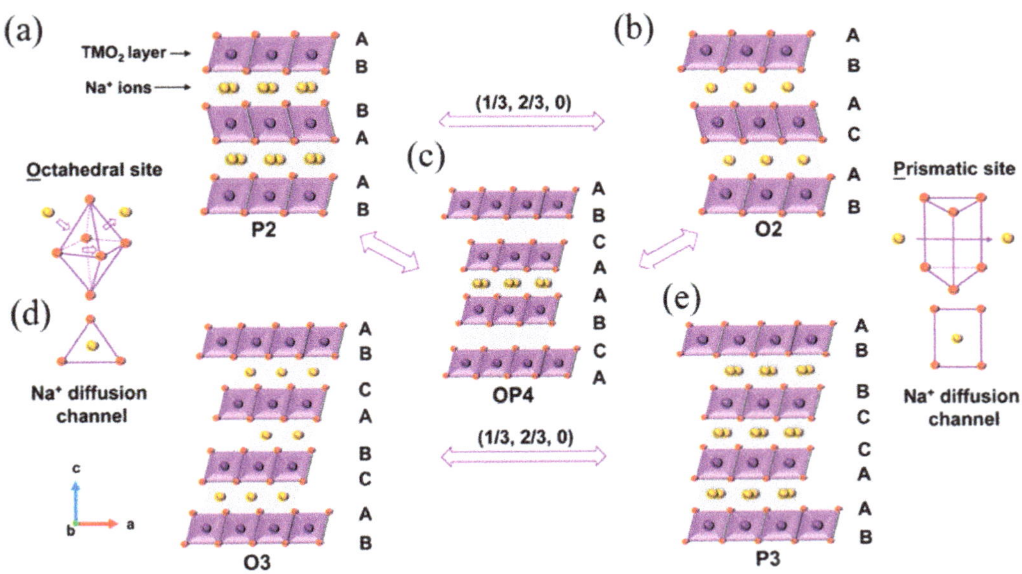

Figure 4. Schematic illustration of the typical crystal structures of Na_xTMO_2, showing (**a**) P2, (**b**) O2, (**c**) OP4, (**d**) O3, and (**e**) P3 phases. Purple spheres represent transition metals, yellow spheres denote Na^+ ions, and red spheres indicate O^{2-} ions. Reprinted from Ref. [42] with permission from Wiley.

2. Materials Classification

2.1. $NaCrO_2$

Among Na_xMO_2 compounds ($0 < x \leq 1$, M = Ni, Co, Mn, Fe, V, and Cr), O3-$NaCrO_2$ has garnered significant attention as a promising cathode material for sodium-ion batteries. It delivers a discharge capacity of 120 mAh g^{-1} with a desirable voltage plateau at ~3 V (vs.

Na/Na^+) and exhibits excellent thermal stability compared to $LiCoO_2$ and $LiFePO_4$. This stability is attributed to the robust chemical stability of Cr^{4+} within CrO_2 slabs at octahedral sites, combined with high-rate capability and excellent coulombic efficiency [43–45].

Interestingly, while $LiCrO_2$, which shares a layered rock salt structure with $LiCoO_2$ and $LiNiO_2$, is electrochemically inactive, $NaCrO_2$ demonstrates high capacity. This discrepancy stems from differences in interstitial spaces between the two compounds. In $LiCrO_2$, the irreversible migration of Cr ions into tetrahedral sites during lithium de-intercalation renders the material inactive after the first charge. In contrast, $NaCrO_2$ avoids this migration when $x \leq 0.5$ due to its larger tetrahedral sites, although further sodium removal ($x > 0.5$) leads to gradual inactivation [46,47]. Structurally, both $LiCrO_2$ and $NaCrO_2$ feature Cr in the trivalent state with a d^3 electronic configuration ($t_{2g}{}^3$, $e_g{}^0$) under octahedral coordination. Upon oxidation, Cr(III) forms the unstable Cr(IV) (d^2), which disproportionate into Cr(III) (d^3) and Cr(VI) (d^0) due to their higher stability. Cr(VI) typically adopts a tetrahedral coordination with four oxygen atoms, as seen in $Cr(VI)O_4{}^{2-}$ of $Li_2CrO_4 \cdot 2H_2O$. The dimensional match between the interstitial tetrahedron and $Cr(VI)O_4{}^{2-}$ plays a crucial role in structural stability [48].

For $LiCrO_2$, the interstitial tetrahedron with an O-O bond length of 2.89 A° and a height of 2.56 A° closely aligns with the $Cr(VI)O_4{}^{2-}$ tetrahedron. However, in $NaCrO_2$, the interstitial tetrahedron within the NaO_2 slab exhibits an O-O bond length of 2.97 A° and a height of 3.15 A°, better accommodating the $Cr(VI)O_4$ structure. Consequently, the bond length mismatch in $NaCrO_2$ prevents irreversible Cr migration, unlike in $LiCrO_2$ (Figure 5). Magnetic measurements of chemically deintercalated $Na_{1-x}CrO_2$ confirm the presence of Cr(IV) during sodium deintercalation.

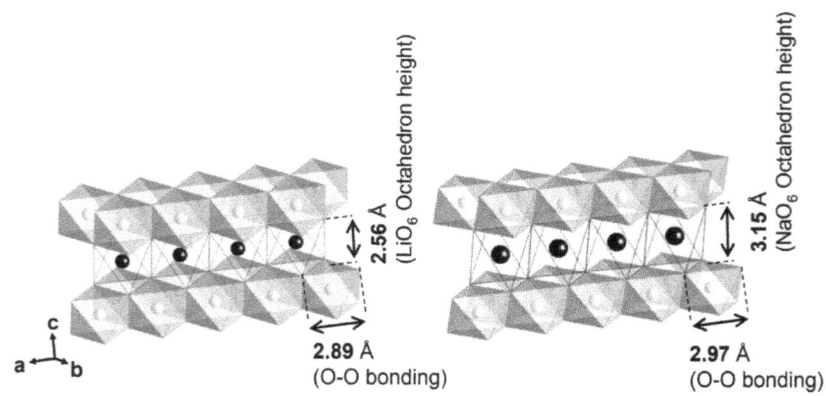

Figure 5. Structural comparison of layered $LiCrO_2$ (**left**) and $NaCrO_2$ (**right**) crystal models. The illustration highlights the differences in interstitial tetrahedral dimensions and their impact on chromium ion migration during electrochemical cycling. Reprinted from Ref. [48] with permission from IOP Publishing.

In $NaCrO_2$, oxygen atoms at the 6c sites are stacked along the c-axis, with Na and Cr ions alternately distributed on both sides of the oxygen layers at the 3a and 3b octahedral sites. This layered structure enables two-dimensional Na^+ ion transport between CrO_2 slabs, enhancing ionic mobility. However, consecutive phase transitions during cycling—from O3 to O′3 and eventually P3—pose significant challenges.

To date, most $NaCrO_2$ materials have been synthesized via solid-state reactions, which suffer from capacity decay and poor power performance due to irreversible phase transitions, low Na^+ diffusivity, and side reactions at the electrode–electrolyte interface [7].

Addressing these issues is critical for realizing the full potential of $NaCrO_2$ as a cathode material in sodium-ion batteries.

The initial charge and discharge curves for $NaCrO_2$ and $LiCrO_2$ are shown in Figure 6a,b, respectively. $LiCrO_2$, upon electrochemical oxidation to 4.5 V vs. Li, demonstrates a capacity of 55 mAh g^{-1}, corresponding to the formation of $Li_{0.81}CrO_2$. This capacity is consistent with the charge compensation mechanism driven by the increased valence of Cr ions during lithium extraction. A slight reversible capacity is observed in the potential range of 3.0–4.5 V; however, subsequent cycles reveal a drastic drop in capacity to less than 10 mAh g^{-1}, signifying the absence of reversible lithium intercalation. This behavior is attributed to the so-called 'freeze effect' of Cr^{6+} ions migrating into tetrahedral sites post-delithiation, as suggested by Komaba et al. [49–51].

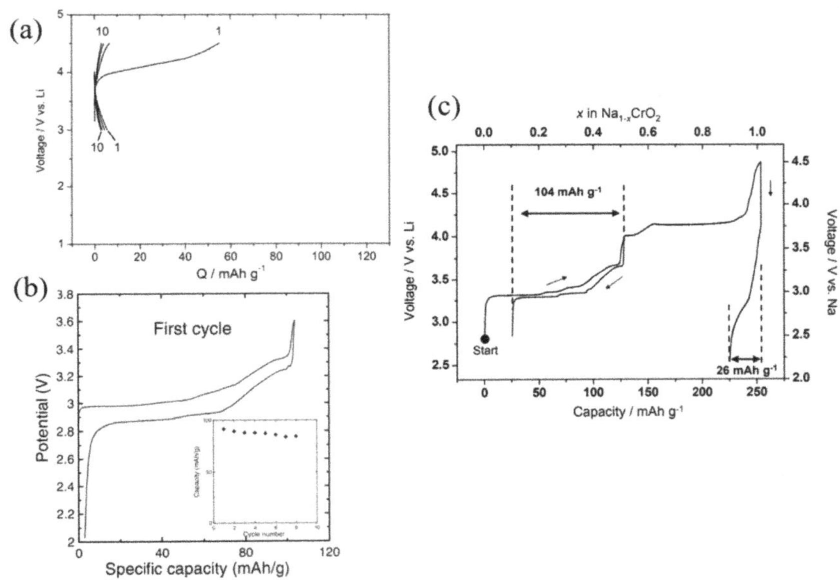

Figure 6. (**a**) Galvanostatic charge and discharge profiles of a $Li/LiCrO_2$ cell cycled at a current density of 20 mA g^{-1} within the voltage range of 3.0–4.5 V. Reprinted from Ref. [49] with permission from Elsevier, (**b**) voltage-specific capacity curve of a $NaCrO_2$ electrode in a $Na/NaCrO_2$ coin cell, cycled between 3.6 and 2.0 V at 25 mA g^{-1}. The inset illustrates the specific capacity versus cycle number. Reprinted from Ref. [52] with permission from IOP Publishing, and (**c**) initial charge and discharge profiles of $NaCrO_2$ in a $Na/NaCrO_2$ cell cycled at 5.0 mA g^{-1}. Reprinted from Ref. [48] with permission from IOP Publishing.

In the case of $NaCrO_2$ synthesized by traditional solid-state reaction of stoichiometric Cr_2O_3 and Na_2CO_3 at 900 °C in Ar environment, sodium deintercalation occurs electrochemically, exhibiting multiple plateaus (Figure 6b) corresponding to phase transitions: hexagonal O3 $\rightarrow$ monoclinic O3 $\rightarrow$ monoclinic P3 [52]. However, when cycled to a high cutoff voltage of 4.5 V, the intercalation process becomes nearly irreversible, resulting in a limited reversible capacity of 26 mAh g^{-1} (Figure 6c). This suggests that complete sodium deintercalation induces an irreversible structural change due to the destabilization of the deintercalated phase, which lacks the sodium pillars necessary to maintain the structural integrity of CrO_2 slabs.

At a lower cutoff voltage of 3.6 V, the charge–discharge curves reveal highly reversible behavior with a capacity of 104 mAh g^{-1}. This highlights that extracting more than

50% of sodium ions from $NaCrO_2$ disrupts crystallinity, transitioning the material into a highly disordered and irreversible phase. The irreversible order-to-disorder transition during deep charging is attributed to Cr migration into vacant Na layers, driven by the disproportionation reaction of Cr(IV) at high potentials ($3Cr^{4+} \rightarrow 2\ Cr^{3+} + Cr^{6+}$). The smaller ionic radius of Cr^{6+} and its preference for tetrahedral coordination facilitate this migration. Importantly, this phenomenon occurs exclusively in O-like phases, as P-like phases lack the necessary tetrahedral sites in their sodium-ion layers.

The disordered bulk structure is believed to comprise a mixture of O3-like $Na_\delta CrO_2$ ($\delta \sim 0$) and rock salt CrO_2, as suggested by prior investigations [43]. These structural transformations underline the importance of optimizing charge cutoff voltages to mitigate irreversible phase changes and enhance the cycling stability of $NaCrO_2$ cathodes.

Figure 7a,b depict the cycling performance of a $Na/NaCrO_2$ cell. The distinct and repeatable plateaus observed in the charge and discharge curves demonstrate the reversible nature of Na^+ intercalation within the $NaCrO_2$ electrode [48].

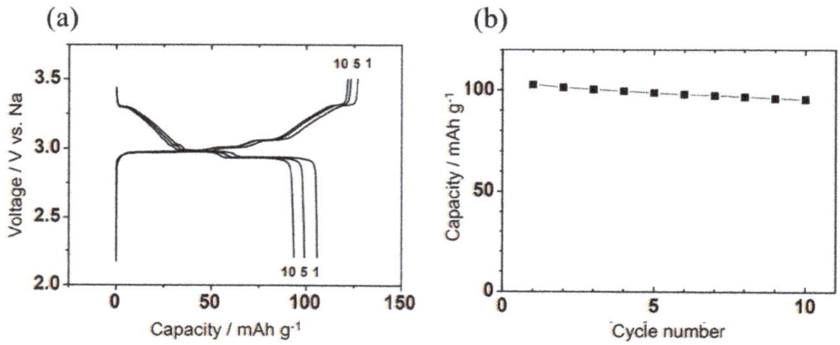

Figure 7. (**a**) Galvanostatic charge and discharge curves of a $Na/NaCrO_2$ cell recorded at a current density of 5.0 mA g^{-1} and (**b**) the corresponding variation in discharge capacity over multiple cycles, illustrating the cycling performance of the $NaCrO_2$ electrode. Reprinted from Ref. [48] with permission from IOP Publishing.

Zhou et al. [51] provided a comprehensive analysis of the reaction mechanism and real-time structural evolution of the layered $NaCrO_2$ cathode material during Na extraction. Utilizing synchrotron-based in situ XRD (Figure 8a), the structural changes in $NaCrO_2$ were examined under a current density of C/12 and a cutoff voltage of 3.6 V.

At the onset of charging, the (003) diffraction peak shifted progressively to lower two-theta angles, indicative of a solid solution reaction accompanied by expansion along the c-axis. When $x = 0.08$, a new (003) peak emerged to the left of the original peak, and its intensity increased with continued charging. This indicated the formation of a distinct phase. In the range $0.08 < x < 0.2$, a two-phase coexistence region was observed, with the intensity of the secondary (003) peak increasing at the expense of the first, while its position continued to shift to lower angles. By $x = 0.25$, the second phase dominated completely, with the original phase disappearing.

As the charge progressed to $x = 0.3$, another (003) peak appeared, and the second phase's peak broadened, marking the formation of a third phase. By $x = 0.4$, only the third phase remained, and its (003) peak shifted further, albeit to a lesser extent than the second phase. This sequence of transitions demonstrated the systematic structural evolution during Na extraction.

Additional insights were obtained by analyzing the (110) peak, initially located at 62.4°, which reflected changes in the a-b lattice plane. As charging advanced, the sharp (110) peak diminished in intensity and broadened, indicating lattice distortion in the a-b plane. Concurrently, its shift to higher two-theta angles suggested a contraction in the a and b lattice parameters.

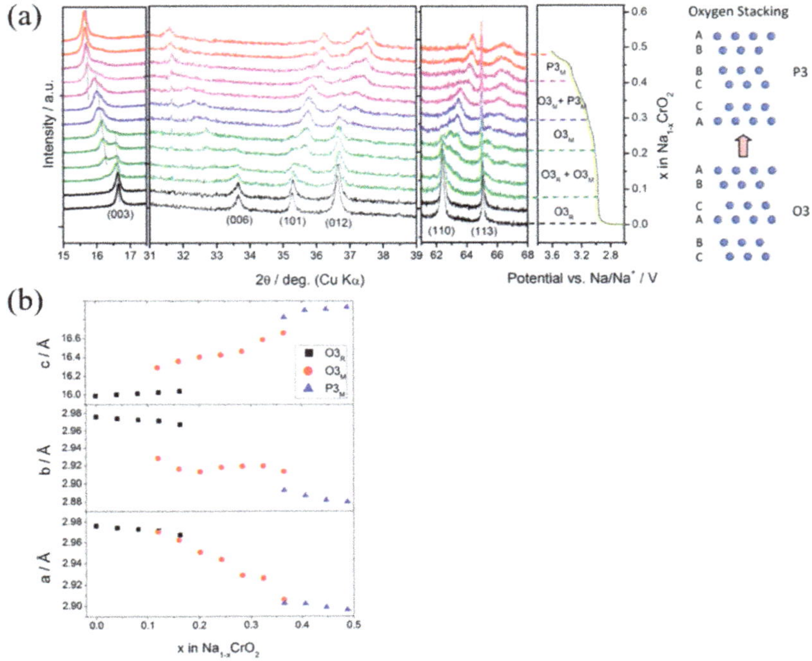

Figure 8. (**a**) In situ X-ray diffraction (XRD) patterns collected during the first charge cycle of a $Na/NaCrO_2$ cell up to 3.6 V at a C/12 rate. The displayed 2θ angles were recalculated to the corresponding values for Cu-Kα radiation (λ = 1.54 A°) from the synchrotron XRD wavelength (λ = 0.7747 A°). The voltage–composition profile is shown adjacent to the XRD patterns. On the far right, the schematic oxygen stacking arrangements for O3 and P3 structures are illustrated and (**b**) the evolution of lattice parameters (a, b, and c) as a function of composition during the first charge, derived from in situ XRD data. For clarity, the monoclinic lattice parameters were converted to their rhombohedral equivalents using the transformation equations: $a_M = \sqrt{3}\, a_R$, $b_M = b_R$, $c_M = c_R/3 \sin\beta$. Reprinted from Ref. [51] with permission from the Royal Society of Chemistry.

Based on the observed Bragg peak variations, the structural evolution of $NaCrO_2$ involved phase transitions among three solid-solution phases, interspersed with two two-phase coexistence regions. Initially, the material adopted an O3-type rhombohedral phase ($O3_R$). After x = 0.08, the second phase emerged, as evidenced by the splitting of the rhombohedral (012) peak, signaling a transition to a monoclinic phase. By x = 0.25, the second phase, identified as the monoclinic $Na_{0.74}CrO_2$ phase ($O3_M$), dominated while maintaining O3 symmetry, albeit with distorted a-b planes and ABCABC oxygen atom stacking.

As x increased from 0.3 to 0.48, the third phase grew to dominate, and its Bragg reflections corresponded to the $Na_{0.52}CrO_2$ phase, identified as a P3-type monoclinic phase ($P3_M$) with ABBCCA oxygen atom stacking. These findings highlight the complexity of the structural transformations occurring in $NaCrO_2$ during electrochemical charging and their implications for its electrochemical performance.

The evolution of the a, b, and c lattice parameters during the initial charge of the $NaCrO_2$ electrode, as derived from in situ XRD analysis, is presented in Figure 8b. During the charging process, the c-axis parameter exhibited an increase, which is attributed to the enhanced electrostatic repulsion between the MO_2 layers as sodium ions were removed.

These sodium ions act as a structural "glue" between the MO_2 layers, and their extraction leads to an expansion of the interlayer spacing.

In contrast, the a and b lattice parameters decreased, reflecting a contraction of the CrO_6 octahedra. This contraction is a direct consequence of the oxidation of Cr^{3+} to higher oxidation states during charging, resulting in shorter bond lengths and a more compact structure within the CrO_2 slabs. These lattice parameter changes collectively highlight the structural adjustments occurring in $NaCrO_2$ during sodium extraction.

The sodium extraction mechanism for the $NaCrO_2$ electrode during the charging process, as elucidated through the authors' experiments, can be summarized as follows:

Open Circuit Voltage (OCV) to 3.05 V:

$$NaCrO_2\ (O3_R) \rightarrow Na_{0.75}CrO_2\ (O3_M) + 0.25\ Na^+ + 0.25\ e^- \quad (1)$$

3.05 V to 3.60 V:

$$Na_{0.75}CrO_2\ (O3_M) \rightarrow Na_{0.5}CrO_2\ (P3_M) + 0.25\ Na^+ + 0.25\ e^- \quad (2)$$

From the OCV to approximately 3.05 V, the $O3_R$-type $NaCrO_2$ undergoes a two-phase transformation into $O3_M$-type $Na_{0.75}CrO_2$ (Reaction 1), facilitated by transitions through two solid-solution regions. Within each region, the 'a' and 'b' lattice parameters decrease, reflecting the contraction of the CrO_6 octahedra, while the 'c' lattice parameter increases due to the reduced electrostatic interactions between the MO_2 layers as sodium ions are extracted. This structural evolution is accompanied by a lattice distortion, transitioning from a rhombohedral to a monoclinic phase. During this process, Cr ions remain confined to octahedral sites, while Na ions migrate from octahedral to pseudo-tetrahedral sites.

Between 3.05 V and 3.60 V, the $O3_M$-type $Na_{0.75}CrO_2$ undergoes another two-phase transformation into the $P3_M$-type $Na_{0.5}CrO_2$ (Reaction 2), also involving two solid-solution regions. In this range, the Cr ions retain their octahedral coordination but decrease in size due to further oxidation. Simultaneously, Na ions transition from pseudo-tetrahedral to prismatic sites. These phase transitions are characterized by the gliding of oxygen planes, which is directly associated with the progressive removal of sodium ions.

Bo et al. [53] explored the structural evolution of $NaCrO_2$ under a current density of C/50 up to 4 V, corroborating the structural changes observed by Zhou et al. [51] up to 3.6 V. As illustrated in Figure 9a, further sodium extraction beyond 3.8 V, corresponding to a $Na_{0.38}CrO_2$ composition, marks the onset of irreversible electrochemical behavior. This is evidenced by the appearance of diffraction peaks associated with new phase(s), notably a broad peak around 8°. In Figure 9b, these unidentified phases are denoted with an 'X', highlighting their ambiguous nature.

The transition from the P′3 phase to the unidentified 'X' phase progresses as the cutoff voltage of 4 V is approached, resulting in a system containing approximately 0.04 Na, as inferred from the capacity measurements. The phase transition near $Na_{0.4}CrO_2$ (around 3.7 V) and the significant decrease in discharge capacity observed beyond 3.8 V imply that this structural transformation is a key contributor to the capacity fading of $NaCrO_2$.

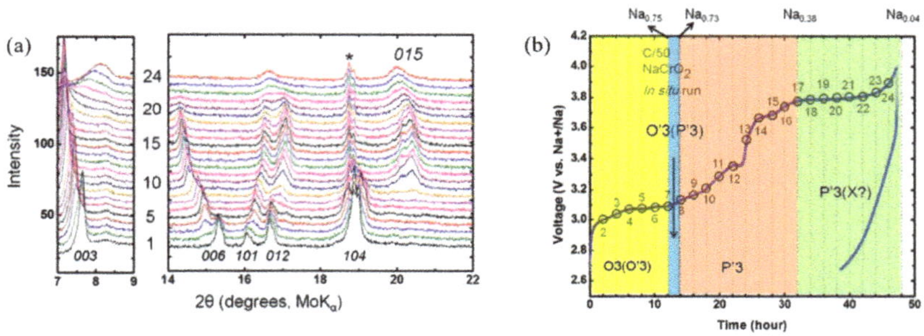

Figure 9. (**a**) In situ XRD patterns collected during the first charge of $NaCrO_2$ at a current density of C/50. A peak originating from the Al current collector overlaps with the 104 reflections of $NaCrO_2$ at approximately 19° (2θ) and is marked with an asterisk and (**b**) the corresponding electrochemical curve, with scan numbers annotated above it. The graph is divided into four color-shaded regions to emphasize the dominant reactions associated with stacking variations during $NaCrO_2$ desodiation. The critical compositions marking transitions between different reaction regimes are also indicated on the upper x-axis. Reprinted from Ref. [53] with permission from the American Chemical Society.

In addition to phase transition characterization techniques, cyclic voltammetry (CV) was employed to investigate the structural changes occurring at both reversible and irreversible voltage ranges. Jakobsen et al. [54] conducted CV experiments in the reversible potential range of 2.5–3.7 V, as depicted in Figure 10a. During the initial charge, two distinct oxidation peaks were observed at 3.07 V (with a shoulder at 3.13 V) and 3.33 V, with corresponding reduction peaks at 2.83 V and 3.28 V during the first discharge. With repeated cycling, the potential hysteresis of the lowest potential event diminished, and after five cycles, the peaks shifted to 3.04 V and 2.87 V. Additionally, the reduction peak revealed a shoulder in the range of 2.9–2.95 V, analogous to the shoulder at 3.13 V during the charging process. Importantly, this shoulder is not associated with irreversibility, as this potential range is characterized by excellent cycling reversibility.

The authors attribute these features to several crystalline-to-crystalline structural transitions, summarized in Figure 10b. Initially, monoclinic distortions result in the formation of the O′3-Na_xCrO_2 phase through a first-order two-phase transition. This phase subsequently transforms into another monoclinic phase, O′3-E-Na_xCrO_2 (where 'E' denotes an expanded interlayer distance), before transitioning to the monoclinic P′3-Na_xCrO_2 phase. Despite the asymmetry between charge and discharge processes, the overall O3 $\leftrightharpoons$ P′3 phase transitions are reversible. Furthermore, O3-$NaCrO_2$ maintains good cycling performance when less than 0.5 Na ions are extracted.

When charged beyond 3.5 V, Jakobsen et al. [43] identified an oxidation peak at 3.67 V in the CV profile. If the charge does not exceed this peak (Figure 10c), the subsequent cycles closely replicate the initial cycle, demonstrating stability in the electrochemical response. Conversely, when the charge surpasses 3.67 V (Figure 10d), subsequent cycles deviate significantly from the initial one, exhibiting broader peaks indicative of increased solid-solution behavior. Despite these changes, Na intercalation remains reversible.

At voltages approaching 4.0 V, corresponding to the complete removal of Na ions, an additional oxidation peak emerges at approximately 3.85 V. As shown in Figure 10e, deep charging to this voltage induces irreversibility in Na_xCrO_2, as evidenced by the absence of reduction peaks during the discharge process. This suggests that the extraction of the final Na ions disrupts the structural integrity of the material, rendering the intercalation mechanism non-reversible.

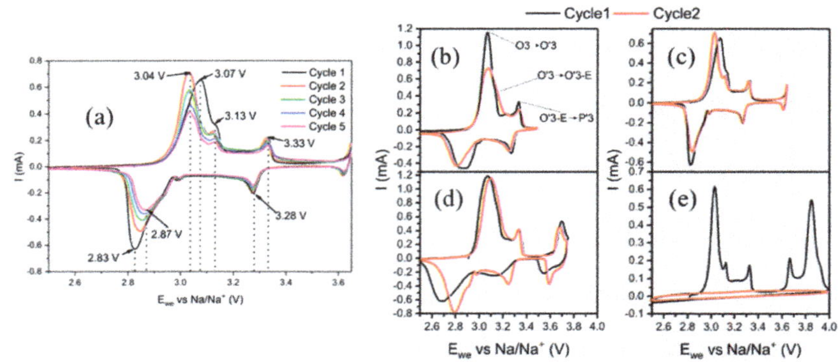

Figure 10. (**a**) Cyclic voltammogram of the first five cycles for the O3-$NaCrO_2$ electrode, ranging from 2.5 V to 3.7 V. The cycles are represented in black (cycle 1), red (cycle 2), green (cycle 3), blue (cycle 4), and magenta (cycle 5). Reprinted from Ref. [54] with permission from Elsevier. The cyclic voltammetry data of O3-$NaCrO_2$ vs. Na/Na^+ are shown for different upper potential limits. Two cycles are presented for each potential window, with the first cycle in black and the second in red. The recorded current is plotted against potential. The potential windows used are (**b**) 2.5–3.5 V, (**c**) 2.5–3.65 V, (**d**) 2.5–3.75 V, and (**e**) 2.5–4.0 V. In (**b**), the well-known crystalline-to-crystalline phase transformations (O3 → O′3, O′3 → O′3-E, and O′3-E → P′3) are noted. Reprinted from Ref. [43] with permission from Elsevier.

Yabuuchi et al. [44] investigated the thermal stability of the $NaCrO_2$ cathode material. Their results showed that $NaCrO_2$ exhibited no phase transitions during heating in the temperature range of 27–527 °C, with only thermal expansion observed. The thermal expansion of $NaCrO_2$ was significantly lower than that of $LiNi_{0.5}Mn_{0.5}O_2$. Specifically, the linear thermal expansion coefficients for $NaCrO_2$ were 0.90×10^{-5} °C^{-1} (a_{hex}) and 1.48×10^{-5} °C^{-1} (c_{hex}), compared to 1.68×10^{-5} °C^{-1} (a_{hex}) and 2.21×10^{-5} °C^{-1} (c_{hex}) for $LiNi_{0.5}Mn_{0.5}O_2$. The thermal expansion of $NaCrO_2$ was anisotropic, with the ratio of a_{hex}/c_{hex} increasing as the temperature rose. Furthermore, oxygen loss in P3-$NaCrO_2$ occurred at significantly higher temperatures (above 500 °C) compared to Ni-based cathode materials in LIBs, which experience oxygen loss between 240 and 260 °C.

In order to improve structural stability of $NaCrO_2$ cathode materials, elemental doping has been an interesting strategy, for example, Ti [55], Sn [56], Ru [57], Nb [58], Mn [59], Al [60], and Ca [46,61]. Ma et al. [62] explored the effect of Sb substitution on the structural stability of $Na_{1-2x}Cr_{1-x}Sb_xO_2$ (x = 0, 0.05, 0.1, and 0.2) as a cathode material for SIBs. The Sb^{5+} ion, a semiconducting metal ion with an ionic radius (0.0600 nm) comparable to Cr (0.0615 nm), enhances structural stability due to its high valence state and the introduction of electron holes, which modify the local electronic environment. Their findings indicated that substituting a suitable amount of Sb, specifically in $Na_{0.90}Cr_{0.95}Sb_{0.05}O_2$ (NCS-0.05), significantly improved the material's cycling and rate performance.

NCS-0.05 demonstrated exceptional cycling stability, retaining 70.9% of its initial capacity after 1000 cycles, in contrast to the 46.8% and 47.8% capacity retention observed for $NaCrO_2$ and $Na_{0.8}Cr_{0.9}Sb_{0.1}O_2$ (NCS-0.1), respectively (Figure 11a). At a low current density of 0.1 C, NCS-0.05 exhibited a reversible capacity of approximately 120 mAh g^{-1}. While the capacity decreased with increasing Sb substitution in the initial cycles, as the current density increased from 0.1 C to 32 C, the capacities of $NaCrO_2$ (NC), NCS-0.05, and NCS-0.1 declined to 31.3, 96.1, and 69.3 mAh g^{-1}, respectively. Upon reverting the current density to 0.1 C, their capacities recovered near their initial values. This comparison highlights that NCS-0.05 exhibited the best rate capability among the samples (Figure 11b). Based on the characterization results, the increase in the c/a ratio due to Sb substitution expands the Na^+ diffusion channel, facilitating enhanced sodium diffusion. This structural adjustment is advantageous for improving the reversibility of Na^+ intercalation and de-intercalation processes.

Ti doping has been reported for both P2- and O3-type Na_xCrO_2 cathode materials in SIBs [55,63,64]. Li et al. [64] investigated the electrochemical performance of O3-$Na_{1-x}Cr_{1-x}Ti_xO_2$ (with x = 0.03, 0.05, and 0.10). Although Ti doping reduces the initial capacity of the material, it significantly enhances the initial coulombic efficiency due to improved structural stabilization and the expansion of the NaO_2 interlayer spacing, attributed to an increase in the c/a lattice ratio. Long-term cycling performance at a rate of 1 C revealed that as x increases, the initial specific discharge capacities of the samples decrease (103.7, 100.6, 96.7, and 88.8 mAh g^{-1} for x = 0, 0.03, 0.05, and 0.10, respectively). However, the corresponding capacity retentions improve substantially (50.0%, 71.6%, 80.1%, and 85.8% after 800 cycles, respectively, Figure 11c). This demonstrates that Ti doping greatly enhances the cycling performance of $NaCrO_2$, further validating the improved structural stability with increasing Ti content. Additionally, the Ti-doped samples exhibit superior rate performance. When tested at a high rate of 30 C (3 A g^{-1}), $Na_{0.95}Cr_{0.9}5Ti_{0.05}O_2$ delivers a reversible capacity of 67 mAh g^{-1}, significantly outperforming the 42 mAh g^{-1} capacity of undoped $NaCrO_2$ (Figure 11d). The improved stability of Ti-doped samples is attributed to the higher bond dissociation energy of the Ti-O bond (661.9 kJ mol^{-1}) compared to the Cr-O bond (427 kJ mol^{-1}) at 298 K. This stronger bond suppresses TMO_2 layer gliding during cycling, delaying the phase transition from the hexagonal O3 structure to the monoclinic P3 structure, thereby enhancing the material's structural integrity and electrochemical performance.

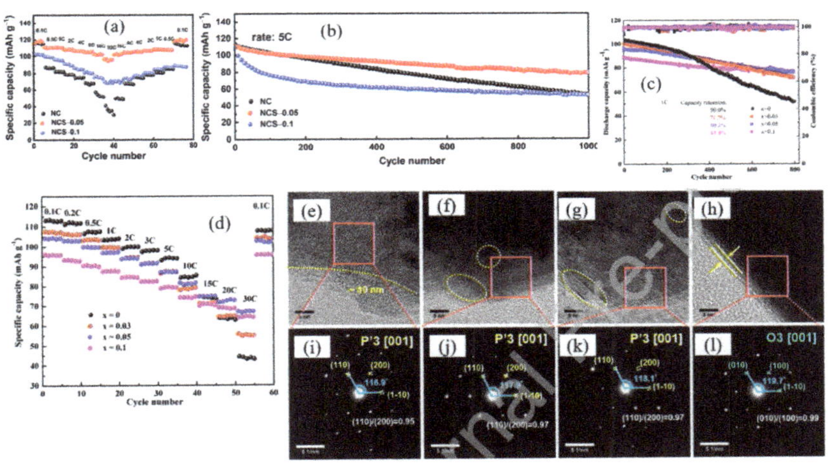

Figure 11. (**a**) Rate capability of NC, NCS-0.05, and NCS-0.1 at rates ranging from 0.1 C to 32 C, highlighting the superior performance of NCS-0.05 under high current densities, (**b**) cycling performance at 5 C for NC, NCS-0.05, and NCS-0.1, with NCS-0.05 exhibiting the highest capacity retention over prolonged cycles. Reprinted from Ref. [62] with permission from Elsevier, (**c**) cycling performance of $Na_{1-x}Cr_{1-x}Ti_xO_2$ at a rate of 1 C, indicating the enhanced capacity retention of Ti-doped samples over prolonged cycling, and (**d**) rate capability of $Na_{1-x}Cr_{1-x}Ti_xO_2$ samples, tested at rates ranging from 0.1 C to 30 C, highlighting the superior rate performance of Ti-doped materials compared to undoped $NaCrO_2$. Reprinted from Ref. [64] with permission from Elsevier. Ex situ HR-TEM images of (**e**) $NaCrO_2$, (**f**) $Na_{0.97}Cr_{0.97}Ti_{0.03}O_2$, (**g**) $Na_{0.93}CrCa_{0.035}O_2$, and (**h**) $Na_{0.9}Ca_{0.035}Cr_{0.97}Ti_{0.03}O_2$ after 1000 cycling at 10 C. SAED pattern of (**i**) $NaCrO_2$, (**j**) $Na_{0.97}Cr_{0.97}Ti_{0.03}O_2$, and (**k**) $Na_{0.93}CrCa_{0.035}O_2$ along the P'3 [001] zone axis, and (**l**) $Na_{0.9}Ca_{0.035}Cr_{0.97}Ti_{0.03}O_2$ along the O3 [001] zone axis, respectively. Reprinted from Ref. [61] with permission from Elsevier.

Lee et al. [61] investigated the effects of Ca and Ti co-substitution in the $NaCrO_2$ structure and found that it significantly improved cycling performance at a high current discharge capacity of 10 C. Based on ex situ TEM images (Figure 11e–h), structural damage and deformation were observed in pure $NaCrO_2$, as well as in samples with single-element substitution (Ca or Ti). In contrast, the co-substituted sample exhibited a smooth surface, indicating minimal distortion in the layered structure, even at the particle edges, which are typically the most sensitive locations.

The structural damage in the other samples was attributed to large variations in unit cell volume and irreversible phase transitions during cycling. Selected area electron diffraction (SAED) analysis (Figure 11i–l) further confirmed the structural stability of $Na_{0.9}Ca_{0.035}Cr_{0.97}Ti_{0.03}O_2$. The angle between the (010) and (1 −10) planes for this sample remained close to that of the O3-$NaCrO_2$ structure (119.7° compared to 120°), demonstrating excellent structural retention over cycling. In contrast, the corresponding angles for cycled $NaCrO_2$, $Na_{0.93}CrCa_{0.035}O_2$, and $Na_{0.97}Cr_{0.97}Ti_{0.03}O_2$ were 116.9°, 117.9°, and 118.1°, respectively, values closer to the P3′-$Na_{0.4}CrO_2$ structure, indicating less structural stability.

Surface modification has been widely recognized as an effective approach to enhance the structural stability of $NaCrO_2$ cathode materials. Techniques such as Cr_2O_3 [60,65] and carbon coatings [66–68] have been frequently employed in various studies. Ding et al. [69] investigated the effects of carbon coating on $NaCrO_2$ cathode materials to improve their stability.

The charge and discharge curves for both uncoated (naked) and carbon-coated $NaCrO_2$ electrodes at different current rates are shown in Figure 12a. While the voltage profiles of the carbon-coated and naked electrodes are similar, the carbon-coated electrode exhibits a lower voltage polarization between the charge and discharge profiles. At a current rate of 5 mA g^{-1}, the carbon-coated $NaCrO_2$ delivers higher charge and discharge capacities of 135 mAh g^{-1} and 116 mAh g^{-1}, respectively, compared to 125 mAh g^{-1} and 106 mAh g^{-1} for the naked electrode. However, an increase in current rate leads to higher polarization and rapid capacity fading for both electrodes.

Figure 12b presents the cycling performance at a current rate of 5 mA g^{-1}. Both the carbon-coated and naked electrodes initially show an increase in discharge capacity, reaching maximum values of 118 mAh g^{-1} and 110 mAh g^{-1} in the third and fourth cycles, respectively. The naked $NaCrO_2$, however, exhibits faster capacity fading during cycling. After 40 cycles, the carbon-coated $NaCrO_2$ retains a discharge capacity of 110 mAh g^{-1}, which is notably higher than that of the uncoated electrode. This highlights the effectiveness of carbon coating in enhancing the cycling stability of $NaCrO_2$ cathode materials.

Yu et al. [19] explored the effect of carbon coating on $NaCrO_2$ cathodes and observed a significant enhancement in conductivity, increasing from 4.4×10^{-3} S cm^{-1} for the bare sample to 4.7×10^{-1} S cm^{-1} for a 14.6 wt% carbon-coated sample (Figure 12c). The initial charge–discharge profiles of bare $NaCrO_2$ and carbon-coated $NaCrO_2$ (C-$NaCrO_2$, 3.4 wt%) were similar, except for the discharge capacities, which were 112 mAh g^{-1} and 121 mAh g^{-1}, respectively. The carbon coating facilitated greater capacity delivery above 3.07 V during charging, attributed to a reversible phase transformation from an O3 to a P3 structure (Figure 12d).

Bare $NaCrO_2$ exhibited significant capacity fading during cycling, while C-$NaCrO_2$ electrodes demonstrated excellent stability, retaining over 96% of their capacity after 50 cycles for coatings ranging from 1.4 to 14.2 wt%. Among these, the 3.4 wt% C-$NaCrO_2$ electrode displayed the best performance (Figure 12e).

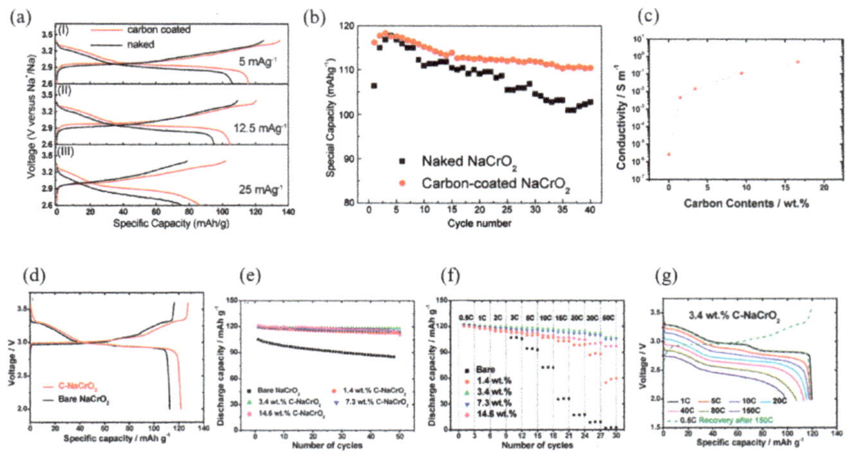

Figure 12. (**a**) Galvanostatic charge–discharge curves of $NaCrO_2/NaClO_4/Na$ cells at current densities of 5 mA g^{-1} (I), 12.5 mA g^{-1} (II), and 25 mA g^{-1} (III), (**b**) specific discharge capacity of $NaCrO_2/NaClO_4/Na$ cells at a current density of 5 mA g^{-1} as a function of cycle number. Reprinted from Ref. [69] with permission from Elsevier, (**c**) electrical conductivity of $NaCrO_2$ as a function of carbon content, (**d**) first charge and discharge curves comparing electrochemical performances of bare and C-$NaCrO_2$, using Na metal as the counter electrode and 0.5 M $NaPF_6$ in PC:FEC as the electrolyte. Cells were tested in the voltage range of 2.0–3.6 V at a charge-discharge rate of 20 mA g^{-1}, (**e**) cycling plots showing capacity as a function of the carbon content, (**f**) rate capability as a function of the carbon content, and (**g**) rate performances of 3.4 wt% C-$NaCrO_2$, which delivered the best results among various carbon-coated samples. Reprinted from Ref. [19] with permission from the Royal Society of Chemistry.

The rate capability of C-$NaCrO_2$ was particularly remarkable (Figure 12f). While bare $NaCrO_2$ delivered only 3 mAh g^{-1} at a high rate of 50 C, the 3.4 wt% and 7.3 wt% C-$NaCrO_2$ electrodes delivered capacities of 106 mAh g^{-1} at 50 C (5.5 A g^{-1}), retaining 87% of the capacity at 0.5 C (55 mA g^{-1}). The performance extended to an extremely high rate of 150 C (16.5 A g^{-1}), with a 27 s discharge delivering 99 mAh g^{-1}, demonstrating that the capacity was solely linked to Na^+ insertion, as evidenced by the flat discharge profile (Figure 12g).

Table 2 compares the electrochemical performance of Cr-based cathode materials in SIBs to provide better clarification of their behavior and capabilities.

Table 2. Electrochemical performance of Cr-based and modified Cr-based cathode for SIBs.

Researchers	Pristine Material	Modified Material	Testing Condition	Pristine Material [mAh g^{-1}]	Modified Material [mAh g^{-1}]
khe et al. [60]	$NaCrO_2$	$NaCr_{0.98}Al_{0.02}O_2$@Cr_2O_3	2.0–3.6 V, RT * 600 mAh g^{-1}	108.5 (initial) 56 (500 cycles)	106.6 (initial) 107 (1000 cycles)
Li et al. [64]	$NaCrO_2$	$Na_{0.95}Cr_{0.95}Ti_{0.05}O_2$	2.3–3.6 V, RT 120 mAh g^{-1}	104 (initial) 52 (800 cycles)	98 (initial) 78 (800 cycles)
Ma et al. [62]	$NaCrO_2$	$Na_{0.9}Cr_{0.95}Sn_{0.05}O_2$	2.5–3.5 V, RT 600 mAh g^{-1}	107 (initial) 50 (1000 cycles)	114 (initial) 81 (1000 cycles)
heng et al. [46]	$NaCrO_2$	$Na_{0.9}Ca_{0.05}CrO_2$	2–3.6 V, RT 120 mAh g^{-1}	110 (initial) 52 (500 cycles)	118 (initial) 90 (500 cycles)
Cai et al. [66]	$NaCrO_2$	$NaCrO_2$@nitrogen doped carbon	2.0–3.6 V, RT 1200 mAh g^{-1}	90 (initial) 47.5 (1000 cycles)	110 (initial) 100.2 (1000 cycles)

Table 2. *Cont.*

Researchers	Pristine Material	Modified Material	Testing Condition	Pristine Material [mAh g^{-1}]	Modified Material [mAh g^{-1}]
Lee et al. [61]	$NaCrO_2$	$Na_{0.9}Ca_{0.035}Cr_{0.97}Ti_{0.03}O_2$	1.5–3.8 V, RT 2400 mAh g^{-1}	34 (initial) 22.1 (500 cycles)	76 (initial) 61.7 (500 cycles)
Wu et al. [70]	$NaCrO_2$	$NaCrO_2$@PVDF-$La_2(CO_3)_3{\cdot}8H_2O$	2.3–3.5 V, RT 240 mAh g^{-1}	119.6 (initial) 56.5 (400 cycles)	111.8 (initial) 103.5 (900 cycles)

* RT stands for room temperature.

2.2. *$NaFeO_2$*

Minor-metal-free $LiFeO_2$ is unsuitable as a cathode material for LIBs due to its electrochemical inactivity. The oxidation of Fe^{3+} to Fe^{4+} is challenging because oxygen evolution precedes Fe^{3+} oxidation, unlike the Fe^{2+} to Fe^{3+} oxidation observed in $LiFePO_4$ [71]. Additionally, the similar ionic radius of Fe^{3+} and Li^+ promote cation mixing. An empirical rule suggests that layered AMO_2 compounds are stabilized when the ionic radius ratio of the alkali metal (r_A) to the transition metal (r_M), r_A/r_M, exceeds 1.23. For $LiFeO_2$ (r_{Li}/r_{Fe} = 1.18), this criterion is unmet, leading to structural instability [72]. In contrast, $NaFeO_2$ (r_{Na}/r_{Fe} = 1.58) avoids such issues due to the larger Na^+ ion. $NaFeO_2$ exists as orthorhombic β-$NaFeO_2$, which is electrochemically inactive, and rhombohedral α-$NaFeO_2$, a cathode with O3-type layered structure per Delmas' notation [73–76].

Okada and co-workers [73,77] were the first to report the reversible sodium cycling of α-$NaFeO_2$, synthesized by solid-state reaction of equal molar ratio Fe_2O_3 and Na_2O_2 at 650 °C in air, in sodium-ion cells using a sodium metal counter electrode. Charging up to 3.6 V vs. Na yielded an initial charge capacity of 103 mAh g^{-1}, corresponding to the formation of $Na_{0.58}FeO_2$, assuming no parasitic reactions. The cell discharged reversibly to 1.5 V vs. Na with an initial discharge capacity of 85 mAh g^{-1} and exhibited a flat operating voltage plateau at ~3.4 V. Sodium iron oxide has garnered interest as a cathode material for SIBs due to its high operating voltage (~3.3 V) based on the Fe^{3+}/Fe^{4+} redox reaction and a capacity of ~80 mAh g^{-1}. Iron's abundance makes it appealing for SIBs; however, challenges arise at high potentials, including structural instability of the cathode and electrolyte decomposition. These issues cause rapid performance degradation due to irreversible structural transitions, particularly the formation of monoclinic $Na_{0.5}FeO_2$ during desodiation beyond x > 0.5 in $Na_{1-x}FeO_2$ [38,73,78,79].

Tetravalent iron (Fe^{4+}) is rare but has been observed in certain perovskite oxides synthesized under high oxygen pressure. Mössbauer spectroscopy of α-$NaFeO_2$ and its charged states reveals that it operates via the Fe^{3+}/Fe^{4+} redox couple. The initial $NaFeO_2$ exhibits high-spin Fe^{3+} characterized by an isomer shift (IS) of 0.36 mm s^{-1} and quadrupole splitting (QS) of 0.50 mm s^{-1}, indicating a uniform Fe^{3+} valence state. Upon charging, an increase in QS suggests distortion of FeO_6 octahedra. Signals for Fe^{4+} (IS = 0.05 mm s^{-1}, QS = 0.55 mm s^{-1}) were observed, with 18% of the iron identified as Fe^{4+} at 3.6 V, demonstrating the feasibility of sodium-ion deintercalation and Fe^{3+} oxidation (Figure 13a) [73,80]. Takeda et al. [81] corroborated these findings through Mössbauer analysis of $NaFeO_2$, confirming Fe^{4+} formation during Na extraction.

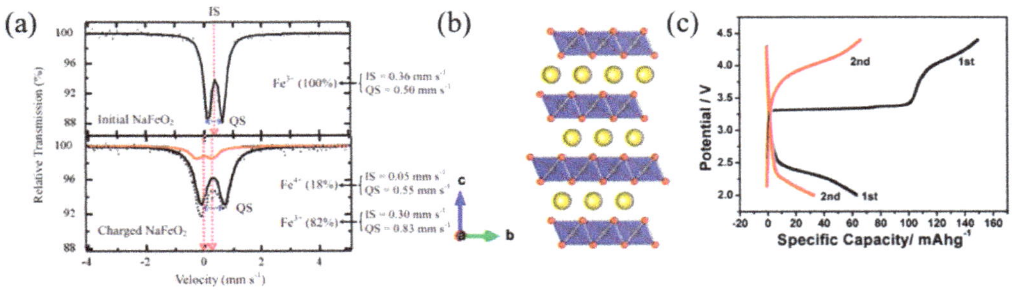

Figure 13. (**a**) Mössbauer spectra of initial and fully charged α-$NaFeO_2$ cathodes. The black line represents Fe^{3+}, while the red line corresponds to Fe^{4+}. Reprinted from Ref. [73] with permission from IOP Publishing, (**b**) schematic illustration of the O3-$NaFeO_2$ structure, where Na and Fe ions are octahedrally coordinated with oxygen atoms, forming a layered structure with Na layers between O-Fe-O slabs, and (**c**) voltage profiles of $NaFeO_2$ during cycling between 2.0 and 4.4 V versus Na^+/Na at a current density of 20 mA g^{-1}, showing potential plateaus and capacity characteristics. Reprinted from Ref. [31] with permission from Elsevier.

The O3-$NaFeO_2$ structure comprises Na and Fe octahedrally coordinated with oxygen, forming Na layers between O-Fe-O slabs (Figure 13b). During the initial charge to 4.4 V, the voltage profile exhibits plateaus at ~3.3 V and ~3.8 V, corresponding to Fe^{3+} oxidation and possible oxygen redox, respectively. Approximately 64% of Na-ion extraction (150 mAh g^{-1}) occurs during the 3.3 V plateau. However, significant polarization (≥1.0 V) and low coulombic efficiency (~42%) were observed during discharge, with voltage plateaus shifting to 2.5 V and 2.3 V. Subsequent cycles show diminished capacity and increased polarization, with the Fe^{3+}/Fe^{4+} redox reaction restricted to the first cycle (Figure 13c). This degradation is attributed to severe structural changes, including Fe migration, highlighting the challenges of using $NaFeO_2$ as a cathode material.

Yabuuchi et al. [82] investigated the influence of cutoff voltage on the structural and electrochemical performance of the α-$NaFeO_2$ cathode in sodium-ion cells. Their findings indicate that the reversibility of the electrode material is significantly affected by the cutoff voltage during charging (sodium extraction), as illustrated in Figure 14a. While a higher cutoff voltage increased the sodium extraction capacity, it simultaneously reduced the reversible capacity. Optimal performance, with minimal polarization, was achieved at a cutoff voltage of 3.4 V. Beyond this value, charging to 4.0 V or higher significantly increased polarization and reduced reversible capacity, rendering the material nearly inactive as an electrode at higher potentials. Approximately 70% of sodium can be extracted from O3-type $NaFeO_2$ by oxidation up to 4.5 V, assuming no side reactions. However, this degree of sodium extraction compromises the electrode's electrochemical activity and structural integrity (Figure 14a). The cyclability of the O3-type $NaFeO_2$ at various cutoff voltages was further examined (Figure 14b,c). Cells charged to different voltages between 3.4 and 4.5 V, then discharged to 2.5 V at a current rate of 12.1 mA g^{-1}, demonstrated varying degrees of performance. A cutoff voltage of 3.4 V maintained good capacity retention, with an initial reversible capacity of 80 mAh g^{-1} and approximately 75% capacity retention after 30 cycles (Figure 14c). Increasing the cutoff voltage to 3.5 V enhanced the first-cycle reversible capacity to 100 mAh g^{-1} but slightly reduced long-term capacity retention compared to the 3.4 V cycle. Further increases in the cutoff voltage to 4.0 V or higher led to significant capacity fade and poor overall cyclability.

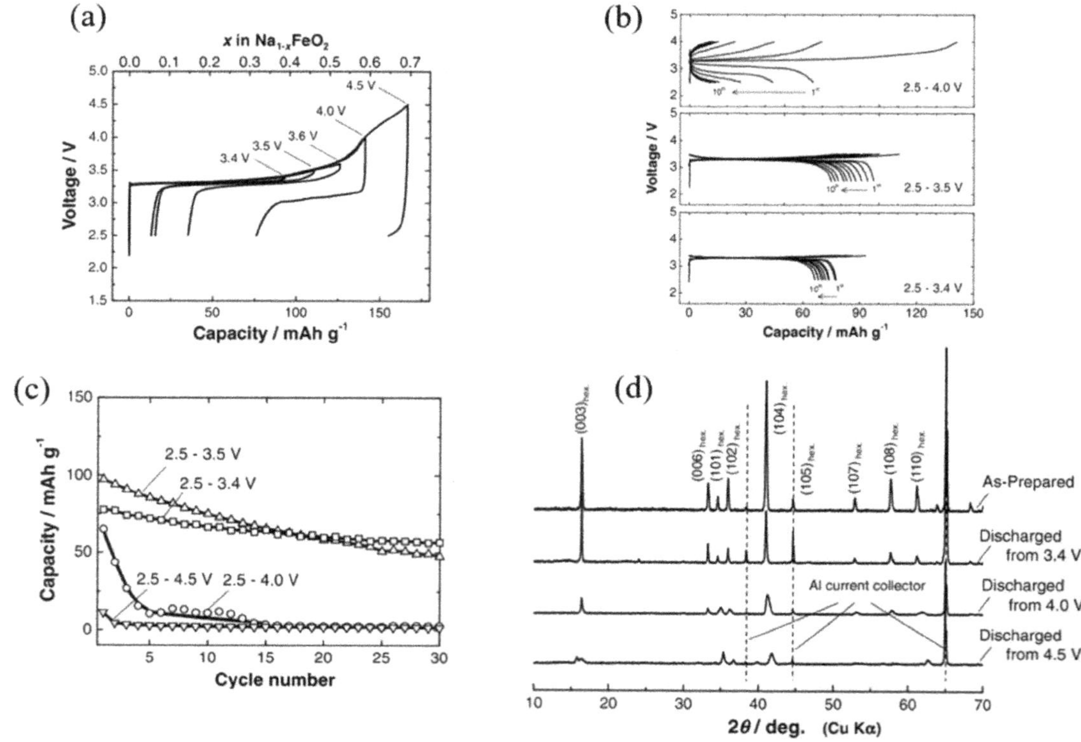

Figure 14. (**a**) Initial charge–discharge profiles of Na/$NaFeO_2$ cells at various cutoff voltages, measured at a rate of 12 mA g^{-1}, (**b**) charge–discharge curves of Na/$NaFeO_2$ cells with different cutoff voltages at the same rate, (**c**) discharge capacity retention over 30 cycles for cells charged to 3.4, 4.0, or 4.5 V and discharged to 2.5 V, and (**d**) XRD patterns of $Na_{1-x}FeO_2$ composite electrodes before and after the initial charge–discharge cycle at different cutoff voltages. The electrodes were charged (oxidized) to 3.4, 4.0, or 4.5 V and discharged to 2.5 V at a rate of 12.1 mA g^{-1}. Reprinted from Ref. [82] with permission from the Electrochemical Society of Japan.

Ex situ XRD analysis was conducted to elucidate the mechanism underlying the poor reversibility observed at higher cutoff voltages (Figure 14d). At a cutoff voltage of 3.4 V, sodium ions were reversibly extracted in a topotactic manner. An increase in the intensity of the $(003)_{hex}$ diffraction line relative to the $(104)_{hex}$ line after cycling between 3.4 and 2.5 V indicates the formation of sodium-ion vacancies at 3b octahedral sites. In O3-type sodium-layered compounds, phase transitions from O3 to P3 are commonly observed during sodium extraction when $x > 0.1$ in $Na_{1-x}MeO_2$. These transitions, caused by sodium/vacancy ordering at 3b sites, lead to stepwise voltage changes during galvanostatic oxidation/reduction. However, in $Na_{1-x}FeO_2$, the cell voltage increases monotonically with sodium extraction.

Irreversible structural changes occur when the charging voltage exceeds 4.0 V ($x > 0.5$ in $Na_{1-x}FeO_2$). At 4.5 V, the irreversible capacity increases significantly, likely due to the formation of additional vacancies at 3b sites. This is accompanied by a reduction in the intensity of the $(003)_{hex}$ diffraction line relative to the $(101)_{hex}$ and $(104)_{hex}$ lines. Such changes suggest possible iron migration from 3a to 6c (face-shared tetrahedral to octahedral sites) or to 3b sites, assuming no oxygen release from the lattice. This migration of iron ions may block sodium conduction pathways, contributing to the observed decline in electrode performance.

The chemical instability of Fe^{4+} in $NaFeO_2$ cathode materials further exacerbates performance issues. Lee et al. [83] demonstrated that the electrochemical performance of $NaFeO_2$ significantly deteriorates when stored in a charged state. In their study, a coin cell charged to 3.6 V, stored for 24 h, and then discharged to 2.0 V exhibited a much lower specific capacity (62.7 mAh g^{-1}) and average cell voltage compared to a cell discharged immediately after charging (80 mAh g^{-1}). This decline is attributed to the chemical instability of Fe^{4+}, which spontaneously reduces Fe^{3+} in the charged state, accompanied by electrolyte decomposition.

Given that batteries are often stored in fully or partially charged states, this instability leads to significant self-discharge and rapid performance degradation. Addressing the chemical instability of Fe^{4+} is thus critical to unlocking the potential of Fe^{3+}/Fe^{4+}-based redox cathodes.

The thermal stability of $NaFeO_2$ is promising compared to cathodes used in LIBs. When $NaFeO_2$ was heated to 450 °C, no exothermic reactions were observed. In contrast, exothermic reactions were reported for $Li_{0.5}CoO_2$—a widely used cathode material in portable devices—at temperatures of 190 °C and 395 °C. However, it is noteworthy that heat generation occurred in the case of charged $NaFeO_2$ ($Na_{0.58}FeO_2$) above 350 °C, as observed in differential scanning calorimetry (DSC) measurements, according to reaction 3 [73]:

$$Na_{0.58}FeO_2 \rightarrow 0.58\ NaFeO_2 + 0.21\ Fe_2O_3 + 0.105\ O_2 \quad (3)$$

Elemental doping has proven to be an effective strategy for enhancing the performance of $NaFeO_2$ cathode materials [84]. Xu et al. [85] incorporated Ru as a doping element into $NaFeO_2$, demonstrating that Ru-doped $NaFeO_2$ (Na_4FeRuO_6) delivered a high reversible capacity of 120 mA h g^{-1} and achieved approximately 80% capacity retention after 100 cycles when charged up to 4 V. Furthermore, Ru doping significantly reduced potential polarization in $NaFeO_2$.

The typical charge–discharge cycles of Na_4FeRuO_6 and $NaFeO_2$ are presented in Figure 15a,b. $NaFeO_2$ delivered about 85 mAh g^{-1} reversible capacity at 0.2 C during the initial charge–discharge cycle, accompanied by a low coulombic efficiency of 81.7%. In contrast, Na_4FeRuO_6 exhibited a reversible capacity of 120 mA h g^{-1} within a voltage range of 2–4 V, with a much higher coulombic efficiency of 96.7%, indicating improved reversibility and reduced polarization.

Na_4FeRuO_6 demonstrated superior rate capability, delivering capacities of 120 mAh g^{-1} at 0.2 C, 96 mAh g^{-1} at 0.5 C, 88 mAh g^{-1} at 1 C, 75 mAh g^{-1} at 2 C, and 55 mAh g^{-1} at 5 C. In contrast, $NaFeO_2$ showed poor rate performance, with negligible capacity at 2 C (Figure 15c,d). The improved performance of Ru-doped $NaFeO_2$ is attributed to the suppression of Fe-ion migration, which reduces internal strain and mitigates particle cracking associated with the structural distortions caused by Fe-ion migration.

Co-doping in $NaFeO_2$ has been shown to significantly enhance its electrochemical performance, as demonstrated by Yoshida et al. [86]. The study revealed that although NaFe0.5$Co_{0.5}O_2$ is considered an intermediate solid solution between $NaFeO_2$ and $NaCoO_2$, it delivers the highest reversible capacity of approximately 160 mAh g^{-1} among the three materials tested. Notably, the average operating voltage of $NaFe_{0.5}Co_{0.5}O_2$ (3.14 V) in Na cells exceeds that of $NaCoO_2$ (2.96 V) (Figure 15e).

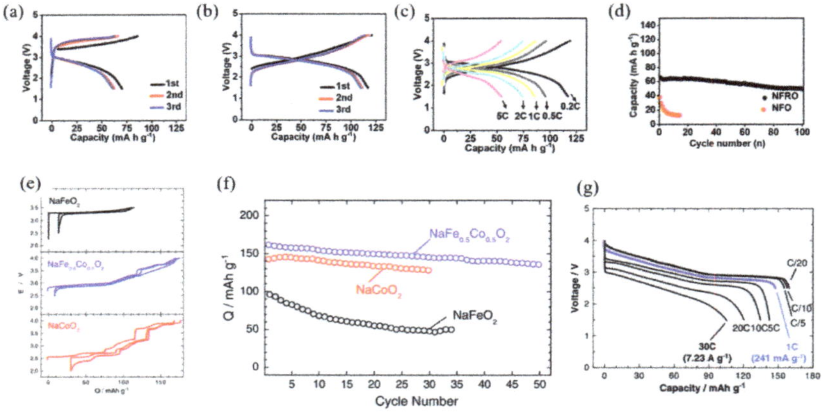

Figure 15. (**a**) Initial charge and discharge curves of Na/$NaFeO_2$ cells at a rate of 0.2 C (where 1 C corresponds to 100 mA g^{-1}), (**b**) initial charge and discharge curves of Na/Na_4FeRuO_6 cells at a rate of 0.2 C, (**c**) rate performance of Na/Na_4FeRuO_6 cells at rates of 0.2 C, 0.5 C, 1 C, and 2 C, and (**d**) cycle performance of Na/Na_4FeRuO_6 and Na/$NaFeO_2$ cells at a rate of 2 C. Reprinted from Ref. [85] with permission from Wiley, (**e**) charge–discharge curves of $NaFeO_2$, $NaFe_{0.5}Co_{0.5}O_2$, and $NaCoO_2$ in Na cells at a rate of 12 mA g^{-1}, (**f**) changes in the discharge capacity over 50 cycles, and (**g**) rate capability of $NaFe_{0.5}Co_{0.5}O_2$. Cells were charged to 4.0 V at a rate of 12 mA g^{-1} and discharged at different rates (C/20 = 12 mA g^{-1} to 30 C = 7260 mA g^{-1}). Reprinted from ref. [86] with permission from Elsevier.

Additionally, both $NaFe_{0.5}Co_{0.5}O_2$ and $NaCoO_2$ exhibit relatively good capacity retention over cycling, as illustrated in Figure 15f. A remarkable finding is the superior rate capability of $NaFe_{0.5}Co_{0.5}O_2$, which significantly outperforms $NaCoO_2$. Even at a high rate of 30 C (7.23 A g^{-1}), micrometer-sized $NaFe_{0.5}Co_{0.5}O_2$ achieves a discharge capacity exceeding 100 mAh g^{-1}, highlighting its excellent fast-charging and high-rate performance.

Shevchenko et al. [87] investigated the effect of Mn and Ni substitution on the performance of $NaFeO_2$-based cathodes, synthesizing $NaFe_{1/3}Mn_{1/3}Ni_{1/3}O_2$ to reduce Fe ion migration into the Na layers. The electrochemical performance of this cathode material was evaluated by charging to 4.0 V and 4.2 V and discharging to 1.9 V. At a C/10 rate, the material delivered specific capacities of approximately 130 mAh g^{-1} and 160 mAh g^{-1}, respectively. However, charging up to 4.2 V introduced a plateau around 4.1 V, which led to capacity degradation over cycling due to structural instability.

2.3. $NaCuO_2$

Copper is abundant, relatively inexpensive, and among the less toxic transition metals. Despite its favorable properties, there have been no reported electrochemical studies on $LiCuO_2$. Efforts to synthesize $LiCuO_2$ via solid-state calcination have been unsuccessful, as the reaction results in the formation of Li_2CuO_2 and CuO, as shown below:

$$2\,CuCO_3 + Li_2CO_3 \rightarrow Li_2CuO_2 + CuO + 3\,CO_2 \tag{4}$$

In contrast, $NaCuO_2$ has been successfully synthesized using the solid-state calcination of Cuo and Na_2O_2 at 450 °C in oxygen environment. This behavior parallels that of nickelates, where $NaNiO_2$ can be readily prepared via solid-state calcination, while stoichiometric $LiNiO_2$ is notably challenging to obtain. Structural studies reveal that $NaCuO_2$ adopts a monoclinic structure with the same space group and elemental arrangement as

$LiCuO_2$, as illustrated in Figure 16a. This structural similarity indicates that $NaCuO_2$ may hold promise as a positive electrode material for sodium-ion batteries.

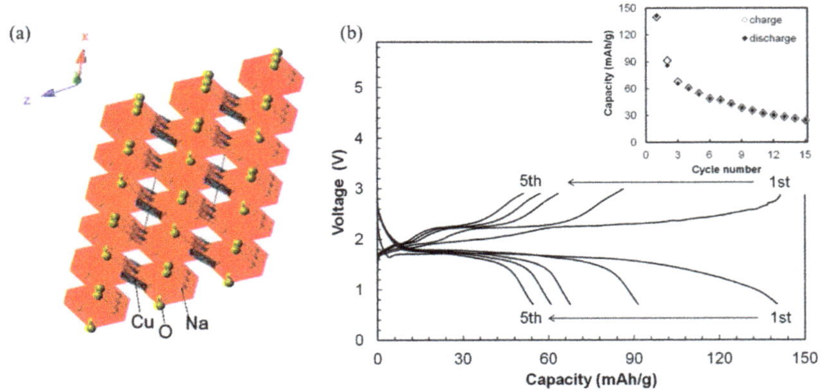

Figure 16. (**a**) Crystal structure of $NaCuO_2$ and (**b**) discharge–charge curves of a Na/$NaCuO_2$ cell cycled between 0.75 and 3.0 V. The inset illustrates the cycling performance. Reprinted from Ref. [88] with permission from IOP Publishing.

Figure 16b presents the discharge–charge profiles of a Na/$NaCuO_2$ cell from the first to the fifth cycle within the voltage range of 0.75 to 3.0 V [88].

Theoretically, sodium-ion insertion and extraction in $NaCuO_2$ correspond to the reduction of Cu^{3+} to Cu^{2+} during discharge and the oxidation of Cu^{3+} to Cu^{4+} during charge, respectively.

$$NaCu^{3+}O_2 + x\,Na^+ + x\,e^- \rightarrow Na_{1+x}Cu^{2+}O_2 \quad (5)$$

$$NaCu^{3+}O_2 \rightarrow Na_{1\text{-}x}Cu^{4+}O_2 + x\,Na^+ + x\,e^- \quad (6)$$

The subsequent charge capacity of $NaCuO_2$ reached 142 mAh g^{-1}, nearly equivalent to its initial discharge capacity. Inductively Coupled Plasma Atomic Emission Spectroscopy (ICP-AES) analysis indicated that sodium-ion insertion and extraction during the first discharge and charge processes corresponded to 0.8 mol and 0.5 mol, respectively. In the case of $LiCuO_2$, lithium-ion insertion occurs at approximately 2.0 V (vs. Li), leading to the formation of $Li_{1.7}CuO_2$.

Despite stable discharge–charge plateaus over the first five cycles, a noticeable decline in discharge capacity was observed with continued cycling. This significant capacity fading suggests that structural changes occur in $NaCuO_2$ during repeated electrochemical cycling, ultimately impacting its performance. Ono et al. [89] investigated the poor cycling performance of $NaCuO_2$ in the voltage range of 1–3 V and identified the underlying mechanism. Sodium-ion insertion into $NaCuO_2$ during the first discharge results in an irreversible conversion of $NaCuO_2$ to an amorphous phase, Na_2CuO_2. Subsequent charge and discharge reactions proceed between these amorphous phases. However, the discharge product, Na_2CuO_2 (more accurately $Na_{1.6}CuO_2$), is unstable and undergoes decomposition, forming copper oxide (CuO) and sodium oxide (Na_2O):

$$NaCuO_2 + Na^+ + e^- \rightarrow Na_2CuO_2 \quad (7)$$

$$Na_2CuO_2 \rightarrow Na_2O + CuO \quad (8)$$

subsequently, CuO reacts with sodium ions, producing Cu_2O and additional Na_2O:

$$CuO + Na^+ + e^- \rightarrow \frac{1}{2}Cu_2O + \frac{1}{2}Na_2O \quad (9)$$

These irreversible reactions contribute to the significant capacity fading observed in $NaCuO_2$ during repeated electrochemical cycling.

2.4. $NaNiO_2$

The structure of $NaNiO_2$ was first identified in 1954 using film methods with a precession camera and later refined through Rietveld analysis with X-ray and neutron scattering in 1977. Subsequent studies in 2000 and 2005 employed neutron and X-ray diffraction, respectively, to further understand its structure [90–93]. $NaNiO_2$ exhibits two polymorphs: a low-temperature monoclinic O3 layered structure and a high-temperature rhombohedral phase. The low-temperature phase features NiO_6 octahedra that are edge-sharing and elongated due to the Jahn–Teller distortion of the Ni^{3+} ions. Sodium ions occupy interlayer positions, forming distorted octahedral coordination with oxygen atoms, and this phase is isostructural with $NaMnO_2$. At approximately 450 K, thermal motion overcomes the static Jahn–Teller distortion, causing a phase transition to the rhombohedral form, which is isostructural with α-$NaFeO_2$ and $LiNiO_2$. The monoclinic $NaNiO_2$ polymorph was synthesized by solid-state reaction of excess amounts of Na_2O and NiO at 650 °C under O_2 flow and tested as a cathode material, exhibiting a stepwise charge–discharge profile with an initial open-circuit voltage of 2.47 V. Vassilaras et al. [94] demonstrated, in contrast to previous studies (which reported during the first electrochemical cycle, only 0.2 sodium ions could be extracted from $NaNiO_2$ [94,95]), a charge capacity of 199 mAh g^{-1} (x = 0.85 in Na_xNiO_2) and a discharge capacity of 147 mAh g^{-1} (x = 0.62 in Na_xNiO_2) within the voltage range of 2.0–4.5 V. Galvanostatic charge and discharge measurements of $NaNiO_2$ at C/10 (1 C = 235 mAh g^{-1}) were conducted over multiple cycles in two different voltage ranges: 1.25–3.75 V and 2.0–4.5 V. The results are illustrated in Figure 17a,b.

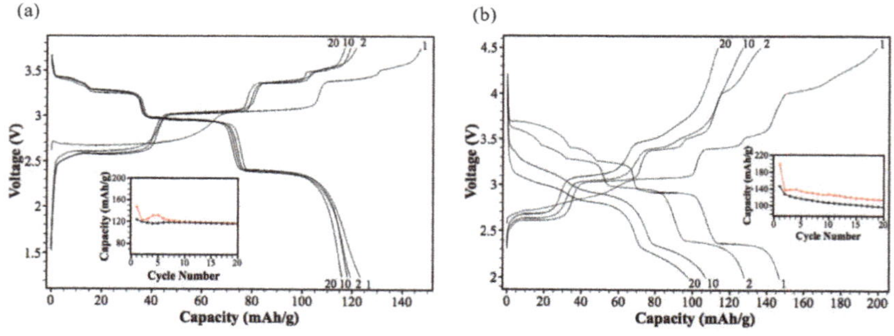

Figure 17. (**a**) Voltage profiles of $NaNiO_2$ during multiple cycles at C/10, with the cell galvanostatically cycled within the voltage range of 1.25–3.75 V and (**b**) voltage profiles of $NaNiO_2$ cycled between 2.0–4.5 V. Reprinted from Ref. [94] with permission from IOP Publishing.

The charge and discharge curves of $NaNiO_2$ exhibit the expected step-like features, indicating distinct phase transitions during sodium extraction and insertion. Except for the first charge cycle, the similarity in the plateaus suggests consistent phase transitions throughout subsequent cycling. In the voltage range of 1.25–3.75 V (Figure 17a), the charge and discharge capacities for the first, second, tenth, and twentieth cycles are 147/123, 121/119, 119/118, and 117/116 mAh g^{-1}, respectively, with corresponding coulombic efficiencies of 83.8%, 98.4%, and 99.2% for the first, second, and subsequent cycles [94].

When the voltage range is extended to 2.0–4.5 V (Figure 17b), the charge and discharge capacities for the first, second, tenth, and twentieth cycles are 199/147, 136/128, 127/107, and 113/98 mAh g^{-1}, respectively. The coulombic efficiencies under this range are 73.7%, 93.7%, 84.6%, and 86.3% for the first, second, tenth, and twentieth cycles, respectively. A significant decline in capacity and coulombic efficiency is observed at higher voltages, indicative of side reactions and structural instability.

The first charge curve deviates from subsequent charge cycles, particularly below 3.0 V. However, above 3.0 V, the charge profiles align with those of later cycles, suggesting that excess capacity in the first cycle is associated with reactions at lower voltages. Cyclic voltammetry analysis confirms the capacity decline and differences between charge and discharge profiles over extended cycling at higher voltages (Figure 18a). The CV curves reveal four reversible oxidation and reduction peaks at 2.7/2.3, 3.1/2.9, 3.4/3.2, and 3.5/3.4 V. Beyond the first cycle, subsequent scans show consistent reversibility within the 2.0–4.0 V range.

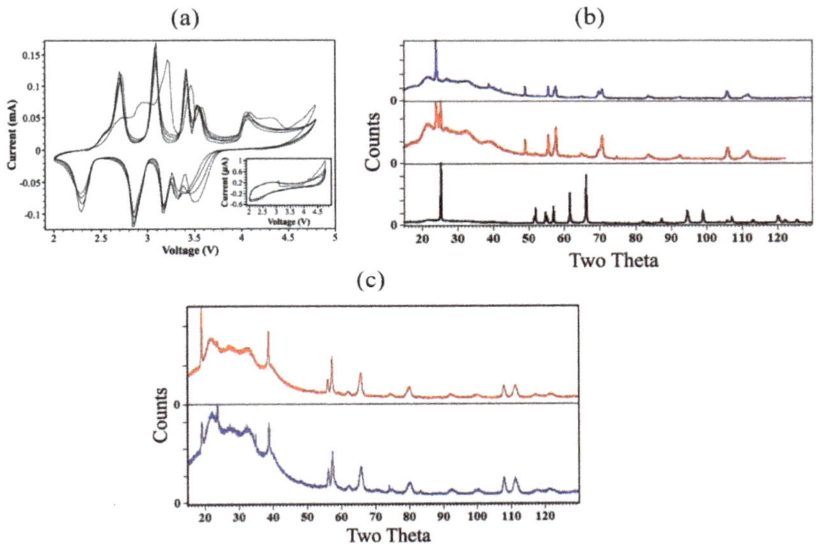

Figure 18. (**a**) Cyclic voltammetry data of $NaNiO_2$ (main plot) compared to the electrolyte (inset). The CV data illustrate the oxidation and reduction processes of $NaNiO_2$ across multiple cycles, highlighting the reversibility of the material in the 2.0–3.75 V range, (**b**) XRD patterns of pristine $NaNiO_2$ (black), $NaNiO_2$ mixed with PTFE and carbon black (red), and the cathode material after the first cycle in the 2.0–3.75 V range (blue). The pristine phase corresponds to $NaNiO_2$, while the mixed electrode includes an additional $Na_{0.91}NiO_2$ phase, attributed to minor sodium loss during electrode preparation. The cathode material post-cycling shows only the $Na_{0.91}NiO_2$ phase, with no evidence of reformation of the fully sodiated $NaNiO_2$ phase. Broad peaks between 10° and 30° are due to the Kapton film used during the measurement, and (**c**) XRD patterns of the cathode after charging to 3.75 V (red) and 4.5 V (blue). Both patterns are largely similar, though the cathode charged to 4.5 V exhibits two additional peaks near 23.7° and 35.0°, indicative of the formation of a new phase. The structure of this new phase remains unidentified. The broad peaks between 10° and 30° are also attributed to the Kapton film. The lattice parameters for $NaNiO_2$ in this study are reported as a = 2.84 A° and c = 20.8 A°. Reprinted from Ref. [94] with permission from IOP Publishing.

The oxidation peaks between 2.5 and 3.0 V during the first cycle differ from those in later cycles, corroborating the galvanostatic cycling data. Above 4.0 V, oxidation processes appear partially reversible, as no corresponding reduction peaks are observed near 4.0 V.

These observations suggest that side reactions at higher voltages contribute to poor cycling performance. Both galvanostatic and CV results highlight distinct reactions occurring in the 2.5–3.0 V range during the initial charge cycle, which are absent in subsequent cycles.

To further understand structural changes after the first cycle, XRD analysis was performed, and the diffraction patterns of three samples are compared in Figure 18b.

The bottom XRD pattern (black) corresponds to the pristine $NaNiO_2$ sample. The middle pattern (red) represents the XRD of the as-prepared electrode, which consists of $NaNiO_2$ mixed with PTFE and carbon black. In addition to the pristine $NaNiO_2$ phase, a secondary phase, identified as $Na_{0.91}NiO_2$, is observed. This indicates a small loss of sodium during electrode preparation, likely caused by air exposure or interactions with the binder and carbon black. The top pattern (blue) shows the XRD of the electrode after discharge following cycling between 2.0 and 3.75 V. Only the $Na_{0.91}NiO_2$ phase is present, with no evidence of the original $NaNiO_2$ phase, suggesting that the discharge process does not fully restore the sodiated $NaNiO_2$ phase.

Further analysis was conducted at higher voltages to investigate the observed capacity reduction. XRD patterns recorded after charging $NaNiO_2$ to 3.75 V and 4.5 V are displayed in Figure 18c. While the majority of the patterns remain similar, the highly charged sample exhibits two additional peaks at approximately 23.7° and 35.0°. These peaks indicate the formation of a new phase, although its structure remains unidentified [94].

Han et al. [96] revisited the phase transitions of $NaNiO_2$ during charge–discharge cycling within a voltage range of 1.5–4.0 V at a rate of C/10 (Figure 19a). During the first charge, four distinct plateaus were observed, reflecting multiple phase transitions. These transitions are attributed to gliding of the oxide layers, variations in structural distortion due to sodium deintercalation, and rearrangement of sodium ions within the interlayer space. The OCVs associated with these plateaus are 2.59 V, 3.04 V, 3.38 V, and 3.53 V, respectively.

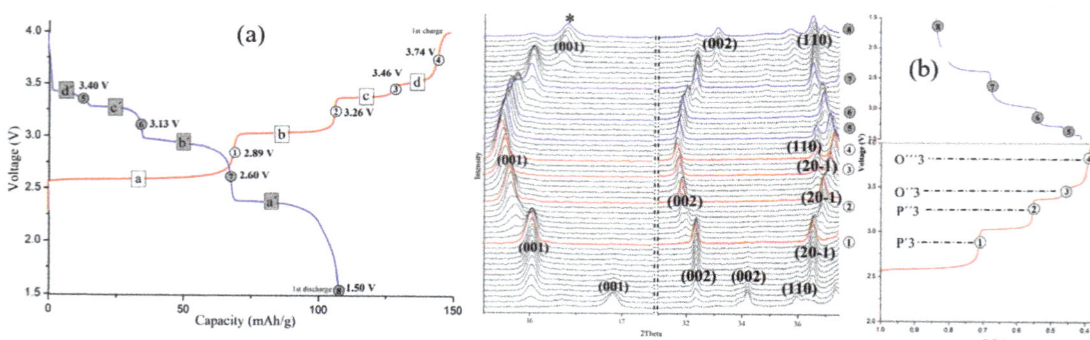

Figure 19. (**a**) First charge–discharge cycle of $NaNiO_2$ within the voltage range of 1.5–4.0 V at a C/10 rate. Numbers 1 to 8 indicate the specific voltage stops at which ex situ and relaxed in situ XRD measurements were performed. The charge–discharge process is divided into phases marked by plateaus a to d′ that correspond to key voltage ranges and phase transitions observed during the cycling and (**b**) in situ XRD patterns of the $NaNiO_2$ electrode during the first charge–discharge process, showing corresponding phase transitions. The voltage–composition curve, combined with the XRD patterns, highlights the structural changes that occur as sodium is extracted and reinserted, corresponding to the different phases during charge and discharge, as detailed in the charge–discharge plateaus. The phase transitions are tracked in real-time, revealing the changes in the crystal structure at various stages of sodium deintercalation and reintercalation. Reprinted from Ref. [96] with permission from Elsevier.

During discharge, four stepwise plateaus similar to those during charge were observed, indicating that the phase transitions are highly reversible throughout the electrochemical process. However, the first discharge capacity is notably lower than the first charge capacity, a well-documented phenomenon attributed to incomplete re-sodiation during the first discharge. This results in significant capacity loss.

Additionally, the lengths of the plateaus during charge and discharge are approximately identical, except for the first plateau in both processes. A pronounced voltage drop below 2.3 V suggests that a highly unfavorable phase transition is necessary to achieve the fully sodiated phase.

Real-time in situ XRD measurements of a $NaNiO_2$ powder electrode were performed within the 2θ ranges of 15–18° and 31–38°, as shown in Figure 19b. The initial scan clearly identifies the pristine material as pure $NaNiO_2$. The initial scan identifies the pristine material as pure $NaNiO_2$. However, during the charging process, the fully sodiated phase diminishes rapidly, and partially desodiated phases emerge, evidenced by a shift of the (001) reflection toward lower 2θ angles. This shift corresponds to an increase in the c-spacing as sodium ions are deintercalated. The removal of positively charged sodium ions enhances electrostatic repulsion between the oxide layers, leading to structural expansion. Concurrently, the (−201) reflection shifts toward a smaller d-spacing, indicating contraction along the a- and b-axes. This shrinkage is attributed to the increasing concentration of smaller Ni^{4+} ions during oxidation.

As the charge progresses, new biphasic regions appear. Between sodium contents of ~1.0 and 0.74 per formula unit, O′3, P′3, and O″″3 phases emerge. The transition between these phases is accompanied by increased distortion, evidenced by changes in lattice parameters. Upon reaching a sodium concentration of 0.55 at 3.26 V, most O′3 and P′3 phases disappear, and the P″3 phase dominates. The a/b ratio increases from 1.73 to 1.76, reflecting enhanced structural distortion, though the β angle remains stable at 121.5°, preserving hexagonal symmetry.

Further oxidation at 3.38 V (sodium content ~0.55 to 0.42) reveals a new plateau corresponding to the coexistence of P″3, P‴3, and O″3 phases. During this stage, the P″3 phase transitions into a more distorted P3-like structure, before the formation of the O″3 phase. The newly identified P‴3 phase exhibits greater distortion than P″3, though it does not conform to the O″3 structural model. As the P‴3 phase transitions to O″3, the a/b ratio remains relatively stable, but the β angle shifts significantly from 119.75° to 106.0°, marking a transition from hexagonal to monoclinic symmetry. This phase evolution, driven by increased interlayer repulsion due to extensive sodium removal, exemplifies the transition from P-type to O-type phases.

A small biphasic region is observed between O″3 and O‴3 when the sodium content decreases from 0.42 to 0.40. This quickly transitions to a solid solution of O‴3. By 3.74 V in the relaxed in situ XRD scans, peaks corresponding to the O′3, P′3, P″3, and O″3 phases are nearly absent, leaving only the O‴3 phase detectable at sodium concentrations as low as 0.37. The lattice parameters 'a' and 'c' represent the diagonal and edge lengths of the MO_6 octahedron, respectively, and relate to the hexagonal lattice parameters as follows:

$$a\ (\text{monoclinic}) = \sqrt{3} \times a\ (\text{hexagonal})$$

$$b\ (\text{monoclinic}) = a\ (\text{hexagonal})$$

Pristine $NaNiO_2$ exhibits an a/b ratio of 1.88 and a β angle of 110.45°, indicative of a significant Jahn–Teller distortion caused by the low-spin $t^6{}_{2g}\ e^1{}_g$ electronic configuration of Ni^{3+} [92]. This distorted monoclinic phase is stable at room temperature, while the ideal rhombohedral phase stabilizes only above 450 K, where thermal energy overcomes the Jahn–Teller effect.

During discharge, the phase transitions and voltage steps mirror those observed during charging. However, as the discharge voltage approaches 2.3 V, polarization increases, indicating that sodium re-intercalation into the host structure becomes kinetically unfavorable near full sodiation. Similar observations were made by Wang et al. [97], who reported irreversible structural changes in $NaNiO_2$ during charge–discharge cycling in the 2–4.5 V range. Two-dimensional phase mapping revealed that these structural changes significantly hinder the electrochemical reversibility of the material.

Notably, a steep voltage drop below 2.3 V and the emergence of an O''''3 phase (marked with an asterisk in Figure 19b) during discharge highlight the difficulty in achieving full sodiation. This behavior underscores the structural and kinetic limitations of $NaNiO_2$ during prolonged cycling.

Kim et al. [98] investigated the thermal stability of $Na_{0.91}NiO_2$ and $Na_{0.5}NiO_2$ in the presence of an electrolyte within the temperature range of 300–600 °C, as shown in Figure 20a–d. Two phase transitions were observed in both samples during heating: the first transition occurred between 247 and 352 °C, and the second between 528 and 600 °C (Figure 20a,c). Both samples exhibited the formation of a Ni metal phase and a structural transition from a layered structure to rock salt (Figure 20b,d).

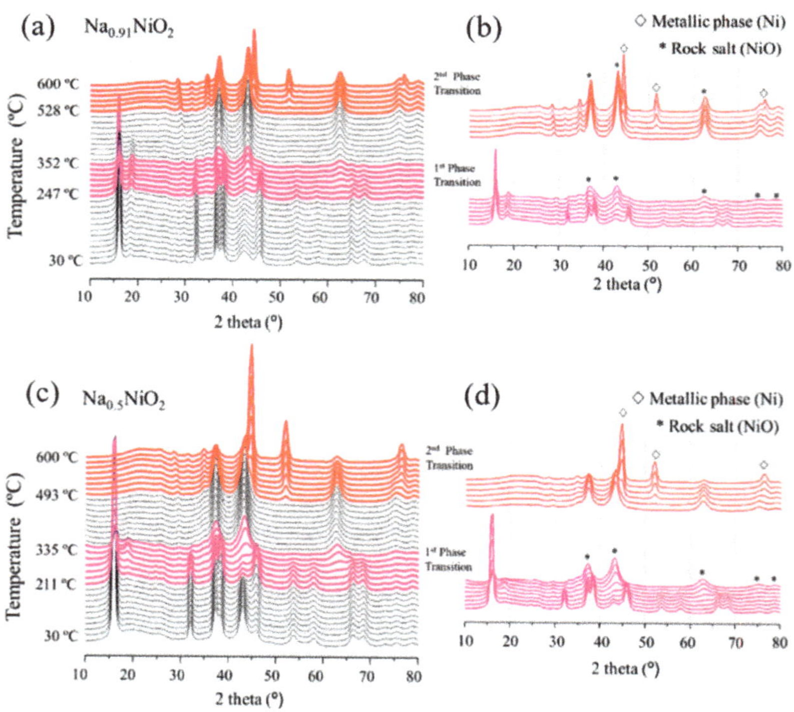

Figure 20. (**a**,**c**) Time-resolved XRD patterns of $Na_{0.91}NiO_2$ and $Na_{0.5}NiO_2$ during heating from 30 to 600 °C, (**b**,**d**) selected regions showing the two phase transition ranges for each material. Reprinted from Ref. [98] with permission from Elsevier.

However, a key difference was identified between the two samples. The onset temperature for phase decomposition was lower for the desodiated sample ($Na_{0.5}NiO_2$), indicating that increased Na extraction from the structure reduces the decomposition temperature.

A substantial number of studies have explored the use of Ni as a primary transition metal in SIB cathode materials, often combined with reinforcing elements such as Ti, Mn, and Co. Various compositions have been successfully synthesized and reported, including: $NaFe_{0.2}Ni_{0.4}Ti_{0.4}O_2$, $NaFe_{0.4}Ni_{0.3}Ti_{0.3}O_2$ [99], $NaNi_{0.5}Mn_{0.5}O_2$ [100], $NaNi_{0.5}Mn_{0.3}Ti_{0.2}O_2$ [101], $NaNi_{0.8}Co_{0.15}Al_{0.05}O_2$ [102], $NaNi_{0.815}Co_{0.15}Al_{0.035}O_2$ [103], $NaNi_xFe_yMn_zO_2$ [104], $NaNi_xMn_{1-x}O_2$ [105], $NaFe_{0.5}Ni_{0.5}O_2$ [106,107], $Na_{0.8}[Ni_{0.3}Co_{0.2}Ti_{0.5}]O_2$ [108], $NaNi_{0.5}Ti_{0.5}O_2$ [107,109], $Na(Ni_{2/3}Sb_{1/3})O_2$ [110], $NaNi_{0.9}Ti_{0.1}O_2$ [111], and $Na_xNi_{0.6}Co_{0.4}O_2$ [112] have been successfully reported as cathode materials in SIBs.

$NaNi_{0.5}Mn_{0.5}O_2$ has been widely studied as a promising cathode material for sodium-ion batteries (SIBs) [113–117]. However, the O3-$NaNi_{0.5}Mn_{0.5}O_2$ phase undergoes multiple complex phase transitions (O3-O3′-P3-P3′) and exhibits multiple voltage platforms during the charge and discharge processes. These transitions result in significant unit cell volume changes, structural collapse, and sluggish kinetic performance [100].

Ti substitution for Mn has been shown to increase interslab distances and suppress phase transformations in the high-voltage region [101,118]. Yuan et al. [117] reported that La doping in the structure of $NaNi_{0.5}Mn_{0.5}O_2$ improves stability and rate capability by reducing particle size, stabilizing the crystal structure, and enhancing Na-ion diffusion. This enhancement is attributed to the formation of strong La-O bonds, which expand the Na interlayer spacing.

Liang et al. [114] developed a core–shell structure to improve the cycling and rate performance of $NaNi_{0.5}Mn_{0.5}O_2$. In this design, a highly stable P2-$Na_{2/3}MnO_2$ coating was deposited on the high-capacity $NaNi_{0.5}Mn_{0.5}O_2$ shell. The reinforced structure demonstrated significant performance improvements, retaining 85.3% of its capacity after 150 cycles at 1C (125 mAh g^{-1}), compared to 74.8% for the bare sample. Additionally, the core–shell structure exhibited excellent high-rate capability, delivering 103 mAh g^{-1} at 15C, which is notably higher than the 82 mAh g^{-1} achieved by the uncoated sample.

Meghnani et al. [119] developed a Na_3PO_4 coating on $NaNi_{0.815}Co_{0.15}Al_{0.035}O_2$ particles to enhance air stability and mitigate the formation of insulating compounds such as NaOH and Na_2CO_3. This modification improved the material's viability as a cathode for SIBs. The Na_3PO_4 coating contributed to a highly stable structure, improved sodium intercalation/deintercalation ability, and facilitated reversible phase transformations during the discharge process. These factors collectively resulted in promising electrochemical performance of Na_3PO_4@$NaNi_{0.815}Co_{0.15}Al_{0.035}O_2$ over extended cycling, as demonstrated in Figure 21a,b.

Li et al. [120] synthesized O3-type Ni-rich $NaNi_{2/3}Mn_{1/6}Fe_{1/6}O_2$ (NNMF) at various calcination temperatures (800 °C, 830 °C, 850 °C, and 880 °C) as a promising cathode material for SIBs. The first-cycle charge–discharge curves of NNMF samples, tested within a voltage range of 1.5–4.2 V at 0.2 C (1 C = 160 mA g^{-1}), are shown in Figure 21c. The first discharge-specific capacities for NNMF-800, NNMF-830, NNMF-850, and NNMF-880 were 113, 117, 226, and 58 mAh g^{-1}, respectively.

The cycling performance after 100 cycles, as shown in Figure 21d, revealed significant differences among the samples. The NNMF-850 electrode exhibited a gradual decrease in discharge capacity from 226 mAh g^{-1} to 158 mAh g^{-1}, corresponding to a capacity retention rate of ~70%, demonstrating excellent cycling stability. In contrast, NNMF-800, NNMF-830, and NNMF-880 showed capacity retention rates of 40%, 46%, and 41.7%, respectively, indicating poor cycling stability.

Figure 21e illustrates the discharge performance of the electrodes at various current densities. The NNMF-850 electrode displayed superior rate performance compared to the other samples. For NNMF-850, the discharge capacities at 0.2 C, 0.5 C, 1 C, and 2 C were 225,

193, 168, and 145 mAh g^{-1}, respectively, and the capacity recovered to 193 mAh g^{-1} upon returning to 0.2 C, demonstrating excellent reversibility. On the other hand, the discharge capacities of NNMF-800, NNMF-830, and NNMF-880 decreased more significantly as the current density increased, indicating inferior rate capabilities.

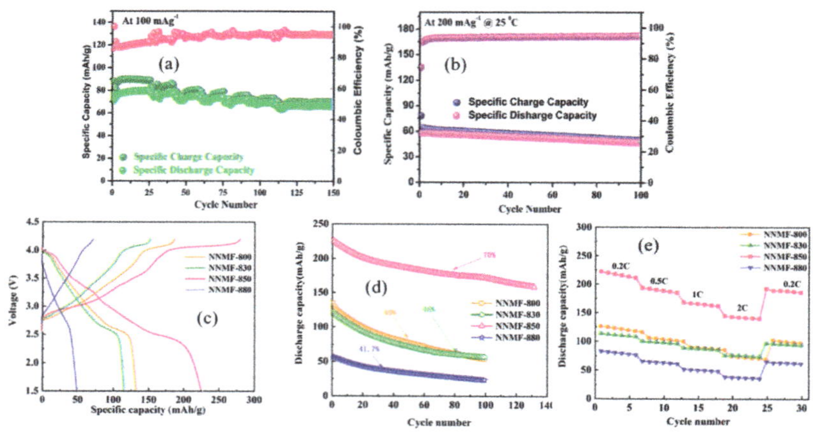

Figure 21. (**a**) Cyclic performance of $Na_3PO_4@NaNi_{0.815}Co_{0.15}Al_{0.035}O_2$ at a current density of 100 mA g^{-1} at room temperature, (**b**) cyclic performance of $Na_3PO_4@NaNi_{0.815}Co_{0.15}Al_{0.035}O_2$ at a current density of 200 mA g^{-1} at room temperature. Reprinted from Ref. [119] with permission from the American Chemical Society, (**c**) first charge–discharge curves of NNMF-800, NNMF-830, NNMF-850, and NNMF-880 at a rate of 0.2C (1 C = 160 mA g^{-1}), (**d**) capacity retention curves of NNMF-800, NNMF-830, NNMF-850, and NNMF-880 at a rate of 0.2 C, and (**e**) rate cycling plots for NNMF-800, NNMF-830, NNMF-850, and NNMF-880. Reprinted from Ref. [120] with permission from Elsevier.

The superior performance of NNMF-850 is attributed to the optimization of its crystallinity and morphology, which enhances its structural stability and Na^+ diffusion, ensuring excellent cycling stability and rate capability.

2.5. $NaCoO_2$

Investigations into Na_xCoO_2 ($0.5 \leq x \leq 1$) have identified four distinct phases: O3 ($0.9 \leq x \leq 1$, α phase), O′3 ($x = 0.75$, α' phase), P3 ($0.55 \leq x \leq 0.6$, β phase), and P2 ($0.64 \leq x \leq 0.74$, γ phase). These phases depend on non-stoichiometry and synthesis conditions (The most common synthesis method is solid-state reaction of (non)stoichiometric Co_3O_4 and Na_2CO_3 at different temperatures, based on phase type). Notably, the P2 phase demonstrates superior reversibility as a cathode material and enhanced thermoelectric power [121–124].

Among these, the O3 phase demonstrates superior energy density at voltages below 3.5 V [125–127]. However, O3 materials are strongly hygroscopic, which poses challenges for processing in air.

P2 compounds, on the other hand, have garnered significant attention due to their higher sodium mobility compared to O3 compounds, as demonstrated in computational and experimental studies. The P2 structure forms in the stoichiometry Na_xMO_2 when $0.6 < x < 0.8$, while the O3 phase is more stable at full sodiation ($x = 1.0$). Consequently, synthesizing P2 compounds requires a higher average transition metal oxidation state ($M^{3.2+}/M^{3.4+}$) than for O3. As a result, only three single-metal P2 compounds (M = V, Mn, and Co) have been synthesized to date, while others are obtained in binary or ternary systems incorporating high-valent ions. However, these high-valence elements often contribute minimally to the electrochemical performance, either being inactive or adding only to the low-voltage capacity [37]. The charge–discharge curves of O3, P3, and P2 sodium cobalt oxide are shown in Figure 22a. Despite differences in the stacking sequences of CoO_2 layers, Na_xCoO_2 phases exhibit multiple voltage plateaus. Large reversible capacities are achieved for all three phases: 140 mAh g^{-1} for the O3 phase, 130 mAh g^{-1} for the P3 phase, and 120 mAh g^{-1} for the P2 phase.

Layered $NaCoO_2$ stands out due to its excellent structural stability and superior cyclability compared to $LiCoO_2$. The sodium-ion diffusion coefficient of Na_xCoO_2 (0.5–1.5 × 10^{-10} cm^2 s^{-1}) is comparable to that of lithium-ion diffusion in $LiCoO_2$ (1 × 10^{-11} cm^2 s^{-1}) [128]. In $LiCoO_2$, excessive lithium-ion removal can lead to a structural transformation from hexagonal to tetragonal, causing capacity degradation. In contrast, $NaCoO_2$ benefits from sodium's larger ionic radius, preventing occupation of tetrahedral sites, which avoids structural deformation and lowers the potential barrier for ion migration.

Among the various phases of Na_xCoO_2 ($0.5 \leq x \leq 1$), the P2 phase ($0.6 \leq x \leq 0.74$) exhibits superior reversibility as a cathode material due to its large trigonal prismatic sites that enhance sodium-ion intercalation and improve cycle life. However, its complex phase transitions—caused by the rotation of MO_6 octahedra and M-O bond breaking—can also enhance its cyclic performance, making the P2 phase a focus of recent research [129].

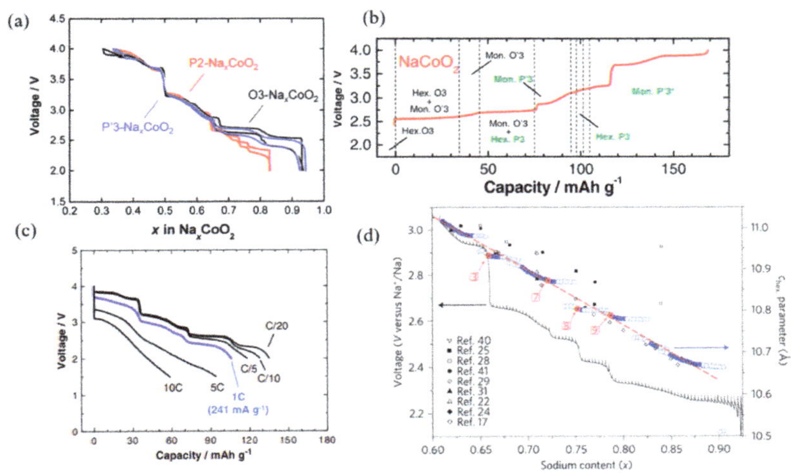

Figure 22. (**a**) Comparison of typical charge–discharge curves of Na cells with layered polymorphs: O3-, P′3-, and P2-type Na_xCoO_2. The curves show similar profiles across the three polymorphs, except in the region with high sodium content ($x > 0.7$), where differences become apparent. Reprinted from Ref. [130] with permission from the American Chemical Society, (**b**) galvanostatic charge curves and structural evolution of $Na_{1-x}CoO_2$, where detailed structural changes have not been reported in the monoclinic P′3 phase above 3.5 V. Reprinted from Ref. [38] with permission from Wiley, (**c**) rate capability of $NaCoO_2$, indicating high capacity at low rates but diminished performance at higher current densities due to phase boundary movements and Na^+/vacancy ordering. Reprinted from Ref. [86] with permission from Elsevier, and (**d**) evolution of the c_{hex} parameter as a function of sodium content in P2-Na_xCoO_2, showing its correlation with the electrochemical potential (red curve) and structural parameters in both single-phase (filled blue circles) and biphasic domains (open circles). In single-phase regions, the c_{hex} parameter shows an approximately linear evolution with sodium content (red dashed line), though deviations occur near phase composition boundaries, especially close to biphasic domains. Key single-phase compositions, corresponding to specific electrochemical transitions (drops no. 3, 7–9), are highlighted as red circles. Reprinted from Ref. [127] with permission from Nature.

Yoshida et al. [86] reported that O3-$NaCoO_2$ delivers a discharge capacity of over 140 mAh g^{-1} within a voltage range of 2.5–4.0 V, featuring a stepwise voltage profile. This profile is attributed to multiple reversible phase transitions involving CoO_2 slab gliding and in-plane vacancy-Na ordering (Figure 22b). The capacity loss during cycling in O3-$NaCoO_2$ is likely due to the degradation of the electrolyte solution, exacerbated by the non-passivated surface of metallic sodium used as the counter electrode. Furthermore, the voltage profile of Na_xCoO_2 suggests a limited rate capability, attributed to phase boundary movements in two-phase regions and Na/vacancy ordering coupled with charge ordering (Figure 22c).

Unlike $LiCoO_2$, the P2-type Na_xCoO_2 exhibits a highly intricate electrochemical charge–discharge profile characterized by several voltage plateaus. These plateaus are attributed to Na^+/vacancy ordering, which arises from in-plane patterning between Na^+ and vacancy sites, a phenomenon absent in the layered structure cathodes of LIBs. Na^+/vacancy ordering results from energy minimization involving three types of cation interactions: Na^+-Na^+, Na^+-$Co^{3+/4+}$, and $Co^{3+/4+}$-$Co^{3+/4+}$, occurring below a critical temperature. This ordering is highly sensitive to the sodium content in the host material and is accompanied by a first-order phase transition at specific sodium concentrations. Such transitions introduce additional activation energy barriers for Na^+ accommodation. Berthelot et al. [127] investigated the structural changes in P2-Na_xCoO_2 cathode materials and observed that the c_{hex} parameter decreases during discharge. This decrease is attributed to enhanced structural cohesiveness upon sodium intercalation. Simultaneously, a slight increase in the a_{hex} parameter-representing the M-M (metal-metal) distance within the slabs-is linked to the reduction of M^{4+} ions (Figure 22d). It can be observed that the voltage plateaus in the electrochemical curve correspond to biphasic transitions, during which the c_{hex} parameter remains constant for each phase. Conversely, in single-phase regions, the c_{hex} parameter evolves continuously with changes in sodium composition.

Similar to other sodium transition metal oxides, the doping of electrochemically (in)active elements has been explored as an effective strategy to enhance the performance of sodium cobalt oxide cathode materials. This approach aims to stabilize the structure, improve cycling stability, and optimize the electrochemical properties by mitigating issues such as phase transitions and structural degradation during sodium intercalation and deintercalation [37,131–133]. Kang et al. [134] investigated the effects of Ti doping in P2-type $Na_{0.67}Co_{1-x}Ti_xO_2$ ($x < 0.2$) cathodes under an extended potential range exceeding 4.4 V (Figure 23a–c). Their findings (Figure 23d) demonstrated that $Na_{0.67}Co_{0.90}Ti_{0.10}O_2$ exhibited excellent capacity retention of 115 mAh g^{-1} even after 100 cycles, while $Na_{0.67}CoO_2$ showed negligible capacity retention (<10 mAh g^{-1}) under a 4.5 V cutoff voltage. Additionally, $Na_{0.67}Co_{0.90}Ti_{0.10}O_2$ displayed exceptional rate capabilities (Figure 23e), achieving 108 mAh g^{-1} at a current density of 1000 mA g^{-1} (7.4 C). These results suggest that the enhanced rate capability is primarily due to reduced kinetic resistance from Ti substitution, rather than the elimination of active sites for Na ions at high potential conditions above 4.4 V.

Ca doping has been identified as an effective way to enhance the cycling stability of P2-type Na_xCoO_2 cathode materials [135,136]. Han et al. [136] explored the cycling and rate performance of $Na_xCa_yCoO_2$ ($0.45 \le x \le 0.64$, $0.02 \le y \le 0.10$) to understand the impact of Ca doping on cathode materials. The ionic radius of Ca^{2+} (1.00 A°) is considerably larger than that of Co^{3+} (0.61 A° for high spin) or Co^{4+} (0.53 A° for high spin), making it unlikely to replace Co directly within the CoO_6 octahedra. Instead, Ca^{2+} ions are positioned between CoO_6 slabs, where they strengthen the electrostatic interactions between these layers. While Ca doping reduces the initial discharge capacity, it significantly enhances the structural stability of $Na_{0.73}CoO_2$. Among the compositions studied, the $Na_{0.60}Ca_{0.07}CoO_2$

cathode exhibited the highest capacity retention over extended cycling (Figure 23f) and superior rate performance (Figure 23g), particularly under high-current conditions.

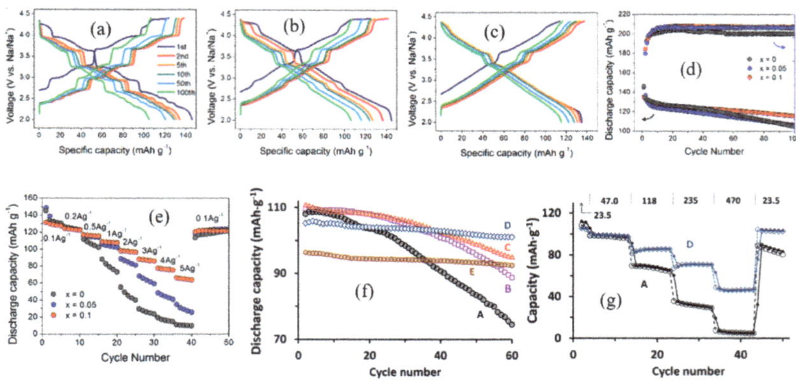

Figure 23. (**a**–**c**) Electrochemical cycling performance of Na/$Na_{0.67}Co_{1-x}Ti_xO_2$ cells with varying titanium substitution levels (x = 0, x = 0.05, and x = 0.1), illustrating specific discharge capacities as a function of cycle number, (**d**) specific capacity and coulombic efficiency over cycles for each composition, highlighting improved capacity retention and efficiency with titanium doping, (**e**) rate performance comparison of Na/$Na_{0.67}Co_{1-x}Ti_xO_2$ cells, emphasizing the enhanced rate capability with x = 0.05 and x = 0.1 substitutions. Reprinted from Ref. [134] with submission from the American Chemical Society, (**f**) discharge capacity retention of $Na_{0.73}CoO_2$ and calcium-doped variants $Na_{0.68}Ca_{0.02}CoO_2$, $Na_{0.63}Ca_{0.05}CoO_2$, $Na_{0.60}Ca_{0.07}CoO_2$, and $Na_{0.52}Ca_{0.10}CoO_2$ at 23.5 mA g $^{-1}$, demonstrating capacity stabilization with optimized calcium doping levels, and (**g**) discharge (open markers) and charge (solid markers) capacities of $Na_{0.73}CoO_2$ (A) and $Na_{0.60}Ca_{0.07}CoO_2$ (D) at varying current densities, where marked current values denote performance trends across rate capabilities. Reprinted from Ref. [136] with permission from Elsevier.

Table 3 provides a summary of the capacities achieved with different phases of sodium cobalt oxide. These materials were all examined using non-aqueous electrolytes, such as organic or ionic solvents, with a sodium metal anode serving as the counter electrode.

Table 3. Specific capacities reported for various phases of Na_xCoO_2.

Compound	Phase Type	Discharge Specific Capacity (mAh g^{-1})	Cycling Rate
$Na_{0.67}CoO_2$ [20]	P2	105.0	2 μA cm^{-2}
$Na_{0.74}CoO_2$ [137]	P2	107.0	0.1 C
$Na_{0.71}CoO_2$ [138]	P2	69.7	0.08 C
$Na_{0.74}CoO_2$ [135]	P2	134.0	11.1 mA g^{-1}
$Na_{0.67}CoO_2$ [139]	P2	146.8	10 mAg^{-1}
$Na_{0.67}CoO_2$ [134]	P2	146.0	100 mA g^{-1}
$Na_{0.7}CoO_2$ [140]	P2	125.0	5 mA g^{-1}
$Na_{0.99}CoO_2$ [20]	O3	118.0	4.3 μA cm^{-2}
$NaCoO_2$ [86]	O3	140.0	12 mA g^{-1}
$Na_{0.57}CoO_2$ [124]	P2	68.0	0.7 C (0.1 A g^{-1})

2.6. $NaVO_2$

Among the various reported Na_xVO_2 compounds, P2- and O3-type compositions have been thoroughly studied. The O3-type Na_xVO_2, obtained by reduction in $NaVO_3$ under H_2 atmosphere at 700 °C, demonstrates the highest initial discharge capacity (~140 mAh g^{-1}) with full reversibility between $Na_{0.5}VO_2$ and Na_1VO_2 compositions, but suffers from poor cycling stability due to significant slab gliding during the charge–discharge process. The intermediate voltage window (1.6–2.6 V), although indicative of decent electrochemical performance, makes it less attractive for practical applications. Moreover, the O3 phase is sensitive to air, where oxygen slightly oxidizes the material, forming a small amount of an oxidized phase on the particle surface following the reaction:

$$NaVO_2 + (1 - x)/4\ O_2 \rightarrow Na_xVO_2 + (1 - x)/2\ Na_2O \quad (10)$$

Figure 24a,b illustrate the electrochemical performance of O3- and P2-Na_xVO_2 phases against Na^+/Na. Figure 24a highlights a highly reversible Na insertion and de-insertion process between 1.2 and 2.4 V. The first charge curve shows a potential plateau indicative of a biphasic domain close to the $Na_{2/3}VO_2$ composition, followed by multiple potential plateaus associated with more complex behavior. Near the $Na_{1/2}VO_2$ composition, another single-phase potential step is observed. Figure 24b expands on this by showcasing the extended cycling behavior and capacity retention, underscoring the material's electrochemical stability and phase transformation reversibility over successive cycles [141–143].

Despite these merits, the cell polarization remains significant, as shown in Figure 24c, even at low rates (C/100). Extending the voltage range beyond 2.5 V results in irreversible transitions, as only a small amount of Na can be reinserted (Figure 24d). The discharge curve shape changes significantly, similar to Na_xTiO_2 systems, attributed to Ti migration into Na layers. The structural changes are evidenced in Figure 24e, where XRPD patterns for the $Na_{2/3}VO_2$ and $Na_{1/2}VO_2$ phases are compared to the pristine $NaVO_2$ phase.

The XRPD analysis reveals that for $Na_{2/3}VO_2$ ($2/3 < x < 1$), a mixture of starting O3-$NaVO_2$ and $Na_{2/3}VO_2$ phases is formed. For smaller x values, Na_xVO_2 phases ($1/2 < x < 2/3$) crystallize in the C2/m space group, representing a distortion of the trigonal lattice in O3-$NaVO_2$. The irreversible decrease in interslab distance suggests vanadium migration into vacancies caused by Na de-intercalation.

The P2-phase Na_xVO_2, while challenging to synthesize in a pure single-phase form, emerges as a more promising candidate. Its cycling behavior within the voltage range of 1.4–2.5 V (Figure 24b) shows minimal polarization and a reversible capacity of ~120 mAh g^{-1}. However, at higher deintercalation voltages (V > 2.5 V), irreversible phase changes occur. The voltage–composition curve resembles the P2-Na_xCoO_2 system (Figure 24f), showing multiple potential plateaus (biphasic domains) and continuous voltage variations (solid-solution phases). Specific potential drops correlate with ordered Na/V ratios (1/2, 5/8, and 2/3), pointing to ordered Na ion arrangements and/or V^{3+}/V^{4+} within the VO_2 slabs [144–147].

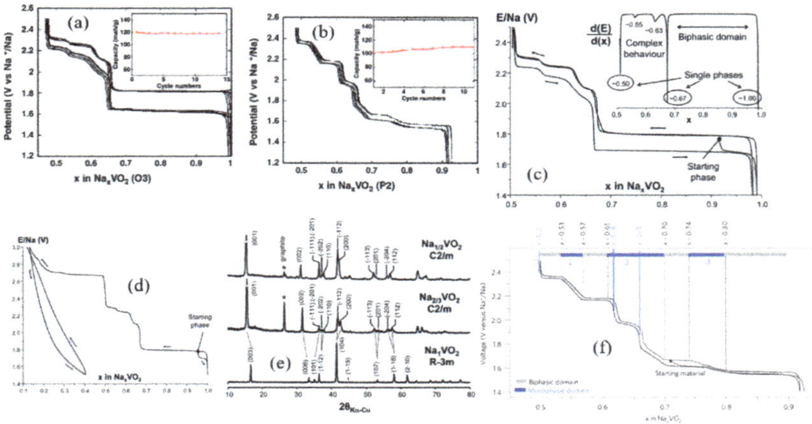

Figure 24. (**a**) Voltage–composition curves for $NaVO_2$ (O3), highlighting the reversible formation of $Na_{1/2}VO_2$ and $Na_{2/3}VO_2$ during Na de-intercalation and the onset of an electrochemically inactive phase at potentials above 2.4 V, (**b**) voltage–composition curves for $Na_{0.7}VO_2$ (P2), demonstrating its electrochemical behavior in the sodium-ion battery system. The inset figures show the delivered discharge capacity during limited cycling. Reprinted from Ref. [145] with permission from Elsevier, (**c**) electrochemical cycling of Na_xVO_2 (x = 0.92, as determined by lever rule) in the range 1.4–2.5 V at rate C/100. The differential curve dE/dx of the first charge is shown in the inset. Na_xVO_2 is a mixture of $NaVO_2$ and $Na_{2/3}VO_2$, (**d**) electrochemical cycling of Na_xVO_2 (x = 0.95, as determined by lever rule) in the range 1.5–3.0 V at rate C/100, (**e**) X-ray powder diffractograms of the as-synthesized Na_1VO_2 phase and the electrochemically obtained $Na_{2/3}VO_2$ and $Na_{1/2}VO_2$ phases. Indexations in the corresponding space groups are shown. (*) is related to graphite peak in the XRD pattern. Reprinted from Ref. [146] with permission from IOP Publishing, (**f**) evolution of cell voltage as a function of sodium content in Na_xVO_2 over the $0.5 \leq x \leq 0.92$ range. The limits of the biphasic domains and the solid solutions are shown by dashed lines. The three single phases for x = 1/2, 5/8, and 2/3 are emphasized by thick blue lines. The numbers 2, 3, and 4 indicate monophasic domains during sodium (de)intercalation. Reprinted from Ref. [144] with permission from Nature.

2.7. *$NaMnO_2$*

$NaMnO_2$ is an intriguing sub-class of lithium transition metal oxides based on manganese, notable for its high theoretical capacity, environmental friendliness, and ease of synthesis. Na_xMnO_2 has been studied since the 1970s, offering a variety of stoichiometries and structures. Depending on sodium content, these materials can form double or triple chains creating 3D tunnel structures (Na/Mn < 0.5) or 2D layered structures (Na/Mn > 0.5). $NaMnO_2$ exists in two main phases: at low temperatures, α-$NaMnO_2$ adopts a distorted O3 layered structure with monoclinic distortion, attributed to the Jahn–Teller effect of Mn^{3+} ions and at high temperatures, it transforms into orthorhombic β-$NaMnO_2$, a layered structure composed of MnO_2 sheets made of edge-sharing MnO_6 octahedra, where sodium ions occupy octahedral sites between adjacent MnO_2 sheets. Density Functional Theory (DFT) calculations estimate similar formation energies for these phases, making it experimentally challenging to synthesize a pure β-$NaMnO_2$ phase [148].

The charge–discharge characteristics of a non-aqueous Na/α-$NaMnO_2$ cell at varying rates, obtained by solid-state reaction of Na_2CO_3 and Mn_2O_3 at 700 °C in air, are shown in Figure 25a,b. Almost 0.9 Na per formula unit is extracted during the first charge, delivering an initial charge capacity of 215 mAh g^{-1}. Monoclinic $NaMnO_2$ provides discharge capacities of 185 and 194 mAh g^{-1} within a voltage range of 2–3.8 V, achieving coulombic efficiencies of 86.4% and 92.8% at C/20 and C/30 rates, respectively (1 C = 244 mAh

g^{-1}). Coulombic efficiency decreases in subsequent cycles due to side reactions, likely involving the electrolyte, particularly at higher voltages. Lower efficiencies at C/30 suggest prolonged exposure to high voltage exacerbates parasitic reactions or degradation [149]. Overcharging beyond 3.8 V (up to 4.2 V at C/10) shows an additional charge capacity of ~90 mAh g^{-1} between 3.8 V and 4.2 V. However, this additional capacity is not recovered during discharge, indicating it arises from electrolyte decomposition rather than reversible Na-ion intercalation (Figure 25c). Monoclinic $NaMnO_2$ exhibits stability issues, as shown in Figure 25d. Structural changes during sodium extraction from O′3-$NaMnO_2$ generate internal stress, leading to capacity fading. This fading is linked to bulk structural changes and crystallite size rather than surface reactions. Gradual and irreversible changes in the charge–discharge profiles during cycling reflect the impact of bulk structural evolution [150].

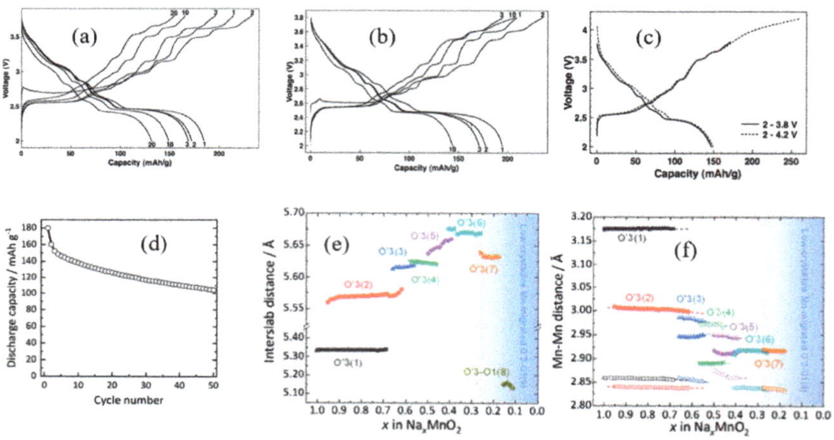

Figure 25. (**a**) Voltage profile of $NaMnO_2$ after multiple cycles at a C/20 rate, (**b**) voltage profile of $NaMnO_2$ after multiple cycles at a C/30 rate. The cell is galvanostatically cycled between 2.0 V and 3.8 V, (**c**) comparison of the voltage profiles of $NaMnO_2$ charged up to 3.8 V and 4.2 V, with the cell cycled at a C/20 rate. Reprinted from Ref. [149] with permission from IOP Publishing, and (**d**) cycling stability of O′3-$NaMnO_2$ over 50 cycles at 12 mA g^{-1} in the voltage range of 2.0–3.8 V at 25 °C. Reprinted from Ref. [150] with permission from the Royal Society of Chemistry, (**e**) refinement results exhibiting the changes in interslab distance, and (**f**) in-plane Mn–Mn distance. Reprinted from Ref. [150] with permission from the Royal Society of Chemistry.

Kubota et al. [150] investigated the phase transitions occurring in monoclinic-$NaMnO_2$ during charging. They identified seven highly crystalline O′3-type phases (O′3(1) to O′3(7)) and less crystalline O′3-O1-like phases. The phase evolution proceeds as follows: In the first plateau around 2.6 V, O′3(2) gradually increases at the expense of O′3(1), undergoing a complete phase transition from O′3(1) to O′3(2). A similar two-phase transition process continues with progressive charging up to 3.8 V, involving seven O′3 phases and one O′3-O1 phase.

With progressive sodium extraction, electrostatic repulsion between adjacent MnO_2 slabs increases, leading to an enlargement of the interslab distance. This trend persists throughout the desodiation process until the emergence of the O′3(7) phase. However, during the transition from O′3(7) to O′3-O1(8) at a voltage plateau around 3.6 V, the interslab distance decreases from 5.63 A° to 5.15 A° (Figure 25e). This transition, associated with decreased crystallinity and high Mn ion migration into the interslab space to form a modulated phase and/or the coexistence of differently oriented domains (Figure 25f), results in the formation of low-crystallinity O′3 phases, causing significant capacity degradation during cycling in a wider voltage range (above 3.55 V).

β-$NaMnO_2$ exhibits a distinctive layered structure that sets it apart from the typical polymorphs of α-$NaMnO_2$. Unlike the planar MnO_6 octahedral layers in α-types, which adjust their stacking sequences to form polymorphs like O3, P2, or P3, β-$NaMnO_2$ is characterized by zigzag layers of edge-sharing MnO_6 octahedra. In this structure, Na^+ ions occupy the octahedral sites between the zigzag MnO_6 layers (Figure 26a). The electrochemical performance of β-$NaMnO_2$, synthesized by firing a mixture of Mn_2O_3 and Na_2CO_3 first at 950 °C for 24 h, after a temperature ramp of 1 °C/min, and second at 950 °C for 24 h, after ramping the temperature at a faster heating rate of 5 °C/min in oxygen atmosphere, is demonstrated through the load curves in Figure 26b, showing Na extraction and reinsertion processes. During the first charge, a distinct plateau appears between the fully sodiated $NaMnO_2$ phase and a desodiated phase close to $Na_{0.57}MnO_2$, with smaller steps observed at compositions around $Na_{0.49}MnO_2$ and $Na_{0.39}MnO_2$. This initial cycle experiences an irreversible capacity loss of approximately 25 mAh g^{-1} (equivalent to 0.1 Na per formula unit). Despite this initial loss, the load curves remain relatively stable over subsequent cycles, with only a slight reduction in capacity. A plateau at 2.7 V corresponds to a phase transition between Jahn–Teller distorted and undistorted structures, accompanied by minimal polarization (below 150 mV). The rate performance of β-$NaMnO_2$ at varying charge–discharge rates (Figure 26c) highlights a gradual but irreversible capacity fade over time. Notably, the material does not recover its initial capacity, indicating some structural degradation during cycling, particularly under higher discharge rates (Figure 26d). These findings suggest that while β-$NaMnO_2$ demonstrates interesting structural and electrochemical features, its cycling stability requires further investigation and optimization [151].

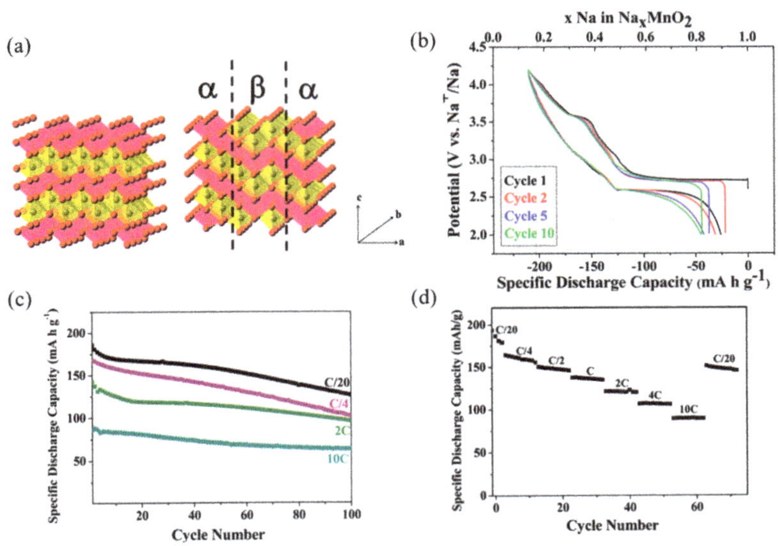

Figure 26. (**a**) Schematic representation of the β-$NaMnO_2$ structure in the Pmnm space group. MnO_6 octahedra are depicted in pink, NaO_6 octahedra in yellow, and oxygen atoms in red. An intergrowth model on the right illustrates the structural relationship between α- and β-$NaMnO_2$. Reprinted from Refs. [151,152] with permission from the American Chemical Society, (**b**) load curves for β-$NaMnO_2$ during Na extraction/reinsertion cycles at a C/20 rate (10 mA g^{-1}), showing the first (black), second (red), fifth (blue), and tenth (green) cycles, (**c**) cycling performance of β-$NaMnO_2$ in the voltage range of 2–4.2 V vs. Na^+/Na at various rates. Na extraction occurs at C/4, while reinsertion is performed

at rates indicated above each dataset. For the C/20 dataset, both extraction and reinsertion occur at C/20, and (**d**) Specific discharge capacities of β-$NaMnO_2$ over 100 cycles between 2 and 4.2 V vs. Na^+/Na at room temperature. The plot illustrates capacity changes with cycling across different rates (C/20 to 10 C). Na extraction occurs at C/4, with reinsertion rates indicated by square markers: black (C/20), purple (C/4), green (C/4), and blue (10 C). Reprinted from Ref. [151] with permission from the American Chemical Society.

To mitigate phase transitions and enhance capacity retention in P2- and O3-type sodium manganese-based layered oxide cathode materials, extensive research has focused on elemental doping and transition metal combinations [107,153–160]. Li et al. [161] synthesized a sodium transition metal oxide, $NaMn_{1/4}Ni_{1/4}Fe_{1/4}Co_{1/4}O_2$, featuring an equimolar ratio of Ni, Mn, Fe, and Co. This cathode material exhibited an initial discharge capacity of 180 mAh g^{-1}, delivered at an average discharge voltage of 3.21 V (Figure 27a). Compared to $NaFe_{0.5}Co_{0.5}O_2$ and $NaMn_{0.5}Ni_{0.5}O_2$, the $NaMn_{1/4}Ni_{1/4}Fe_{1/4}Co_{1/}4O_2$ cathode demonstrated superior cycling performance across both low and high upper cutoff voltages (Figure 27b,c). The improvements in capacity retention were attributed to the absence of multiple phase transitions (for example, from hexagonal to monoclinic structures) and the reduction in Jahn–Teller distortion of Mn ions.

Further advancements have been made by investigating Fe substitution within the $NaMn_{1\text{-}x}Ni_{1\text{-}x}Fe_xO_2$ system. Yuan et al. [162] explored the effect of Fe substitution in $NaMn_{0.5}Ni_{0.5}O_2$ cathode materials, finding that Fe doping successfully suppressed capacity decay during charge–discharge cycling. This enhancement was linked to improved structural stability, specifically the inhibition of MO_2 layer gliding and phase conversion.

Additionally, Li et al. [163] developed a MoS_2 coating for $NaMn_{0.4}Ni_{0.4}Fe_{0.2}O_2$ particles and studied its impact on the material's electrochemical performance. The coating not only improved the material's electrical conductivity but also enhanced its structural stability, even in air. After exposure to air for 10 days, the cathode retained a specific discharge capacity of 128.9 mAh g^{-1} (initially 148.4 mAh g^{-1} at 0.1 C within a 2–4.2 V voltage window). These improvements were attributed to the increased conductivity and reduced side reactions occurring in the electrolyte.

While significant research has been conducted on P2- and O3-type Na_xMnO_2 cathode materials, making them strong candidates for high capacity and stability, relatively few studies have addressed modifications of β-Na_xMnO_2. For instance, Jiang et al. [164] investigated the effects of Cu doping on the electrochemical performance of β-$NaMnO_2$, specifically examining $NaCu_{0.1}Mn_{0.9}O_2$. The charge–discharge curves, capacity retention, and rate performance are presented in Figure 27d–g. As shown in Figure 27d, $NaCu_{0.1}Mn_{0.9}O_2$ exhibited an initial discharge capacity corresponding to 0.6 Na reversible intercalation/deintercalation with a coulombic efficiency of 94.1%, surpassing that of undoped β-$NaMnO_2$. Remarkably, $NaCu_{0.1}Mn_{0.9}O_2$ demonstrated excellent reversibility at the voltage plateau of ~3.5 V, unlike the irreversible behavior observed in β-$NaMnO_2$.

Moreover, $NaCu_{0.1}Mn_{0.9}O_2$ exhibited improved cycling stability, retaining 80.1% of its initial capacity after 100 cycles, compared to 68.4% retention in β-$NaMnO_2$ (Figure 27f). It also showed superior rate performance (Figure 27e,g). This enhanced performance may stem from the absence of local structural rearrangements associated with planar defects in β-$NaMnO_2$, which are mitigated by Cu doping.

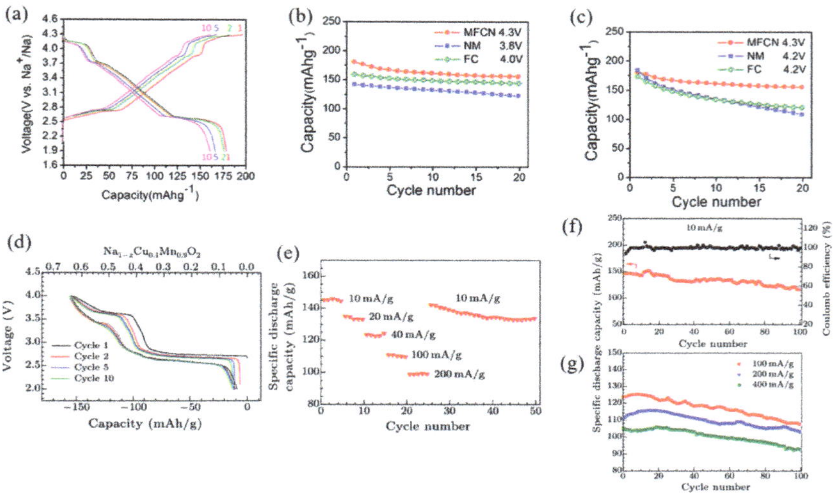

Figure 27. (**a**) Galvanostatic charge and discharge profiles at the 1st, 2nd, 5th, and 10th cycles for $NaMn_{1/4}Fe_{1/4}Co_{1/4}Ni_{1/4}O_2$ (MFCN), (**b**) discharge capacity of MFCN cycled between 1.9 and 4.3 V at different cycles compared with $NaMn_{0.5}Ni_{0.5}O_2$ (NM) cycled between 1.9 and 3.8 V and $NaFe_{0.5}Co_{0.5}O_2$ (FC) cycled between 1.9 and 4.0 V, all at a C/10 rate. (**c**) Discharge capacity of MFCN cycled between 1.9 and 4.3 V at different cycles compared with NM and FC cycled between 1.9 and 4.2 V, all at a C/10 rate. Reprinted from Ref. [161] with permission from Elsevier, (**d**) charge–discharge curves for $NaCu_{0.1}Mn_{0.9}O_2$ at a rate of C/20 (10 mA/g) in the voltage range of 2.0–4.0 V versus Na^+/Na. The 1st, 2nd, 5th, and 10th Na extraction/reinsertion cycles are represented in black, red, blue, and green, respectively, (**e**) rate capability of the $NaCu_{0.1}Mn_{0.9}O_2$ electrode material, (**f**) the capacity and coulombic efficiency versus cycle number at 10 mA g^{-1} for $NaCu_{0.1}Mn_{0.9}O_2$, (**g**) the capacity versus cycle number at 100 mA g^{-1}, 200 mA g^{-1}, and 400 mA g^{-1} for $NaCu_{0.1}Mn_{0.9}O_2$. Reprinted from Ref. [164] with permission from IOP Publishing.

Based on the discussion of transition metal oxide layered materials for SIBs, it is evident that Mn- and Ni-based compounds are excellent candidates for cathode materials. Wang et al. [116] investigated the substitution of both Li and Ti for Mn and Ni in the structure of $NaMn_{0.5}Ni_{0.5}O_2$. The electrochemical performance of $NaLi_{1/9}Ni_{1/3}Mn_{4/9}Ti_{1/9}O_2$ and $NaMn_{0.5}Ni_{0.5}O_2$ is presented in Figure 28a,b.

Unlike the charge–discharge curves of $NaMn_{0.5}Ni_{0.5}O_2$, which exhibit multiple voltage plateaus due to complex phase transitions—identified as the primary cause of capacity decay during cycling—the curves for $NaLi_{1/9}Ni_{1/3}Mn_{4/9}Ti_{1/9}O_2$ are much smoother. This indicates reduced phase transitions, enabling more stable material operation.

Elemental substitution significantly improved the cycling stability of $NaMn_{0.5}Ni_{0.5}O_2$. After 100 cycles at 1 C, the capacity retention increased from 46.97% for the pristine material to 80% for the substituted material, as shown in Figure 28c. The primary reason for this improvement is the expanded Na layer spacing, which enhances Na-ion diffusion and effectively suppresses complex phase transitions (Figure 28d,e).

Beyond conventional elemental doping strategies, structural modifications such as single crystalline and binary structures have been explored as an effective approach to improve transition metal layered oxide cathode materials [165,166]. Single-crystal structural synthesis has emerged as a promising solution to enhance capacity retention during cycling [167–171].

Darga et al. [113] reported that single-crystalline $NaMn_{0.5}Ni_{0.5}O_2$ demonstrated significantly higher capacity retention compared to its polycrystalline counterpart, with 69%

retention versus 35% after 200 cycles at 0.2C. The superior stability of the single-crystalline structure is attributed to its enhanced mechanical stability, as illustrated in Figure 28f,g.

As shown, the polycrystalline cathode particles experience pronounced microcracking due to volume expansion, exposing newly formed surfaces of primary particles to parasitic reactions with the electrolyte. This accelerates capacity fading. In contrast, the monocrystalline structured sample exhibited reduced cracking and minimal intragranular fractures, effectively mitigating these degradation mechanisms.

Compared to single-phase materials, heterostructured materials can leverage the multifunctional advantages of multiple phase structures. This design minimizes phase transformation in the material while simultaneously delivering exceptional electrochemical properties [172–177].

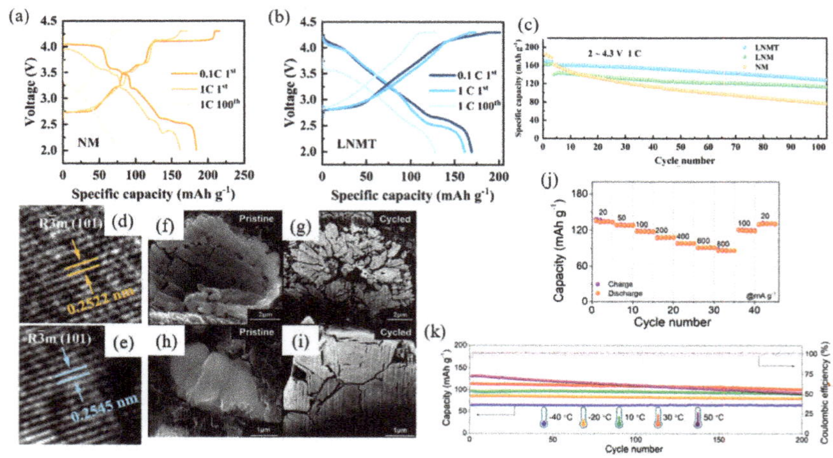

Figure 28. Charge–discharge curves of (**a**) $NaMn_{0.5}Ni_{0.5}O_2$ and (**b**) $NaLi_{1/9}Ni_{1/3}Mn_{4/9}Ti_{1/9}O_2$ during cycling, (**c**) cycling performance of $NaMn_{0.5}Ni_{0.5}O_2$, $NaLi_{1/9}Ni_{1/3}Mn_{5/9}O_2$, and $NaLi_{1/9}Ni_{1/3}Mn_{4/9}Ti_{1/9}O_2$ cathodes over 100 cycles at 1 C in the voltage range of 2.0–4.3 V. Lattice fringes of (**d**) $NaMn_{0.5}Ni_{0.5}O_2$ and (**e**) $NaLi_{1/9}Ni_{1/3}Mn_{4/9}Ti_{1/9}O_2$. Reprinted from Ref. [116] with permission from the American Chemical Society. Cross-sectional SEM image of (**f**) pristine polycrystalline particles dispersed in carbon/PVdF and (**g**) FIB-SEM image of polycrystalline particles after 200 cycles. (**h**) Cross-sectional SEM image of a pristine single-crystalline particle and (**i**) FIB-SEM image of single-crystalline particles after 200 cycles. Reprinted from Ref. [113] with permission from the American Chemical Society. (**j**) Rate capability, and (**k**) cycling performance of P2/O3-$Na_{0.7}Mn_{0.4}Ni_{0.3}Cu_{0.1}Fe_{0.1}Ti_{0.1}O_{1.95}F_{0.1}$ at different temperatures. Reprinted from Ref. [178] with permission from Elsevier.

Zhou et al. [178] reported a high-entropy biphasic P2/O3 material with a phase ratio of 23:77, $Na_{0.7}Mn_{0.4}Ni_{0.3}Cu_{0.1}Fe_{0.1}Ti_{0.1}O_{1.95}F_{0.1}$, as an excellent cathode material exhibiting high rate capability and cycling performance (Figure 28j,k) across a wide temperature range of −40 to 50 °C and at high current densities.

The key factor in optimizing the performance of biphasic materials is achieving an ideal phase ratio. An imbalanced ratio can degrade the cycling stability of the multiphase structure, making it less stable than its single-phase counterparts, especially during prolonged cycling.

3. Conclusions and Perspectives

Sodium-ion batteries (SIBs) are gaining significant attention as a promising alternative to lithium-ion batteries (LIBs), particularly for large-scale applications. The primary focus has been on improving the cathode materials to enhance their capacity density, operating voltage, and overall energy density. Layered oxide transition metal compounds, widely used as cathode materials in LIBs, have emerged as strong candidates for SIBs.

This review has outlined the key challenges that hinder the widespread application of layered oxide cathode materials in SIBs. The most significant obstacle is the structural instability of these materials, which leads to multiple phase transitions that degrade capacity retention over cycling and under high current conditions. Understanding these phase transitions, especially at higher cutoff voltages, is crucial, as the transition behavior in SIBs differs from that of LIBs due to the distinct sizes of Na and Li ions. Consequently, different phase transition mechanisms are at play, affecting the long-term performance of the cathodes.

As layered oxide materials for SIBs are still in the early stages of development, it is crucial to investigate the side reactions occurring at the cathode–electrolyte interface. Furthermore, detailed research into the thermal stability of battery cells, particularly at elevated temperatures, is essential to ensure their reliability under real-world operating conditions.

Based on the materials discussed, layered oxide compounds containing Mn, Fe, and Ni show great promise as cathode materials for SIBs. These materials can be further optimized through strategies such as introducing electrochemically inactive elements like Nb, Al, and Ti, as well as structural modifications, including single-crystalline designs and biphasic structures combining P2 and O3 phases.

A key focus for new cathode materials is reducing production costs and eliminating the need for expensive elements like Co in their composition. Additionally, the development of coatings is necessary to enhance material conductivity during cycling and high-rate operation, as well as to mitigate parasitic reactions with the electrolyte.

Thus, continued research is imperative to advance the understanding of capacity degradation mechanisms and to refine modification techniques. Ultimately, further exploration into these aspects is critical to enabling the widespread adoption of layered oxide materials as cathodes in SIBs, especially for large-scale energy storage systems.

Author Contributions: Conceptualization, writing, methodology, M.A.; review and editing, M.Z.; supervision, H.L. All authors have read and agreed to the published version of the manuscript.

Funding: The work is funded by the National Science Foundations: NSF-OIA-2119688, NSF-OIA-2429592, NSF-CMMI-2215982, NSF-CHE-2216473; and Department of Energy: DE-FE-0032475.

Data Availability Statement: No new data were created or analyzed in this study. Data sharing is not applicable to this article.

Conflicts of Interest: The authors declare no conflicts of interest.

ferences

Yabuuchi, N.; Ikeuchi, I.; Kubota, K.; Komaba, S. Thermal Stability of $NaCrO_2$ for Rechargeable Sodium Batteries: Studies By Differential Scanning Calorimetry, and High-Temperature X-Ray Diffraction. In Proceedings of the 224th ECS Meeting, San Francisco, CA, USA, 27 October–1 November 2013.

Wable, M.; Bal, B.; Capraz, Ö.Ö. Probing electrochemical strain generation in sodium chromium oxide ($NaCrO_2$) cathode in Na-ion batteries during charge/discharge. *Energy Adv.* **2024**, *3*, 601–608. [CrossRef]

Li, G.; Yue, X.; Luo, G.; Zhao, J. Electrode potential and activation energy of sodium transition-metal oxides as cathode materials for sodium batteries: A first-principles investigation. *Comput. Mater. Sci.* **2015**, *106*, 15–22. [CrossRef]

Wang, S.; Chen, F.; He, X.-D.; Zhang, L.-M.; Chen, F.; Wang, J.-R.; Dong, J.-M.; Chen, C.-H. Self-template synthesis of $NaCrO_2$ submicrospheres for stable sodium storage. *ACS Appl. Mater. Interfaces* **2021**, *13*, 12203–12210. [CrossRef]

5. Wang, Y.; Li, W.; Hu, G.; Peng, Z.; Cao, Y.; Gao, H.; Du, K.; Goodenough, J.B. Electrochemical performance of large-grai $NaCrO_2$ cathode materials for Na-ion batteries synthesized by decomposition of $Na_2Cr_2O_7{\cdot}2H_2O$. *Chem. Mater.* **2019**, 5214–5223. [CrossRef]
6. Fukunaga, A.; Nohira, T.; Hagiwara, R.; Numata, K.; Itani, E.; Sakai, S.; Nitta, K. Performance validation of sodium-ion batte using an ionic liquid electrolyte. *J. Appl. Electrochem.* **2016**, *46*, 487–496. [CrossRef]
7. Liang, J.; Liu, L.; Liu, X.; Meng, X.; Zeng, L.; Liu, J.; Li, J.; Shi, Z.; Yang, Y. O3-type $NaCrO_2$ as a superior cathode material sodium/potassium-ion batteries ensured by high structural reversibility. *ACS Appl. Mater. Interfaces* **2021**, *13*, 22635–22 [CrossRef]
8. Doi, T.; Zhao, L.; Zhou, M.; Okada, S.; Yamaki, J.-I. Quantitative studies on the thermal stability of the interface between grap electrode and electrolyte. *J. Power Sources* **2008**, *185*, 1380–1385. [CrossRef]
9. Klimpel, M.; Kovalenko, M.V.; Kravchyk, K.V. Advances and challenges of aluminum–sulfur batteries. *Commun. Chem.* **202** 77. [CrossRef] [PubMed]
10. Mărgulescu, S.; Mărgulescu, E. Lithium Market Trends. *Glob. Econ. Observ.* **2016**, *4*, 117.
11. Azevedo, M.; Baczyńska, M.; Hoffman, K.; Krauze, A. *Lithium Mining: How New Production Technologies Could Fuel the Global Revolution*; McKinsey & Company: New York, NY, USA, 2022; Volume 12.
12. Gao, Y.; Wang, Z.; Lu, G. Atomistic understanding of structural evolution, ion transport and oxygen stability in layered NaFe *J. Mater. Chem. A.* **2019**, *7*, 2619–2625. [CrossRef]
13. Kim, J.-K.; Lim, Y.J.; Kim, H.; Cho, G.-B.; Kim, Y. A hybrid solid electrolyte for flexible solid-state sodium batteries. *Energy Envi Sci.* **2015**, *8*, 3589–3596. [CrossRef]
14. Lu, Y.; Alonso, J.A.; Yi, Q.; Lu, L.; Wang, Z.L.; Sun, C. A high-performance monolithic solid-state sodium battery with Ca^{2+}-do $Na_3Zr_2Si_2PO_{12}$ electrolyte. *Adv. Energy Mater.* **2019**, *9*, 1901205. [CrossRef]
15. Xue, Z.; He, D.; Xie, X. Poly(ethylene oxide)-based electrolytes for lithium-ion batteries. *J. Mater. Chem. A* **2015**, *3*, 19218–19 [CrossRef]
16. Faglioni, F.; Merinov, B.V.; Goddard, W.A., III; Kozinsky, B. Factors affecting cyclic durability of all-solid-state lithium batte using poly(ethylene oxide)-based polymer electrolytes and recommendations to achieve improved performance. *Phys. Ch Chem. Phys.* **2018**, *20*, 26098–26104. [CrossRef] [PubMed]
17. Long, L.; Wang, S.; Xiao, M.; Meng, Y. Polymer electrolytes for lithium polymer batteries. *J. Mater. Chem. A* **2016**, *4*, 10038–10 [CrossRef]
18. Manthiram, A.; Yu, X.; Wang, S. Lithium battery chemistries enabled by solid-state electrolytes. *Nat. Rev. Mater.* **2017**, *2*, 16 [CrossRef]
19. Yu, C.-Y.; Park, J.-S.; Jung, H.-G.; Chung, K.-Y.; Aurbach, D.; Sun, Y.-K.; Myung, S.-T. $NaCrO_2$ cathode for high-rate sodium- batteries. *Energy Environ. Sci.* **2015**, *8*, 2019–2026. [CrossRef]
20. Shibata, T.; Fukuzumi, Y.; Kobayashi, W.; Moritomo, Y. Fast discharge process of layered cobalt oxides due to high Na^+ diffus *Sci. Rep.* **2015**, *5*, 9006. [CrossRef] [PubMed]
21. L Li, Y.; Lu, Y.; Zhao, C.; Hu, Y.-S.; Titirici, M.-M.; Li, H.; Huang, X.; Chen, L. Recent advances of electrode materials for low-c sodium-ion batteries towards practical application for grid energy storage. *Energy Storage Mater.* **2017**, *7*, 130–151. [CrossRef
22. Hirsh, H.S.; Li, Y.; Tan, D.H.S.; Zhang, M.; Zhao, E.; Meng, Y.S. Sodium-ion batteries paving the way for grid energy storage. *A Energy Mater.* **2020**, *10*, 2001274. [CrossRef]
23. Chen, C.-Y.; Matsumoto, K.; Nohira, T.; Hagiwara, R.; Fukunaga, A.; Sakai, S.; Nitta, K.; Inazawa, S. Electrochemical and structu investigation of $NaCrO_2$ as a positive electrode for sodium secondary battery using inorganic ionic liquid NaFSA–KFSA. *J. Po Sources* **2013**, *237*, 52–57. [CrossRef]
24. Shen, B.; Zhan, R.; Dai, C.; Li, Y.; Hu, L.; Niu, Y.; Jiang, J.; Wang, Q.; Xu, M. Manipulating irreversible phase transition of NaC towards an effective sodium compensation additive for superior sodium-ion full cells. *J. Colloid Interface Sci.* **2019**, *553*, 524– [CrossRef] [PubMed]
25. Su, J.; Pei, Y.; Yang, Z.; Wang, X. First-principles investigation on the structural, electronic properties and diffusion barriers Mg/Al doped $NaCoO_2$ as the cathode material of rechargeable sodium batteries. *RSC Adv.* **2015**, *5*, 27229–27234. [CrossRef]
26. Guo, S.; Yu, H.; Jian, Z.; Liu, P.; Zhu, Y.; Guo, X.; Chen, M.; Ishida, M.; Zhou, H. A High-Capacity, Low-Cost Layered Sodi Manganese Oxide Material as Cathode for Sodium-Ion Batteries. *ChemSusChem* **2014**, *7*, 2115–2119. [CrossRef]
27. Yabuuchi, N.; Kajiyama, M.; Iwatate, J.; Nishikawa, H.; Hitomi, S.; Okuyama, R.; Usui, R.; Yamada, Y.; Komaba, S. P2-ty $Na_x[Fe_{1/2}Mn_{1/2}]O_2$ made from earth-abundant elements for rechargeable Na batteries. *Nat. Mater.* **2012**, *11*, 512–517. [CrossR [PubMed]
28. Ponrouch, A.; Marchante, E.; Courty, M.; Tarascon, J.-M.; Palacín, M.R. In search of an optimized electrolyte for Na-ion batter *Energy Environ. Sci.* **2012**, *5*, 8572–8583. [CrossRef]

Dahbi, M.; Yabuuchi, N.; Kubota, K.; Tokiwa, K.; Komaba, S. Negative electrodes for Na-ion batteries. *Phys. Chem. Chem. Phys.* **2014**, *16*, 15007–15028. [CrossRef] [PubMed]

Kim, Y.; Kim, Y.; Choi, A.; Woo, S.; Mok, D.; Choi, N.S.; Jung, Y.S.; Ryu, J.H.; Oh, S.M.; Lee, K.T. Tin phosphide as a promising anode material for Na-ion batteries. *Adv. Mater.* **2014**, *26*, 4139–4144. [CrossRef] [PubMed]

Li, Y.; Gao, Y.; Wang, X.; Shen, X.; Kong, Q.; Yu, R.; Lu, G.; Wang, Z.; Chen, L. Iron migration and oxygen oxidation during sodium extraction from $NaFeO_2$. *Nano Energy* **2018**, *47*, 519–526. [CrossRef]

Wei, J.; Shaw, L.; Chen, W. First-principles comparative study of Cr migration in O3 and O/P hybrid-phased $NaCrO_2$. *Phys. Rev. Mater.* **2022**, *6*, 095403. [CrossRef]

Li, G.; Zhu, W. First-principles investigation on the crystal, electronic structures and diffusion barriers of F-doped $NaMO_2$ (M= V, Cr, Co and Ni) for rechargeable Na-ion batteries. *J. Solid State Chem.* **2021**, *302*, 122440. [CrossRef]

Sawicki, M.; Ortiz, A.L.; Luo, M.; Shaw, L.L. Structural-Defect-Controlled Electrochemical Performance of Sodium Ion Batteries with $NaCrO_2$ Cathodes. *ChemElectroChem* **2017**, *4*, 3222–3230. [CrossRef]

Wei, J.; Shaw, L.; Chen, W. First-principles prediction of Na diffusivity in doped $NaCrO_2$ layered cathode materials with van der Waals interactions. *J. Phys. Chem. C.* **2020**, *124*, 12239–12248. [CrossRef]

Zhao, C.; Yao, Z.; Wang, Q.; Li, H.; Wang, J.; Liu, M.; Ganapathy, S.; Lu, Y.; Cabana, J.; Li, B.; et al. Revealing high Na-content P2-type layered oxides as advanced sodium-ion cathodes. *J. Am. Chem. Soc.* **2020**, *142*, 5742–5750. [CrossRef]

Bianchini, M.; Wang, J.; Clément, R.; Cede, G. A first-principles and experimental investigation of nickel solubility into the P2 Na_xCoO_2 sodium-ion cathode. *Adv. Energy Mater.* **2018**, *8*, 1801446. [CrossRef]

Kubota, K.; Asari, T.; Yoshida, H.; Yabuuchi, N.; Shiiba, H.; Nakayama, M.; Komaba, S. Understanding the structural evolution and redox mechanism of a $NaFeO_2$–$NaCoO_2$ solid solution for sodium-ion batteries. *Adv. Funct. Mater.* **2016**, *26*, 6047–6059. [CrossRef]

Yan, P.; Zheng, J.; Zhang, J.-G.; Wang, C. Atomic resolution structural and chemical imaging revealing the sequential migration of Ni, Co, and Mn upon the battery cycling of layered cathode. *Nano Lett.* **2017**, *17*, 3946–3951. [CrossRef]

Tian, M.; Gao, Y.; Xiao, R.; Wang, Z.; Chen, L. Structural stability and stabilization of Li_2MoO_3. *Phys. Chem. Chem. Phys.* **2017**, *19*, 17538–17543. [CrossRef] [PubMed]

Gao, Y.; Ma, J.; Wang, Z.; Lu, G.; Chen, L. Vacancy-induced MnO_6 distortion and its impacts on structural transition of Li_2MnO_3. *Phys. Chem. Chem. Phys.* **2017**, *19*, 7025–7031. [CrossRef]

Liu, Y.-F.; Han, K.; Peng, D.-N.; Kong, L.-Y.; Su, Y.; Li, H.-W.; Hu, H.-Y.; Li, J.-Y.; Wang, H.-R.; Fu, Z.-Q.; et al. Layered oxide cathodes for sodium-ion batteries: From air stability, interface chemistry to phase transition. *InfoMat* **2023**, *5*, 12422. [CrossRef]

Jakobsen, C.L.; Andersen, B.P.; Johansen, M.; Christensen, C.K.; Drejer, A.Ø.; Karlsen, M.A.; Ravnsbæk. Structure and evolution of disordered deep-charge phases in Na_xCrO_2 Na-ion battery electrodes. *J. Power Sources* **2024**, *591*, 233875. [CrossRef]

Yabuuchi, N.; Ikeuchi, I.; Kubota, K.; Komaba, S. Thermal stability of Na_xCrO_2 for rechargeable sodium batteries; studies by high-temperature synchrotron X-ray diffraction. *ACS Appl. Mater. Interfaces* **2016**, *8*, 32292–32299. [CrossRef]

Luo, M.; Ortiz, A.L.; Shaw, L.L. Enhancing the electrochemical performance of $NaCrO_2$ through structural defect control. *ACS Appl. Energy Mater.* **2020**, *3*, 7216–7227. [CrossRef]

Zheng, L.; Bennett, J.C.; Obrovac, M.N. Stabilizing $NaCrO_2$ by sodium site doping with calcium. *J. Electrochem. Soc.* **2019**, *166*, A2058. [CrossRef]

Kubota, K.; Ikeuchi, I.; Nakayama, T.; Takei, C.; Yabuuchi, N.; Shiiba, H.; Nakayama, M.; Komaba, S. New insight into structural evolution in layered $NaCrO_2$ during electrochemical sodium extraction. *J. Phys. Chem. C* **2015**, *119*, 166–175. [CrossRef]

Komaba, S.; Nakayama, T.; Ogata, A.; Shimizu, T.; Takei, C.; Takada, S.; Hokura, A.; Nakai, I. Electrochemically reversible sodium intercalation of layered $NaNi_{0.5}Mn_{0.5}O_2$ and $NaCrO_2$. *ECS Trans.* **2009**, *16*, 43. [CrossRef]

Komaba, S.; Takei, C.; Nakayama, T.; Ogata, A.; Yabuuchi, N. Electrochemical intercalation activity of layered $NaCrO_2$ vs. $LiCrO_2$. *Electrochem. Commun.* **2010**, *12*, 355–358. [CrossRef]

Myung, S.-T.; Komaba, S.; Hirosaki, N.; Kumagai, N.; Arai, K.; Kodama, R.; Nakai, I. Structural Investigation of Layered $Li_{1-\delta}Mn_xCr_{1-x}O_2$ by XANES and In Situ XRD Measurements. *J. Electrochem. Soc.* **2003**, *150*, A1560. [CrossRef]

Zhou, Y.-N.; Ding, J.-J.; Nam, K.-W.; Yu, X.; Bak, S.-M.; Hu, E.; Liu, J.; Bai, J.; Li, H.; Fu, Z.-W.; et al. Phase transition behavior of $NaCrO_2$ during sodium extraction studied by synchrotron-based X-ray diffraction and absorption spectroscopy. *J. Mater. Chem. A.* **2013**, *1*, 11130–11134. [CrossRef]

Xia, X.; Dahn, J.R. $NaCrO_2$ is a fundamentally safe positive electrode material for sodium-ion batteries with liquid electrolytes. *Electrochem. Solid-State Lett.* **2011**, *15*, A1. [CrossRef]

Bo, S.-H.; Li, X.; Toumar, A.J.; Ceder, G. Layered-to-rock-salt transformation in desodiated Na_xCrO_2 (x 0.4). *Chem. Mater.* **2016**, *28*, 1419–1429. [CrossRef]

Jakobsen, C.L.; Brighi, M.; Andersen, B.P.; Ducrest, G.; Černý, R.; Ravnsbæk, D.B. Expanded solid-solution behavior and charge-discharge asymmetry in Na_xCrO_2 Na-ion battery electrodes. *J. Power Sources.* **2022**, *535*, 231317. [CrossRef]

55. Tsuchiya, Y.; Takanashi, K.; Nishinobo, T.; Hokura, A.; Yonemura, M.; Matsukawa, T.; Ishigaki, T.; Yamanaka, K.; Ohta Yabuuchi, N. Layered $Na_xCr_xTi_{1-x}O_2$ as Bifunctional Electrode Materials for Rechargeable Sodium Batteries. *Chem. Mater.* 2 *28*, 7006–7016. [CrossRef]
56. Park, W.B.; Nathan, M.G.T.; Han, S.C.; Lee, J.-W.; Sohn, K.-S.; Pyo, M. Aliovalent-doped sodium chromium oxide $_{0.9}Cr_{0.9}Sn_{0.1}O_2$ and $Na_{0.8}Cr_{0.9}Sb_{0.1}O_2$) for sodium-ion battery cathodes with high-voltage characteristics. *RSC Adv.* 2 *10*, 43273–43281. [CrossRef]
57. Xi, K.; Chu, S.; Zhang, X.; Zhang, X.; Zhang, H.; Xu, H.; Bian, J.; Fang, T.; Guo, S.; Liu, P.; et al. A high-performance laye Cr-based cathode for sodium-ion batteries. *Nano Energy* **2020**, *67*, 104215. [CrossRef]
58. Wang, S.; Chen, F.; He, H.-Y.; Zhu, Y.-R.; Liu, H.-B.; Chen, C.-H. Enhancing structural stability of $NaCrO_2$ by Nb-substituting sodium-ion battery. *J. Alloys Compd.* **2022**, *925*, 166690. [CrossRef]
59. Wang, Y.; Cui, P.; Zhu, W.; Feng, Z.; Vigeant, M.-J.; Demers, H.; Guerfi, A.; Zaghib, K. Enhancing the electrochemical performa of an O3-$NaCrO_2$ cathode in sodium-ion batteries by cation substitution. *J. Power Sources* **2019**, *435*, 226760. [CrossRef]
60. Ikhe, A.B.; Park, W.B.; Manasi, M.; Ahn, D.; Sohn, K.-S.; Pyo, M. Unprecedented cyclability and moisture durability of NaC sodium-ion battery cathode via simultaneous Al doping and Cr_2O_3 coating. *ACS Appl. Mater. Interfaces* **2023**, *15*, 14958–14 [CrossRef]
61. Lee, I.; Oh, G.; Lee, S.; Yu, T.-Y.; Alfaruqi, M.H.; Mathew, V.; Sambandam, B.; Sun, Y.-K.; Hwang, J.-Y.; Kim, J. Cationic transition metal co-substitution strategy of O3-type $NaCrO_2$ cathode for high-energy sodium-ion batteries. *Energy Storage M* **2021**, *41*, 183–195. [CrossRef]
62. Ma, C.; Li, X.-L.; Yue, X.-Y.; Bao, J.; Luo, R.-J.; Zhou, Y.-N. Suppressing O3- O'3 phase transition in $NaCrO_2$ cathode enabling h rate capability for sodium-ion batteries by Sb substitution. *Chem. Eng. J.* **2022**, *432*, 134305. [CrossRef]
63. Wang, Y.; Xiao, R.; Hu, Y.-S.; Avdeev, M.; Chen, L. P2-$Na_{0.6}[Cr_{0.6}Ti_{0.4}]O_2$ cation-disordered electrode for high-rate symme rechargeable sodium-ion batteries. *Nat. Commun.* **2015**, *6*, 6954. [CrossRef]
64. Li, W.; Wang, Y.; Hu, G.; Peng, Z.; Cao, Y.; Zeng, Y.; Du, K. Ti-doped $NaCrO_2$ as cathode materials for sodium-ion batteries w excellent long cycle life. *J. Alloys Compd.* **2019**, *779*, 147–155. [CrossRef]
65. Wang, S.; Chen, F.; Zhu, T.-Y.; He, X.-D.; Liao, J.-Y.; Zhang, L.-M.; Ding, X.; Hu, Q.; Chen, C.-H. In situ-formed Cr_2O_3 coating $NaCrO_2$ with improved sodium storage performance. *ACS Appl. Mater. Interfaces.* **2020**, *12*, 44671–44678. [CrossRef]
66. Cai, J.; Wei, Y.; Zhu, Y.; Tian, L.; Zhang, J.; Shen, Y.; Wang, M. Nitrogen-doped carbon-coating enables high-rate capability long-cycle stability of $NaCrO_2$ cathode for sodium-ion battery. *Appl. Surf. Sci.* **2025**, *683*, 161804. [CrossRef]
67. Bhardwaj, A.; Panwar, A.K. Effect of carbon shell over $NaCrO_2$ core by C_2H_2 decomposition to enhance electrochemical proper for rechargeable Sodium-ion batteries. *Appl. Surf. Sci.* **2022**, *573*, 151449. [CrossRef]
68. Shi, Z.; Wang, Z.; Shaw, L.L.; Ashuri, M. On the Electrochemical Properties of Carbon-Coated $NaCrO_2$ for Na-Ion Batte *Batteries* **2023**, *9*, 433. [CrossRef]
69. Ding, J.-J.; Zhou, Y.-N.; Sun, Q.; Fu, Z.-W. Cycle performance improvement of $NaCrO_2$ cathode by carbon coating for sodium batteries. *Electrochem. Commun.* **2012**, *22*, 85–88. [CrossRef]
70. Wu, J.; Hu, G.; Du, K.; Peng, Z.; Huang, M.; Fan, J.; Gong, Y.; Guan, D.; Shi, Y.; Liu, R.; et al. Inhibiting electrochemical ph transition of $NaCrO_2$ with long-cycle stability by surface fluorination treatment. *Electrochim. Acta* **2022**, *403*, 139641. [CrossR
71. Yabuuchi, N.; Yano, M.; Yoshida, H.; Kuze, S.; Komaba, S. Synthesis and electrode performance of O3-type NaFe $NaNi_{1/2}Mn_{1/2}O_2$ solid solution for rechargeable sodium batteries. *J. Electrochem. Soc.* **2013**, *160*, A3131. [CrossRef]
72. Okada, S.; Yamaki, J. Iron-Based Rare-Metal-Free Cathodes. In *Lithium Ion Rechargeable Batteries*; John Wiley & Sons: Hoboken, USA, 2010.
73. Zhao, J.; Zhao, L.; Dimov, N.; Okada, S.; Nishida, T. Electrochemical and thermal properties of α-$NaFeO_2$ cathode for Na- batteries. *J. Electrochem. Soc.* **2013**, *160*, A3077. [CrossRef]
74. Watanabe, H.; Usui, H.; Domi, Y.; Nishida, T.; Uetake, K.; Tanaka, T.; Kurokawa, H.; Sakaguchi, H. $NaFeO_2$: Possible Materials Anode and Cathode of Na-Ion Batteries. *ACS Electrochem.* **2025**, *1*, 73–81. [CrossRef]
75. McAuliffe, R.D.; Kamm, G.E.; McDermott, M.J.; Hermann, R.P.; Vasquez-Garcia, N.; Sacci, R.L.; Persson, K.A.; Chapman, K Veith, G.M. Direct Mechanochemical Synthesis, Phase Stability, and Electrochemical Performance of α-$NaFeO_2$. *Inorg. Chem.* 2 *62*, 3358–3367. [CrossRef]
76. Blesa, M.C.; Moran, E.; León, C.; Santamaria, J.; Tornero, J.D.; Menéndez, N. α-$NaFeO_2$: Ionic conductivity and sodium extract *Solid State Ion.* **1999**, *126*, 81–87. [CrossRef]
77. Okada, S.; Takahashi, Y.; Kiyabu, T.; Doi, T.; Yamaki, J.-I.; Nishida, T. Layered transition metal oxides as cathodes for sodi secondary battery. *ECS Meet. Abstr.* **2006**, *MA2006-02*, 201. [CrossRef]
78. Tabuchi, M.; Kataoka, R. Structure and electrochemical properties of α-$NaFeO_2$ obtained under various hydrothermal conditi *J. Electrochem. Soc.* **2019**, *166*, A2209. [CrossRef]

Kuganathan, N.; Kelaidis, N.; Chroneos, A. Defect chemistry, sodium diffusion and doping behavior in $NaFeO_2$ polymorphs as cathode materials for Na-ion batteries: A computational study. *Materials* **2019**, *12*, 3243. [CrossRef]
Kikkawa, S.; Miyazaki, S.; Koizumi, M. Sodium deintercalation from α-$NaFeO_2$. *Mater. Res. Bull.* **1985**, *20*, 373–377. [CrossRef]
Takeda, Y.; Nakahara, K.; Nishijima, M.; Imanishi, N.; Yamamoto, O.; Takano, M.; Kanno, R. Sodium deintercalation from sodium iron oxide. *Mater. Res. Bull.* **1994**, *29*, 659–666. [CrossRef]
Yabuuchi, N.; Yoshida, H.; Komaba, S. Crystal structures and electrode performance of alpha-$NaFeO_2$ for rechargeable sodium batteries. *Electrochem* **2012**, *80*, 716–719. [CrossRef]
Lee, E.; Brown, D.E.; Alp, E.E.; Ren, Y.; Lu, J.; Woo, J.-J.; Johnson, C.S. New insights into the performance degradation of Fe-based layered oxides in sodium-ion batteries: Instability of Fe^{3+}/Fe^{4+} redox in α-$NaFeO_2$. *Chem. Mater.* **2015**, *27*, 6755–6764. [CrossRef]
Thorne, J.S.; Chowdhury, S.; Dunlap, R.A.; Obrovac, M.N. Structure and electrochemistry of $Na_xFe_xTi_{1-x}O_2$ ($1.0 \geq x \geq 0.75$) for Na-Ion battery positive electrodes. *J. Electrochem. Soc.* **2014**, *161*, A1801. [CrossRef]
Xu, J.; Han, Z.; Jiang, K.; Bai, P.; Liang, Y.; Zhang, X.; Wang, P.; Guo, S.; Zhou, H. suppressing cation migration and reducing particle cracks in a layered Fe-based cathode for advanced sodium-ion batteries. *Small* **2020**, *16*, 1904388. [CrossRef] [PubMed]
Yoshida, H.; Yabuuchi, N.; Komaba, S. $NaFe_{0.5}Co_{0.5}O_2$ as high energy and power positive electrode for Na-ion batteries. *Electrochem. Commun.* **2013**, *34*, 60–63. [CrossRef]
Shevchenko, V.A.; Glazkova, I.S.; Novichkov, D.A.; Skvortsova, I.; Sobolev, A.V.; Abakumov, A.M.; Presniakov, I.A.; Drozhzhin, O.A.; Antipov, E.V. Competition between the Ni and Fe redox in the O3-$NaNi_{1/3}Fe_{1/3}Mn_{1/3}O_2$ cathode material for na-ion batteries. *Chem. Mater.* **2023**, *35*, 4015–4025. [CrossRef]
Ono, Y.; Yui, Y.; Hayashi, M.; Asakura, K.; Kitabayashi, H.; Takahashi, K.I. Electrochemical properties of $NaCuO_2$ for sodium-ion secondary batteries. *ECS Trans.* **2014**, *58*, 33. [CrossRef]
Ono, Y.; Yui, Y.; Asakura, K.; Nakamura, J.; Hayashi, M.; Takahashi, K.I. Characterization of electrochemical cycling-induced new products of $NaCuO_2$ cathode material for sodium secondary batteries. *Am. J. Phys. Chem.* **2014**, *3*, 61–66. [CrossRef]
Dyer, L.D.; Borie, B.S., Jr.; Smith, G.P. Alkali metal-nickel oxides of the type $MNiO_2$. *J. Am. Chem. Soc.* **1954**, *76*, 1499–1503. [CrossRef]
Dick, S.; Müller, M.; Preissinger, F.; Zeiske, T. The structure of monoclinic $NaNiO_2$ as determined by powder X-ray and neutron scattering. *Powder Diffr.* **1997**, *12*, 239–241. [CrossRef]
Chappel, E.; Núñez-Regueiro, M.D.; Chouteau, G.; Isnard, O.; Darie, C. Study of the ferrodistorsive orbital ordering in $NaNiO_2$ by neutron diffraction and submillimeter wave ESR. *Eur. Phys. J. B* **2000**, *17*, 615–622. [CrossRef]
Sofin, M.; Jansen, M. New route of preparation and properties of $NaNiO_2$. *Z. Naturforsch. B* **2005**, *60*, 701–704. [CrossRef]
Vassilaras, P.; Ma, X.; Li, X.; Ceder, G. Electrochemical properties of monoclinic $NaNiO_2$. *J. Electrochem. Soc.* **2012**, *160*, A207. [CrossRef]
Miyazaki, S.; Kikkawa, S.; Koizumi, M. Chemical and electrochemical deintercalations of the layered compounds $LiMO_2$ (M = Cr, Co) and $NaM'O_2$ (M' = Cr, Fe, Co, Ni). *Synth. Met.* **1983**, *6*, 211–217. [CrossRef]
Han, M.H.; Gonzalo, E.; Casas-Cabanas, M.; Rojo, T. Structural evolution and electrochemistry of monoclinic $NaNiO_2$ upon the first cycling process. *J. Power Sources* **2014**, *258*, 266–271. [CrossRef]
Wang, L.; Wang, J.; Zhang, X.; Ren, Y.; Zuo, P.; Yin, G.; Wang, J. Unravelling the origin of irreversible capacity loss in $NaNiO_2$ for high voltage sodium ion batteries. *Nano Energy* **2017**, *34*, 215–223. [CrossRef]
Kim, D.H.; Kim, J.-Y.; Park, J.-H.; Kim, S.-O.; Kim, H.-S.; Kim, K.-B.; Chung, K.Y. RT-XAMF and TR-XRD studies of solid-state synthesis and thermal stability of $NaNiO_2$ as cathode material for sodium-ion batteries. *Ceram. Int.* **2022**, *48*, 19675–19680. [CrossRef]
Singh, G.; Aguesse, F.; Otaegui, L.; Goikolea, E.; Gonzalo, E.; Segalini, J.; Rojo, T. Electrochemical performance of $NaFe_x(Ni_{0.5}Ti_{0.5})_{1-x}O_2$ (x = 0.2 and x = 0.4) cathode for sodium-ion battery. *J. Power Sources* **2015**, *273*, 333–339. [CrossRef]
Chang, Y.; Zhou, Y.; Wang, Z.; Li, X.; Wang, D.; Duan, J.; Wang, J.; Yan, G. Synthesis of $NaNi_{0.5}Mn_{0.5}O_2$ cathode materials for sodium-ion batteries via spray pyrolysis method. *J. Alloys Compd.* **2022**, *922*, 166283. [CrossRef]
Wang, H.; Gu, M.; Jiang, J.; Lai, C.; Ai, X. An O3-type $NaNi_{0.5}Mn_{0.3}Ti_{0.2}O_2$ compound as new cathode material for room-temperature sodium-ion batteries. *J. Power Sources* **2016**, *327*, 653–657. [CrossRef]
Zheng, L.; Fielden, R.; Bennett, J.C.; Obrovac, M.N. Hexagonal and monoclinic $NaNi_{0.8}Co_{0.15}Al_{0.05}O_2$ (Na-NCA) for sodium ion batteries. *J. Power Sources* **2019**, *433*, 226698. [CrossRef]
Meghnani, D.; Singh, R.K. Influence of synthesis route on the structure and electrochemical performance of biphasic (O'3/O3) $NaNi_{0.815}Co_{0.15}Al_{0.035}O_2$ cathode for sodium-ion batteries. *Electrochim. Acta* **2022**, *419*, 140403. [CrossRef]
Deng, C.; Gabriel, E.; Skinner, P.; Lee, S.; Barnes, P.; Ma, C.; Gim, J.; Lau, M.L.; Lee, E.; Xiong, H. Origins of Irreversibility in Layered $NaNi_xFe_yMn_zO_2$ Cathode Materials for Sodium Ion Batteries. *ACS Appl. Mater. Interfaces* **2020**, *12*, 51397–51408. [CrossRef]

105. Fielden, R.; Obrovac, M.N. Investigation of the $NaNi_xMn_{1-x}O_2$ ($0 \leq x \leq 1$) System for Na-ion Battery Cathode Materials (Prepr *J. Electrochem. Soc.* **2015**, *162*, A453. [CrossRef]
106. Moeez, I.; Susanto, D.; Ali, G.; Jung, H.-G.; Lim, H.-D.; Chung, K.Y. Effect of the interfacial protective layer on the $NaFe_{0.5}Ni_{0.5}$ cathode for rechargeable sodium-ion batteries. *J. Mater. Chem. A* **2020**, *8*, 13964–13970. [CrossRef]
107. Nanba, Y.; Iwao, T.; Mortemard de Boisse, B.; Zhao, W.; Hosono, E.; Asakura, D.; Niwa, H.; Kiuchi, H.; Miyawaki, J.; Harada et al. Redox potential paradox in Na_xMO_2 for sodium-ion battery cathodes. *Chem. Mater.* **2016**, *28*, 1058–1065. [CrossRef]
108. Guo, H.; Wang, Y.; Han, W.; Yu, Z.; Qi, X.; Sun, K.; Hu, Y.-S.; Liu, Y.; Chen, D.; Chen, L. Na-deficient O3-type cathode mate $Na_{0.8}[Ni_{0.3}Co_{0.2}Ti_{0.5}]O_2$ for room-temperature sodium-ion batteries. *Electrochim. Acta* **2015**, *158*, 258–263. [CrossRef]
109. Maletti, S.; Sarapulova, A.; Schökel, A.; Mikhailova, D. Operando studies on the $NaNi_{0.5}Ti_{0.5}O_2$ cathode for Na-ion batter Elucidating titanium as a structure stabilizer. *ACS Appl. Mater. Interfaces* **2019**, *11*, 33923–33930. [CrossRef]
110. Ma, J.; Bo, S.-H.; Wu, L.; Zhu, Y.; Grey, C.P.; Khalifah, P.G. Ordered and disordered polymorphs of $Na(Ni_{2/3}Sb_{1/3})O_2$: Honeyco ordered cathodes for Na-ion batteries. *Chem. Mater.* **2015**, *27*, 2387–2399. [CrossRef]
111. An, S.; Karger, L.; Dreyer, S.L.; Hu, Y.; Barbosa, E.; Zhang, R.; Lin, J.; Fichtner, M.; Kondrakov, A.; Janek, J. Improving cycli performance of the $NaNiO_2$ cathode in sodium-ion batteries by titanium substitution. *Mater. Futures* **2024**, *3*, 035103. [CrossR
112. Saadoune, I.; Maazaz, A.; Ménétrier, M.; Delmas, C. On the $Na_xNi_{0.6}Co_{0.4}O_2$ system: Physical and electrochemical studies *Solid State Chem.* **1996**, *122*, 111–117. [CrossRef]
113. Darga, J.; Manthiram, A. Facile synthesis of O3-type $NaNi_{0.5}Mn_{0.5}O_2$ single crystals with improved performance in sodium-i batteries. *ACS Appl. Mater. Interfaces* **2022**, *14*, 52729–52737. [CrossRef]
114. Liang, X.; Yu, T.-Y.; Ryu, H.-H.; Sun, Y.-K. Hierarchical O3/P2 heterostructured cathode materials for advanced sodium- batteries. *Energy Storage Mater.* **2022**, *47*, 515–525. [CrossRef]
115. Zhang, S.; Li, X.; Su, Y.; Yang, Y.; Yu, H.; Wang, H.; Zhang, Q.; Lu, Y.; Song, D.; Wang, S.; et al. Four-in-one strategy to boost performance of $Na_x[Ni, Mn]O_2$. *Adv. Funct. Mater.* **2023**, *33*, 2301568. [CrossRef]
116. Wang, Q.; Yu, G.; Luo, B.; Ji, W.; Liu, Z.; Li, M.; Nong, Y.; Tian, Y.; Wang, X.; Zhang, J.; et al. Suppression of Adverse Ph Transition of Layered Oxide Cathode via Local Electronic Structure Regulation for High-Capacity Sodium-Ion Batteries. *A Nano* **2024**, *18*, 18622–18634. [CrossRef]
117. Yuan, T.; Li, P.; Sun, Y.; Che, H.; Zheng, Q.; Zhang, Y.; Huang, S.; Qiu, J.; Pang, Y.; Yang, J.; et al. Refining O3-Type Ni/Mn-Bas Sodium-Ion Battery Cathodes via "Atomic Knife" Achieving High Capacity and Stability. *Adv. Funct. Mater.* **2024**, *34*, 2414 [CrossRef]
118. Wang, P.F.; Yao, H.R.; Liu, X.Y.; Zhang, J.N.; Gu, L.; Yu, X.Q.; Yin, Y.X.; Guo, Y.G. Ti-substituted $NaNi_{0.5}Mn_{0.5-x}Ti_xO_2$ catho with reversible O3−P3 phase transition for high-performance sodium-ion batteries. *Adv. Mater.* **2017**, *29*, 1700210. [CrossR [PubMed]
119. Meghnani, D.; Srivastava, N.; Tiwari, R.K.; Mishra, R.; Patel, A.; Tiwari, A.; Samriddhi; Singh, S.P.; Yadav, V.; Singh, R Surface-Coated $NaNi_{0.815}Co_{0.15}Al_{0.035}O_2$ Cathode-Based Sodium-Ion Batteries with Enhanced Performance. *Energy Fuels* **20** *38*, 694–706. [CrossRef]
120. Li, M.; Qiu, X.; Yin, Y.; Wei, T.; Dai, Z. O3-Type Ni-Rich $NaNi_{2/3}Mn_{1/6}Fe_{1/6}O_2$: A high-performance cathode material sodium-ion batteries. *J. Alloys Compd.* **2023**, *969*, 172406. [CrossRef]
121. Bhide, A.; Hariharan, K. Physicochemical properties of Na_xCoO_2 as a cathode for solid state sodium battery. *Solid State Ion.* **20** *192*, 360–363. [CrossRef]
122. Shibata, A.; Kobayashi, W.; Moritomo, Y. Sodium ion diffusion in layered Na_xMnO_2 ($0.49 \leq x \leq 0.75$): Comparison with Na_xCo *Appl. Phys. Express* **2014**, *7*, 067101. [CrossRef]
123. Zandbergen, H.W.; Foo, M.; Xu, Q.; Kumar, V.; Cava, R.J. Sodium ion ordering in Na_xCoO_2: Electron diffraction study. *Phys. R B* **2004**, *70*, 024101. [CrossRef]
124. Nwanya, A.C.; Ndipingwi, M.M.; Ezema, F.I.; Iwuoha, E.I.; Maaza, M. Bio-synthesized P2-$Na_{0.57}CoO_2$ nanoparticles as catho for aqueous sodium ion battery. *J. Electroanal. Chem.* **2020**, *878*, 114600. [CrossRef]
125. Su, J.-C.; Zhou, G.; Pei, Y.; Yang, Z.-H.; Wang, X.-Y. First-principles investigation on structural and electrochemical properties $NaCoO_2$ for rechargeable Na-ion batteries. *J. Cent. South Univ.* **2015**, *22*, 2036–2042. [CrossRef]
126. Takahashi, Y.; Gotoh, Y.; Akimoto, J. Single-crystal growth, crystal and electronic structure of $NaCoO_2$. *J. Solid State Chem.* **20** *172*, 22–26. [CrossRef]
127. Berthelot, R.; Carlier, D.; Delmas, C. Electrochemical investigation of the P2-Na_xCoO_2 phase diagram. *Nat. Mater.* **2011**, *10*, 74– [CrossRef] [PubMed]
128. Gao, L.; Chen, S.; Zhang, L.; Yang, X. High Areal Capacity $Na_{0.67}CoO_2$ Bundle Array Cathode Tailored for High-Performar Sodium-Ion Batteries. *ChemElectroChem* **2019**, *6*, 947–952. [CrossRef]
129. Shiprath, K.; Manjunatha, H.; Nadh, R.V.; Babu, T.A.; Ramesh, S.; Naidu, K.C.B.; Ramesha, M. Synthesis and Electrochemi Characterization of $NaCoO_2$ as Cathode Material in 2 M NaOH Aqueous Electrolyte. *ChemistrySelect* **2021**, *6*, 1874–1881. [CrossR

Yabuuchi, N.; Kubota, K.; Dahbi, M.; Komaba, S. Research development on sodium-ion batteries. *Chem. Rev.* **2014**, *114*, 11636–11682. [CrossRef] [PubMed]

Clément, R.J.; Bruce, P.G.; Grey, C.P. manganese-based P2-type transition metal oxides as sodium-ion battery cathode materials. *J. Electrochem. Soc.* **2015**, *162*, A2589. [CrossRef]

Carlier, D.; Cheng, J.H.; Berthelot, R.; Guignard, M.; Yoncheva, M.; Stoyanova, R.; Hwang, B.J.; Delmas, C. The P2-$Na_{2/3}Co_{2/3}Mn_{1/3}O_2$ phase: Structure, physical properties and electrochemical behavior as positive electrode in sodium battery. *Dalton Trans.* **2011**, *40*, 9306–9312. [CrossRef]

Beck, F.R.; Cheng, Y.Q.; Bi, Z.; Feygenson, M.; Bridges, C.A.; Moorhead-Rosenberg, Z.; Manthiram, A.; Goodenough, J.B.; Paranthaman, M.P.; Manivannan, A. Neutron diffraction and electrochemical studies of $Na_{0.79}CoO_2$ and $Na_{0.79}Co_{0.7}Mn_{0.3}O_2$ cathodes for sodium-ion batteries. *J. Electrochem. Soc.* **2014**, *161*, A961. [CrossRef]

Kang, S.M.; Park, J.-H.; Jin, A.; Jung, Y.H.; Mun, J.; Sung, Y.-E. Na^+/Vacancy Disordered P2-$Na_{0.67}Co_{1-x}Ti_xO_2$: High-Energy and High-Power Cathode Materials for Sodium Ion Batteries. *ACS Appl. Mater. Interfaces* **2018**, *10*, 3562–3570. [CrossRef] [PubMed]

Matsui, M.; Mizukoshi, F.; Imanishi, N. Improved cycling performance of P2-type layered sodium cobalt oxide by calcium substitution. *J. Power Sources* **2015**, *280*, 205–209. [CrossRef]

Han, S.C.; Lim, H.; Jeong, J.; Ahn, D.; Park, W.B.; Sohn, K.-S.; Pyo, M. Ca-doped Na_xCoO_2 for improved cyclability in sodium ion batteries. *J. Power Sources* **2015**, *277*, 9–16. [CrossRef]

Ding, J.J.; Zhou, Y.N.; Sun, Q.; Yu, X.Q.; Yang, X.Q.; Fu, Z.W. Electrochemical properties of P2-phase $Na_{0.74}CoO_2$ compounds as cathode material for rechargeable sodium-ion batteries. *Electrochim. Acta* **2013**, *87*, 388–393. [CrossRef]

Rai, A.K.; Anh, L.T.; Gim, J.; Mathew, V.; Kim, J. Electrochemical properties of Na_xCoO_2 (x~ 0.71) cathode for rechargeable sodium-ion batteries. *Ceram. Int.* **2014**, *40*, 2411–2417. [CrossRef]

Hwang, S.; Lee, Y.; Jo, E.; Chung, K.Y.; Choi, W.; Kim, S.M.; Chang, W. Investigation of Thermal Stability of P2-Na_xCoO_2 Cathode Materials for Sodium Ion Batteries Using Real-Time Electron Microscopy. *ACS Appl. Mater. Interfaces* **2017**, *9*, 18883–18888. [CrossRef]

Fang, Y.; Yu, X.-Y.; Lou, X.W. (David). A practical high-energy cathode for sodium-ion batteries based on uniform P2-$Na_{0.7}CoO_2$ microspheres. *Angew. Chem. Int. Ed.* **2017**, *56*, 5801–5805. [CrossRef]

Onoda, M. Geometrically frustrated triangular lattice system Na_xVO_2: Super paramagnetism in x = 1 and trimerization in $x \approx 0.7$. *J. Phys. Condens. Matter.* **2008**, *20*, 145205. [CrossRef]

Fielden, R.; Cole, L.; Obrovac, M.N. Successive orbital ordering transitions in $NaVO_2$. *Phys. Rev. Lett.* **2008**, *101*, 166402. [CrossRef]

Fielden, R.; Cole, L.; Obrovac, M.N. Low voltage sodium intercalation in $Na_xV_xTi_{1-x}O_2$ ($2/3 \leq x \leq 1$). *J. Electrochem. Soc.* **2017**, *164*, A490. [CrossRef]

Guignard, M.; Didier, C.; Darriet, J.; Bordet, P.; Elkaïm, E.; Delmas, C. P2-Na_xVO_2 system as electrodes for batteries and electron-correlated materials. *Nat. Mater.* **2013**, *12*, 74–80. [CrossRef]

Hamani, D.; Ati, M.; Tarascon, J.-M.; Rozier, P. Na_xVO_2 as possible electrode for Na-ion batteries. *Electrochem. Commun.* **2011**, *13*, 938–941. [CrossRef]

Didier, C.; Guignard, M.; Denage, C.; Szajwaj, O.; Ito, S.; Saadoune, I.; Darriet, J.; Delmas, C. Electrochemical Na-Deintercalation from $NaVO_2$. *Electrochem. Solid-State Lett.* **2011**, *14*, A75. [CrossRef]

Didier, C.; Guignard, M.; Darriet, J.; Delmas, C. O′3-Na_xVO_2 System: A Superstructure for $Na_{1/2}VO_2$. *Inorg. Chem.* **2012**, *51*, 11007–11016. [CrossRef] [PubMed]

Shishkin, M.; Kumakura, S.; Sato, S.; Kubota, K.; Komaba, S.; Sato, H. Unraveling the role of doping in selective stabilization of $NaMnO_2$ polymorphs: Combined theoretical and experimental study. *Chem. Mater.* **2018**, *30*, 1257–1264. [CrossRef]

Ma, X.; Chen, H.; Ceder, G. Electrochemical properties of monoclinic $NaMnO_2$. *J. Electrochem. Soc.* **2011**, *158*, A1307. [CrossRef]

Kubota, K.; Miyazaki, M.; Kim, E.J.; Yoshida, H.; Barpanda, P.; Komaba, S. Structural change induced by electrochemical sodium extraction from layered O′3-$NaMnO_2$. *J. Mater. Chem. A* **2021**, *9*, 26810–26819. [CrossRef]

Billaud, J.; Clément, R.J.; Armstrong, A.R.; Canales-Vázquez, J.; Rozier, P.; Grey, C.P.; Bruce, P.G. β-$NaMnO_2$: A high-performance cathode for sodium-ion batteries. *J. Am. Chem. Soc.* **2014**, *136*, 17243–17248. [CrossRef] [PubMed]

Abakumov, A.M.; Tsirlin, A.A.; Bakaimi, I.; Van Tendeloo, G.; Lappas, A. Multiple twinning as a structure directing mechanism in layered rock-salt-type oxides: $NaMnO_2$ polymorphism, redox potentials, and magnetism. *Chem. Mater.* **2014**, *26*, 3306–3315. [CrossRef]

Yu, H.; Walsh, M.; Liang, X. Improving the comprehensive performance of $Na_{0.7}MnO_2$ for sodium ion batteries by ZrO_2 atomic layer deposition. *ACS Appl. Mater. Interfaces* **2021**, *13*, 54884–54893. [CrossRef] [PubMed]

Wen, Y.; Wang, B.; Zeng, G.; Nogita, K.; Ye, D.; Wang, L. Electrochemical and structural study of layered P2-type $Na_{2/3}Ni_{1/3}Mn_{2/3}O_2$ as cathode material for sodium-ion battery. *Chem. Asian J.* **2015**, *10*, 661–666. [CrossRef] [PubMed]

155. Thorne, J.S.; Dunlap, R.A.; Obrovac, M.N. Structure and electrochemistry of $Na_xFe_xMn_{1-x}O_2$ ($1.0 \leq x \leq 0.5$) for Na-ion bat positive electrodes. *J. Electrochem. Soc.* **2012**, *160*, A361. [CrossRef]
156. Li, Z.-Y.; Zhang, J.; Gao, R.; Zhang, H.; Hu, Z.; Liu, X. Unveiling the Role of Co in Improving the High-Rate Capability Cycling Performance of Layered $Na_{0.7}Mn_{0.7}Ni_{0.3-x}Co_xO_2$ Cathode Materials for Sodium-Ion Batteries. *ACS Appl. Mater. Interf* **2016**, *8*, 15439–15448. [CrossRef] [PubMed]
157. Han, M.H.; Gonzalo, E.; Sharma, N.; López del Amo, J.M.; Armand, M.; Avdeev, M.; Saiz Garitaonandia, J.J.; Rojo, T. H performance P2-phase $Na_{2/3}Mn_{0.8}Fe_{0.1}Ti_{0.1}O_2$ cathode material for ambient-temperature sodium-ion batteries. *Chem. M* **2016**, *28*, 106–116. [CrossRef]
158. Gao, X.; Chen, J.; Liu, H.; Yin, S.; Tian, Y.; Cao, X.; Zou, G.; Hou, H.; Wei, W.; Chen, L.; et al. Copper-substituted Na_xMO_2 (M= Mn) cathodes for sodium ion batteries: Enhanced cycling stability through suppression of Mn (III) formation. *Chem. Eng. J.* **2** *406*, 126830. [CrossRef]
159. Bucher, N.; Hartung, S.; Franklin, J.B.; Wise, A.M.; Lim, L.Y.; Chen, H.-Y.; Nelson Weker, J.; Toney, M.F.; Srinivasan, M. $Na_xCo_yMn_{1-y}O_2$ (y = 0, 0.1) as Cathode Materials in Sodium-Ion Batteries-Effects of Doping and Morphology To Enhance Cyc Stability. *Chem. Mater.* **2016**, *28*, 2041–2051. [CrossRef]
160. Xu, S.; Chen, H.; Li, C.; Nie, R.; Yang, Y.; Zhou, M.; Zhang, X.; Zhou, H. Effect of Ti doping on the structural and electrochem performance of O3-type Na $(Ni_{0.3}Fe_{0.2}Mn_{0.5})_{1-x}Ti_xO_2$ cathode materials for sodium-ion batteries. *J. Alloys Compd.* **2023**, 171199. [CrossRef]
161. Li, X.; Wu, D.; Zhou, Y.-N.; Liu, L.; Yang, X.-Q.; Ceder, G. O3-type Na $(Mn_{0.25}Fe_{0.25}Co_{0.25}Ni_{0.25})O_2$: A quaternary layered cath compound for rechargeable Na ion batteries. *Electrochem. Commun.* **2014**, *49*, 51–54. [CrossRef]
162. Yuan, D.D.; Wang, Y.X.; Cao, Y.L.; Ai, X.P.; Yang, H.X. Improved electrochemical performance of Fe-substituted $NaNi_{0.5}Mn_{0.}$ cathode materials for sodium-ion batteries. *ACS Appl. Mater. Interfaces* **2015**, *7*, 8585–8591. [CrossRef] [PubMed]
163. Li, X.; Ge, M.; Zhang, M.; Tang, X.; Liu, X.; Cui, Y.; Zhang, H.; Yang, Y.; Yin, Y.; Yang, S.-T. Prilling and Coating Strateg Synthesize High-Performance Spherical $NaNi_{0.4}Fe_{0.2}Mn_{0.4}O_2$ Cathode Materials for Sodium Ion Batteries. *Langmuir* **2024**, 18610–18618. [CrossRef]
164. Jiang, L.-W.; Lu, Y.-X.; Wang, Y.-S.; Liu, L.-L.; Qi, X.-G.; Zhao, C.-L.; Chen, L.-Q.; Hu, Y.-S. A high-temperature β-phase NaMr stabilized by Cu doping and its Na storage properties. *Chin. Phys. Lett.* **2018**, *35*, 048801. [CrossRef]
165. Wang, X.-Z.; Zuo, Y.; Qin, Y.; Zhu, X.; Xu, S.-W.; Guo, Y.-J.; Yan, T.; Zhang, L.; Gao, Z.; Yu, L.; et al. Fast Na^+ Kinetics Suppressed Voltage Hysteresis Enabled by a High-Entropy Strategy for Sodium Oxide Cathodes. *Adv. Mater.* **2024**, *36*, 2312 [CrossRef] [PubMed]
166. Ding, F.; Zhao, C.; Xiao, D.; Rong, X.; Wang, H.; Li, Y.; Yang, Y.; Lu, Y.-X.; Hu, Y.-S. Using high-entropy configuration strateg design Na-ion layered oxide cathodes with superior electrochemical performance and thermal stability. *J. Am. Chem. Soc.* **2** *18*, 8286–8295. [CrossRef]
167. Lamb, J.; Jarvis, K.; Manthiram, A. Molten-salt synthesis of O3-type layered oxide single crystal cathodes with contro morphology towards long-life sodium-ion batteries. *Small* **2022**, *43*, 2106927. [CrossRef]
168. Zhang, L.; Huang, J.; Song, M.; Lu, C.; Wu, W.; Wu, X. Single-crystal growth of P2-type layered oxides with increased exposur {010} planes for high-performance sodium-ion batteries. *ACS Appl. Mater. Interfaces* **2023**, *40*, 47037–47048. [CrossRef]
169. Yang, D.; Long, Y.; Gao, X.-W.; Zhao, Z.; Chen, H.; Lai, Q.; Li, C.; Niu, R.; Liu, Z.; Gu, Q.; et al. Single Crystal Sodium Laye Oxide Achieves Superior Cyclability at High Voltage. *Adv. Energy Mater.* **2024**, *14*, 2404999. [CrossRef]
170. Xu, L.; Song, M.; Xie, J.; Chen, M.; Wu, W.; Tan, Z.; Qiu, S.; Wu, X. Suppressing the P2-OP4 phase transition of single-cry P2-type Ni/Zn/Mn-based layered oxide for advanced sodium-ion batteries. *Powder Technol.* **2024**, *448*, 120314. [CrossRef]
171. Zhang, F.; Lu, Y.; Guo, Y.; Li, C.; Liu, Y.; Yang, M.; Zhao, B.; Wu, W.; Wu, X. Highly stabilized single-crystal P2-type layered oxi obtained via rational crystal orientation modulation for sodium-ion batteries. *Chem. Eng. J.* **2023**, *458*, 141515. [CrossRef]
172. Xiao, Y.; Wang, H.-R.; Hu, H.-Y.; Zhu, Y.-F.; Li, S.; Li, J.-Y.; Wu, X.-W.; Chou, S.-L. Formulating high-rate and long-cy heterostructured layered oxide cathodes by local chemistry and orbital hybridization modulation for sodium-ion batteries. *A Mater.* **2022**, *33*, 2202695. [CrossRef]
173. Yang, L.; López del Amo, J.M.; Shadike, Z.; Bak, S.-M.; Bonilla, F.; Galceran, M.; Nayak, P.K.; Buchheim, J.R.; Yang, X.-Q.; Rojo et al. A Co- and Ni-free P2/O3 biphasic lithium stabilized layered oxide for sodium-ion batteries and its cycling behavior. *A Funct. Mater.* **2020**, *42*, 2003364. [CrossRef]
174. Zhou, Y.-N.; Wang, P.-F.; Niu, Y.-B.; Li, Q.; Yu, X.; Yin, Y.-X.; Xu, S.; Guo, Y.-G. A P2/P3 composite layered cathode high-performance Na-ion full batteries. *Nano Energy* **2019**, *55*, 143–150. [CrossRef]
175. Zhou, D.; Huang, W.; Lv, X.; Zhao, F. A novel P2/O3 biphase $Na_{0.67}Fe_{0.425}Mn_{0.425}Mg_{0.15}O_2$ as cathode for high-performa sodium-ion batteries. *J. Power Sources* **2019**, *421*, 147–155. [CrossRef]
176. Liang, X.; Sun, Y.-K. A novel pentanary metal oxide cathode with P2/O3 biphasic structure for high-performance sodium- batteries. *Adv. Funct. Mater.* **2022**, *44*, 2206154. [CrossRef]

Liu, R.; Huang, W.; Liu, J.; Li, Y.; Wang, J.; Liu, Q.; Ma, L.; Kwon, G.; Ehrlich, S.N.; Wu, Y.; et al. Revealing the Nature of Binary-Phase on Structural Stability of Sodium Layered Oxide Cathodes. *Adv. Mater.* **2024**, *36*, 2401048. [CrossRef]
Zhou, P.; Che, Z.; Liu, J.; Zhou, J.; Wu, X.; Weng, J.; Zhao, J.; Cao, H.; Zhou, J.; Cheng, F. High-entropy P2/O3 biphasic cathode materials for wide-temperature rechargeable sodium-ion batteries. *Energy Storage Mater.* **2023**, *57*, 618–627. [CrossRef]

micromachines

Review

Advances in Host-Free White Organic Light-Emitting Diodes Utilizing Thermally Activated Delayed Fluorescence: A Comprehensive Review

Wenxin Zhang [1,2], Yaxin Li [1,2], Gang Zhang [1,2,*], Xiaotian Yang [3], Xi Chang [1,2], Guoliang Xing [4], He Dong [1,2], Jin Wang [1,2,*], Dandan Wang [5], Zhihong Mai [5] and Xin Jiang [6]

1 College of Information Technology, Jilin Engineering Research Center of Optoelectronic Materials and Devices, Jilin Normal University, Siping 136000, China; zwx00719@163.com (W.Z.); 13944698385@163.com (Y.L.); cx02008@126.com (X.C.); donghe@jlnu.edu.cn (H.D.)
2 Key Laboratory of Functional Materials Physics and Chemistry of Ministry of Education, Jilin Normal University, Siping 136000, China
3 Key Laboratory of Preparation and Applications of Environmental Friendly Material of the Ministry of Education, College of Chemistry, Jilin Normal University, Changchun 130103, China; hanyxt@163.com
4 Jilin Special Equipment Inspection Center, Jilin Special Equipment Accident Investigation Service Center, No. 866 Huadan Street, Longtan District, Jilin 132013, China; guoliangxingjlsei@163.com
5 Hubei Jiufengshan Laboratory, Wuhan 430206, China; wangdandan@jfslab.com.cn (D.W.); maizhihong@jfslab.com.cn (Z.M.)
6 State Key Laboratory of Integrated Optoelectronics, College of Electronic Science and Engineering, Jilin University, Changchun 130012, China; jiangxin22@mails.jlu.edu.cn
* Correspondence: zg982@163.com (G.Z.); jwang@jlnu.edu.cn (J.W.)

Abstract: The ever-growing prominence and widespread acceptance of organic light-emitting diodes (OLEDs), particularly those employing thermally activated delayed fluorescence (TADF), have firmly established them as formidable contenders in the field of lighting technology. TADF enables achieving a 100% utilization rate and efficient luminescence through reverse intersystem crossing (RISC). However, the effectiveness of TADF-OLEDs is influenced by their high current density and limited device lifetime, which result in a significant reduction in efficiency. This comprehensive review introduces the TADF mechanism and provides a detailed overview of recent advancements in the development of host-free white OLEDs (WOLEDs) utilizing TADF. This review specifically scrutinizes advancements from three distinct perspectives: TADF fluorescence, TADF phosphorescence and all-TADF materials in host-free WOLEDs. By presenting the latest research findings, this review contributes to the understanding of the current state of host-free WOLEDs, employing TADF and underscoring promising avenues for future investigations. It aims to serve as a valuable resource for newcomers seeking an entry point into the field as well as for established members of the WOLEDs community, offering them insightful perspectives on imminent advancements.

Keywords: WOLED; intrinsically host-free engineering; thermally activated delayed fluorescence (TADF); phosphorescence

1. Introduction

In contemporary society, electronic displays have become an indispensable component of various industries. The growing demand for display devices has led scientists to explore new and efficient light-emitting materials and structures of devices. Among the proposed alternatives, organic light emitting diodes (OLEDs) have become the most promising display technology, experiencing rapid and dynamic development [1–3]. Compared with traditional LED materials, OLEDs have significant advantages, such as vibrant color reproduction, flexibility, ultrathin form factor, and low energy consumption. These characteristics make OLEDs a convincing innovative solution to meet the requirements of ultra-lightweight, ultrathin, energy-efficient, and high-color-saturation displays [4–7].

Various types of WOLEDs have been developed, including all-phosphorescent white light [8,9], all-fluorescent white light [10,11], TADF white light [12,13], hybrid white light [14–16], and others. The first generation of OLED luminescent materials, namely fluorescent materials, were unable to effectively utilize the generated triplet excitons, resulting in an internal quantum efficiency (IQE) [17] of only 25%. Therefore, devices using these materials have difficulty in achieving an external quantum efficiency (EQE) of over 5%.

In 1998, Forrest and Professor Ma Yuguang from Jilin University introduced the second generation of luminescent materials [18]: phosphorescent materials. In 2012, Adachi et al. discovered a class of organic materials capable of achieving 100% IQE by leveraging the phenomenon of TADF [19]. By reducing the energy gap between singlet and triplet excitons, triplet excitons can be converted to singlet excitons under the condition of thermal activation, resulting in radiative recombination [20]. Although fluorescent materials can achieve complete utilization of excitons, the commercial feasibility is still limited due to their high cost [21–23].

Organic luminescent materials can be categorized based on their distinct luminescent mechanisms, such as fluorescent materials [24,25], phosphorescent materials [26], TADF materials [27,28], and others. The exciton decay process of organic materials can be divided into radiative transition, intersystem crossing (ISC), reverse intersystem crossing (RISC), and internal conversion (IC). The triplet state of exciton transitions back to the excited singlet state through the anti-system crossing process, where the fluorescence generated by the deactivation transition to the ground state is called delayed fluorescence, and the T_1 of TADF transforms into the luminescent S_1 through RISC [29–31], thus achieving 100% theoretical quantum effect. High-quality white light emission can be achieved based on exciplex and electroplates emission [32,33].

We listed four common types of TADF-WOLEDs, as shown in Figure 1, and summarized the key data from their representative literature in Table 1. Most research works concerning OLEDs are carried out based on doping and excitable compounds, but at the same time, there are problems such as complex device structure [34,35], strict doping ratio [36,37], and easy contamination during the evaporation process [38]. However, ultrathin WOLEDs (the luminous layer is usually a few nanometers or even less than 1nm of the device) can make it easier to conduct experiments by virtue of its simple device structure. Therefore, this work summarizes the research progress of host-free TADF-WOLEDs in terms of compound synthesis, device structure, and device photoelectric properties.

Table 1. Device performance of TADF based WOLEDs.

Devices	CE (cd/A)	PE (lm/w)	CRI	CIE (x,y)	EQE (%)	Ref.
TADF Ultrathin WOLED	30.3	28.8	88	(0.47,0.44)	18.7	[39]
	34.14	33	–	(0.369,0.460)	–	[40]
	18.9	21.1	86	(0.375,0.410)	8.7	[41]
TADF Doped WOLED	28.83	20.83	–	(0.45,0.46)	11.5	[42]
	16.3	13.3	–	(0.37,0.39)	8.39	[35]
	26.04	23.83	90	(0.490,0.473)	12.79	[43]
TADF-phosphor sensitization WOLED	60.0	42.8	64	(0.33,0.44)	22.8	[36]
	46.8	56.5	–	(0.44,0.48)	15.2	[44]
	26.9	22.3	–	(0.51,0.41)	12.8	[45]
TADF interfacial exciplex WOLED	52.8	33.2	80	(0.35,0.39)	22.9	[46]
	69	69.9	–	(0.40,0.57)	20	[47]
	57.8	62.5	–	–	22.7	[48]

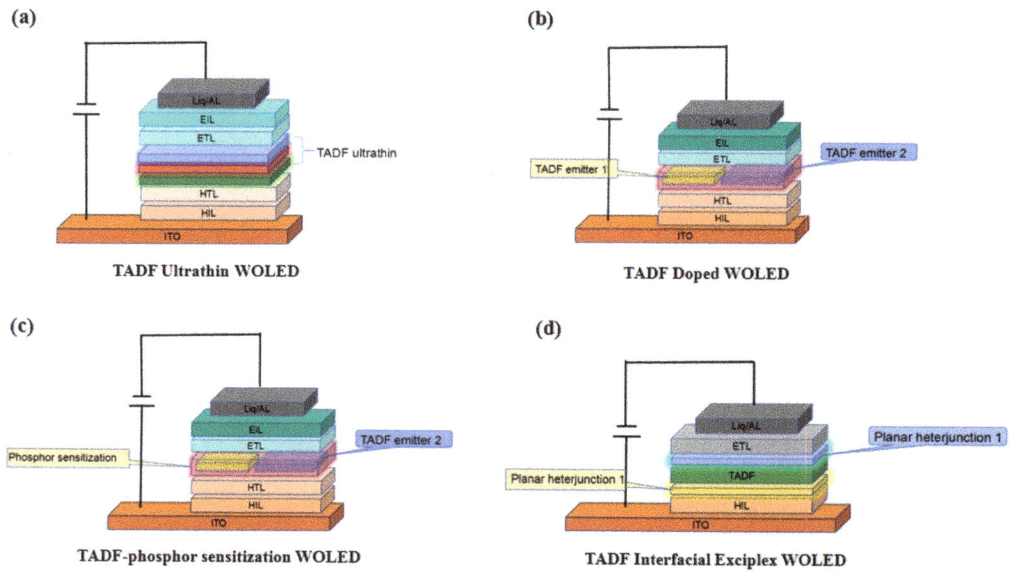

Figure 1. (**a**) TADF ultrathin WOLED. (**b**) TADF doped WOLED. (**c**) TADF-phosphor sensitization WOLED. (**d**) TADF-interfacial exciplex WOLED.

2. WOLEDs Based on Host-Free TADF-Conventional Fluorescence Materials

The pursuit of high electroluminescence efficiency in WOLEDs has led to extensive research on all-phosphorescent systems and fluorescent–phosphorescent systems. These systems utilize phosphorescent materials, which exhibit high efficiency in exciton utilization. However, the widespread application of WOLEDs has been hindered by the limited availability of strong blue phosphors [49–51]. In contrast, the traditional emitting layer of WOLEDs can only harness singlet excitons. When doped into TADF materials, the electroluminescence (EL) efficiency is significantly enhanced compared to common host materials [52]. Moreover, full-fluorescent WOLEDs utilizing TADF materials show outstanding competitiveness and advantages as hosts and sensitizers in terms of high colorimetric purity and the stability of fluorescent emitters. These advancements highlight the potential and desirability of full-fluorescent TADF-based WOLEDs over traditional full-fluorescent WOLEDs [53,54].

In a recent study by Miao et al. in 2021 [55], a multifunctional compound incorporating tetra(4-bromophenyl)anthracene (TBAN) was designed and synthesized. Through the utilization of hybrid local electron and charge transmission mechanisms, as well as aggregation-induced luminescence, a novel and highly simplified two-color WOLED was proposed. The WOLED comprised solely three host-free organic emitting layers, with the molecular formula and structural diagram depicted in Figure 2. TBAN acts as the yellow emission layer and hole transport layer, while $Bepp_2$ functions as the electron transport layer and a blue emission layer, ultimately producing white light. To achieve a more balanced white emission, a charge-regulating layer was incorporated between the two transport layers. This adjustment facilitated the controlled distribution of carriers and excitons in the transport layer, enhancing the overall performance and color balance of the WOLED.

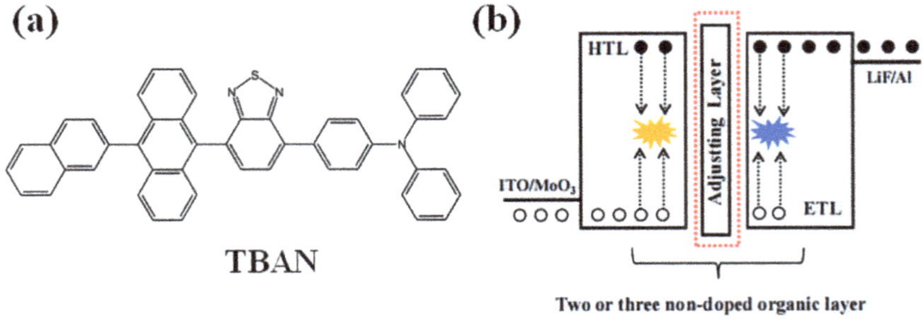

Figure 2. (**a**) Molecular formula and (**b**) structural formula representation.

In their experimental study, Miao et al. aimed to develop WOLEDs by combining the simultaneous emission of TBAN and bis[2-(2-pyridinyl)phenolato]berylliuM ($Bepp_2$). Initially, the authors tested the EL performance of yellow and blue OLEDs to assess the potential for achieving white emission. The connection lines of the two best-performing devices' Commission Internationale de l'Eclairage (CIE) coordinates intersected within the pure white light region, indicating the possibility of achieving high-quality white emission by balancing the intensity of TBAN and $Bepp_2$ emissions [56,57]. Considering the significant influence of $Bepp_2$ thickness on device performance, Miao et al. designed a series of two-color WOLEDs with varying $Bepp_2$ thicknesses. The aim was to adjust the emission spectrum by optimizing the $Bepp_2$ thickness. To balance the distribution of excitons in the yellow and blue emission transport layers, a high-level charge-regulating layer composed of TCTA was inserted between the two respective transport layers [58]. The three-layer host-free structure of TBAN/4,4′,4″-Tris(carbazol-9-yl)-triphenylamine(TCTA)/$Bepp_2$ was optimized in WOLEDs, and 86 white-light devices with one of the highest CRI reported in the literature were obtained. Additionally, the maximum brightness achieved was measured at 43,460 cd/m^2. The EL spectrum's change in CIE coordinates exhibited a minimal shift of (0.00, −0.01) when the turn-on voltage increased from 5 V to 7 V. This change corresponded to a brightness increase from 1011 cd/m^2 to 9310 cd/m^2.

In a separate study conducted by Zhao et al. in 2023 [50], the all-fluorescence device consisted of an orange-light-emitting layer sandwiched between two sky-blue-light-emitting layers. The design achieves a record power efficiency (PE) of 130.7 lm/W and 31.1% EQE. For a WOLED composed of all-TADF, triplet-triplet and singlet-triplet annihilation would be more serious due to the longer lifetime of triplet excitons in TADF materials [31]. Therefore, the hybrid WOLED structure composed of TADF luminescent layers is very important as it can prevent excitons from reaching the non-radiative lowest triplet excited state of the traditional luminescent layer.

In 2018, Zhao et al. [59] reported a yellow TADF emitter with strong emission bis(2-(diphenylphosphine)phenyl)ether oxide (OPDPO), a blue emitter 9,9′-((2-(4″-(9H-carbazol-9-yl)-[1,1′-biphenyl]-4-yl)ethene-1,1-dial)bis(4,1-phenylene))bis(9H-carbazole)(2CzTPEpCz) with high aggregation-induced emission (AIE)-active fluorescent materials in the T_1 state. Similarly, in order to achieve the AIE effect, Wang et al. [60] added donor/acceptor pairs to the polymer to form a double space charge transfer channel. The white light device achieved a 14% EQE.

It is envisaged that a blue AIE-active fluorophore will be combined to realize completely host-free high-efficiency WOLEDs. As shown in Figure 3, the OPDPO and 2CzTPEpCz peaked at 577 nm and 470 nm, respectively. The S_1 and T_1 states of OPDPO were reached at 2.59 and 2.46 eV. Due to the small difference in the excitation energy between the two states, the non-emissive T_1 state of 2CzTPEpCz is higher than that of OPDPO. When the two emissive layers (EMLs) simply stack together without any intermediate layer, the triplet exciton quenching of OPDPO can be effectively inhibited by 2CzTPEPCz.

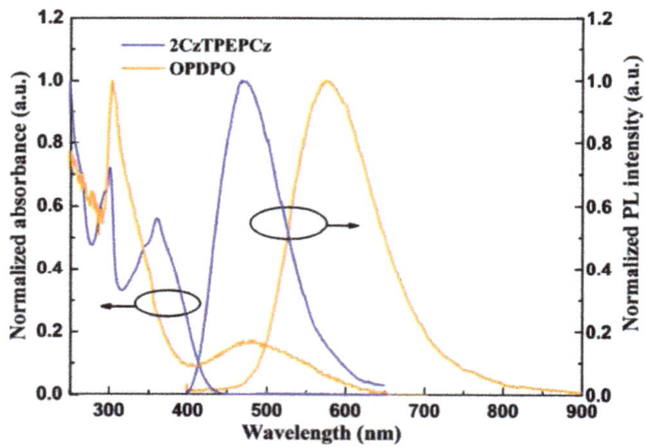

Figure 3. PL spectra and UV-vis absorption of OPDPO and 2CzTPEPCz films.

Most reports on high-efficiency TADF devices for use in host-free OLEDs focus on blue and green emitters rather than yellow-light-emitting layers [61,62], and solution-treated OLEDs have the potential to be printed into complex structures and luminous regions [63,64]. Thus, Zhao et al. also prepared a host-free device for OPDPO: indium-tin-oxide (ITO)/PEDOT: PSS/4,4′-bis(9H-carbazol-9-yl)biphenyl (CBP) (33 nm)/OPDPO (7 nm)/1,3,5-tris(N-phenylbenzimidazol-2-yl)benzene (TPBI) (40 nm)/Mg: Ag. The maximum current efficiency (CE), PE, and EQE are 37.6 CD/A, 14.8 lm/W, and 16.6%, respectively. The maximum brightness is 3600 cd/m^2, which has an EL performance suitable for practical application. Host-free devices based on 2CzTPEpCz: ITO/PEDOT: PSS (40 nm)/CBP (20 nm)/2CzTPEPCZ (15 nm)/TPBI (40 nm)/Mg: Ag. The maximum EQE, CE, and PE are 5.6%, 11.3 cd/A, and 7.1 lm/W. The EQE of the device reaches 3.3% at a brightness of 1000 lm/W, which has a rather low efficiency roll-off. It is one of the most efficient host-free blue OLED devices using traditional AIE fluorescent layers without any external coupling method [65–67].

It is shown that the TADF-fluorescent hybrid has the advantage of low rolling efficiency compared to all-TADF WOLEDs. Based on an important advantage of traditional fluorescent light-emitting layers, blue OLEDs generally have lower efficiency roll-off compared to yellow TADF OLEDs. Based on this, a completely host-free two-color WOLED has the following components: ITO/poly(3,4-methylenedioxy-thiophene): poly(styrene sulfonate) (PEDOT: PSS) (40 nm)/CBP (20 nm)/OPDPO (4 nm)/2CzTPEPCZ (15 nm)/TPBI (40 nm)/Mg: Ag with dual EMLs without any intermediate layer was constructed, and its structure and energy level diagrams are shown in Figure 4.

The tightly stacked light-emitting layer devices make charge carrier recombination more efficient into the 2CzTPEPCz layer from the TPBi light-emitting layer and then into the OPDPO light-emitting layer, where the energy barrier is avoided. It is also easy for holes to enter the OPDPO layer from the CBP luminescent layer. Due to the low HOMO energy level barrier, some holes enter the 2CzTPEPCz through the OPDPO luminescent layer. Most excitons compound in the OPDPO luminescent layer, and only a small part of them are at the 2CzTPEPCz interface, resulting in yellow light accounting for a greater proportion of the emission. OPDPO has a wide bandwidth, so the device ends up with warm white light. In summary, most excitons can be used for radioluminescence by timely or delayed fluorescence of OPDPO, thus obtaining higher EQE.

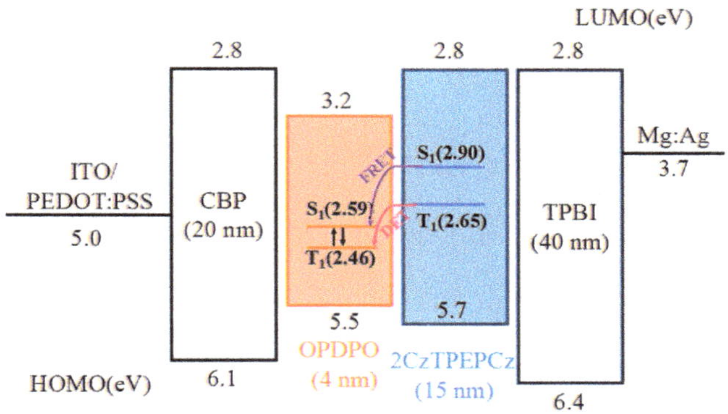

Figure 4. The energy diagram of the WOLED device and the diagram of the triplet harvesting mechanism.

From the characteristic curve in Figure 5, the maximum PE, CE, and EQE are 18.0 lm/W, 45.9 cd/A, and 20.8%. At a brightness of 1000 cd/m^2, the EQE achieved by the device still reaches 13.7%, which is much higher than the EQE of most traditional host-free fluorescent WOLEDs reported in the previous literature. As the brightness increased from 400 to 10,000 cd/m^2, the device CIE coordinates and CCT varied slightly from (0.45, 0.44) to (0.43, 0.44) and 3049 to 3392K, respectively, with the color rendering index (CRI) increasing from 73 to 76. A completely warm WOLED of the host-free two-color device, based on a traditional fluorescence-TADF hybrid emitter, has been successfully fabricated.

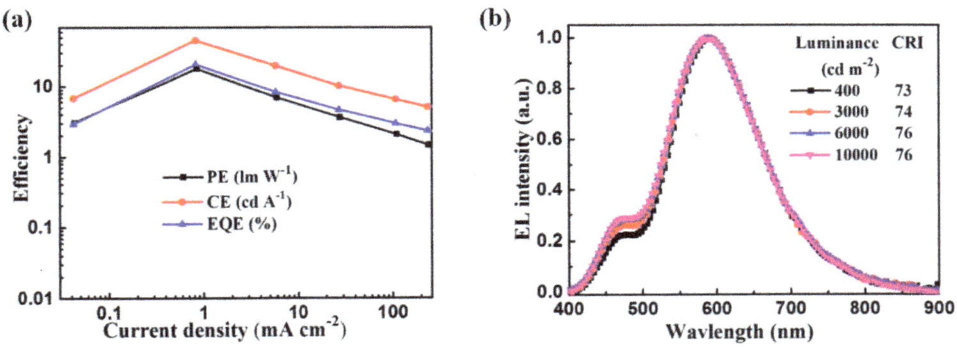

Figure 5. (**a**) PE-CE-EQE-current density curves and (**b**) normalized EL spectra at different luminance of the WOLED device.

3. WOLEDs Based on Host-Free Phosphorescence-TADF Materials

Traditional fluorescent WOLEDs have limitations in achieving high efficiency, with only 25% maximum IQE [23]. However, the phosphorescent introduction of light-emitting layer molecules has significantly improved OLED efficiency [68,69]. These phosphorescent materials typically contain heavy metal atoms, such as Os, Pd, Pt, and Ir, in their organic metal complexes [70]. Compared to traditional luminescent materials, the singlet S_1 and triplet T_1 excitons can be effectively utilized by the phosphor through the heavy atom effect, resulting in a potential IQE of 100% and enabling the realization of high-performance OLEDs [71,72]. Furthermore, due to their high cost, fluorescence-based materials are less feasible for mass production, making phosphorescent materials the primary focus of experimental research [73,74]. In 2020, Huo et al. [75] demonstrated for the first time that bi-TADF-TADF /room-temperature phosphorescence (RTP) emission can be utilized to obtain WOLED with a high CRI.

In 2022, Wu et al. [76] conducted a study in which a series of compounds with AIE properties were designed and synthesized. These compounds belong to the donor-acceptor-donor (D-A-D) type and utilize the donor-acceptor-donor′ (D-A-D′) framework, as depicted in Figure 6.

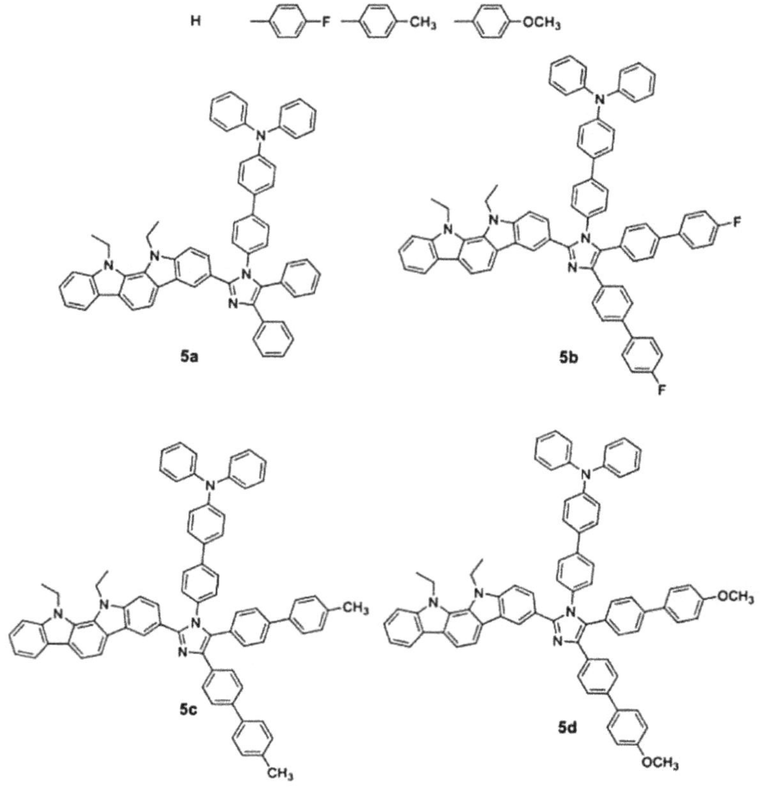

Figure 6. Molecular formula representation of the donor-acceptor-donor (D-A-D) compounds.

A high-efficiency WOLED was studied, in which the $Irppy_3$ and Ir-$(MDQ)_2$(acac)-doped compound was used as the double-blue-emission layer. The WOLED maximum brightness was 66,424 cd/m^2 and it had a CE of 67.6 cd/A. Notably, the device demonstrated excellent chromatic stability with minimal deviation ($\Delta x = \pm 0.003$, $\Delta y = \pm 0.002$) when the turn-on voltage changed from 5 V to 12 V. The color stability was improved by optimizing the structure of the double-blue-emission layer. High blue-fluorescence emission intensity was better matched with green and red phosphor emission intensity. The harmonious balance among the emission colors contributed to the device's color stability and excellent electroluminescent properties. These findings underscore the significant potential of the newly developed wide band gap materials for applications in white lighting devices.

Compound **5b**, used as a host-free blue OLED, exhibits excellent luminous performance with the 10,770 cd/m^2 maximum luminance and a CE of 6.93 cd/A. The device structure of the hybrid WOLED is illustrated in Figure 7. In this configuration, the fluorescent double emission layer uses compound 5b's material, while compound **5d** is doped with $Irppy_3$ (with an energy transfer (ET) value of 2.4 eV) and IR-$(MDQ)_2$ACAC (with an ET value of 2.0 eV) to function as the green and red phosphorescent emission layers, respectively. The electron transport layer material uses TPBi. The hybrid WOLED demonstrated impressive performance metrics, including the 66,424 cd/m^2 maximum luminance, the

67.6 cd/A maximum CE, the 56.6 lm/W maximum PE, and the 24.3% maximum external quantum efficiency. The device was (0.337,0.341) in CIE coordinates, the CRI reached 85, and the pure white light emission was realized.

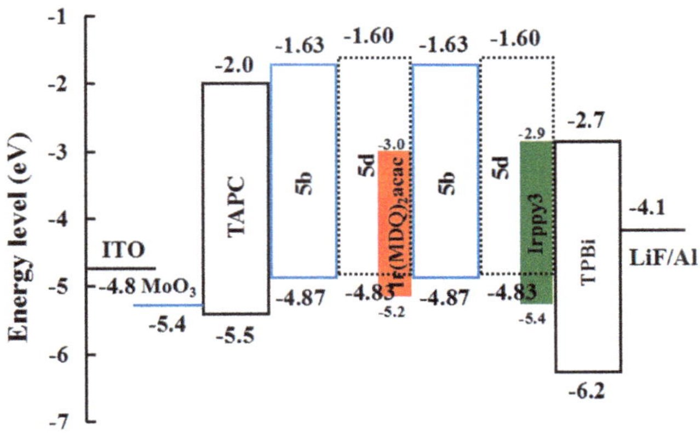

Figure 7. Schematic representation of the device structure for a hybrid WOLED.

3.1. Phosphorescent Ultrathin Emitting Layer WOLED

Although controlling ultrathin EML is more difficult to control in experiments than traditional host-free EML, it can avoid intramolecular aggregation, thereby building highly efficient host-free WOLED. In 2019, Li et al. [77] proposed a novel concept interlayer (IL) to insert sky-blue ultrathin emission layers (Ph-UEMLs) between the yellow and blue-light-emitting layers, effectively reducing the efficiency roll down of TADF. The device finally achieved 24.47% EQE. In 2023, Miao et al. [78] proposed a straightforward, host-free approach that uses single or complementary phosphorescent Ph-UEMLs directly in the interface polymers of TADF OLEDs. By implementing this method, all monochromatic OLEDs achieved the 2.4–2.5 V low turn-on voltage and exhibited remarkable device performance, with EQE max surpassing 20%. bis(4,6-difluorophenylpyridinato-N, C2)picolinate iridium (FIrpic) and bis(2-phenylbenzothiazolato)(acetylacetonate)iridium(III)(IR(BT_2(ACAC))) were used as complementary Ph-UEMLs and directly stacked onto the interface of the m-bis(N-carbazolyl)benzene(mCP)/2,4,6-tris[3-(diphenylphosphine)phenyl]-1,3,5-triazine (PO-T2T)) exciton system. In order to study the TADF characteristics of exciton formation at the MCP/PO-T2T interface, the photoluminescence spectra of co-deposited MCP: PO-T2T films were studied at low temperatures, as depicted in Figure 8. The results indicate an energy level difference of 0.03 eV, satisfying the thermodynamic conditions for the RISC of triplet excitons. This finding suggests that the excitons formed within the mCP/PO-T2T system possess TADF characteristics.

The optimized device structure consists of ITO (180 nm)/MoO_3 (3 nm)/TAPC (40 nm)/mCP (10 nm)/FIrpic (0.40 nm)/Ir$(BT)_2$(acac) (0.01, 0.02, 0.03, 0.04 nm)/PO-T2T (50 nm)/LiF (1 nm)/Al (100 nm), forming a two-color white-light-emitting-layer device, as illustrated in Figure 9. Remarkably, the device achieved a 23.48% maximum EQE, surpassing the EQE values of most doped dichroic white devices. Notably, the white light emission spectrum can be easily controlled from cold to warm by simply adjusting the thickness of the Ph-UEMLs.

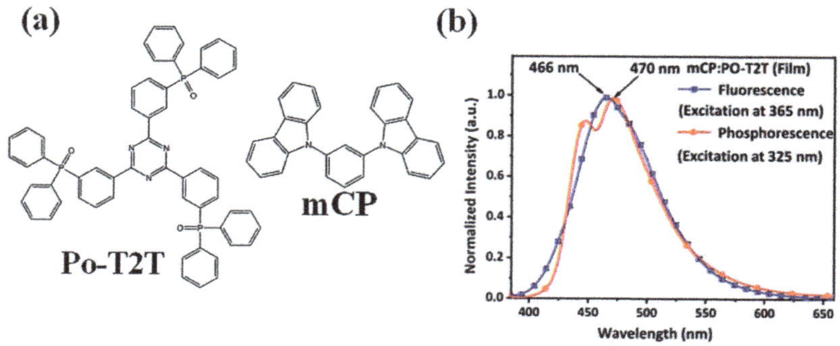

Figure 8. (**a**) Molecular structure formula representation; (**b**) low-temperature photoluminescence spectra of MCP:PO-T2T (1:1) films.

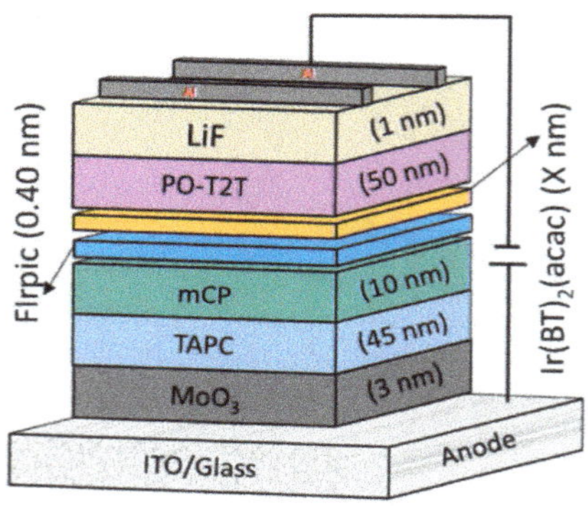

Figure 9. Schematic diagram of two-color white-light-emitting-layer device.

In the experimental investigation, four devices were fabricated and tested, each incorporating phosphorescence. Ph-UEMLs with different thicknesses denoted as 0.01, 0.02, 0.03, and 0.04 nm correspond to the devices labeled as W1, W2, W3, and W4, respectively. Table 2 summarizes the key performance parameters of WOLED.

Table 2. Key performance parameters of white components W1, W2, W3, and W4.

Devices	Turn-On Voltage (v)	Maximum			
		CE (cd/A)	PE (lm/W)	EQE (%)	Luminance (cd/m^2)
W1	2.5	53.96	64.77	20.50	12,140
W2	2.5	64.93	78.80	23.48	13,220
W3	2.5	59.96	75.35	21.27	16,050
W4	2.5	57.50	71.02	20.29	16,870

The effect of the thickness of IR(BT)$_2$(ACAC)-UEML on the emission characteristics of WOLEDs has been studied systematically. As the IR(BT)$_2$(ACAC)-UEML thickness increases, the relative intensity of blue emission decreases gradually, while that of yellow emission increases, eventually dominating the emission spectrum. At an IR(BT)$_2$(ACAC)-UEML thickness of 0.02 nm, a well-balanced intensity between the two emission bands was achieved, resulting in the W2 device displaying a more balanced white light. By simply adjusting the PhUEML thickness, it is possible to achieve controlled white light emission with a tunable color temperature ranging from cold to warm. Remarkably, the W2 device demonstrated a 23.48% max EQE, surpassing most doped two-color WOLEDs, while maintaining the 2.5 V low turn-on voltage.

In order to obtain high-performance WOLEDs, the combination of TADF and phosphorescent materials has attracted great attention. However, the lack of stable and efficient blue phosphors has hindered the development of phosphorescent WOLEDs [79,80]. To address this challenge, various strategies have been proposed to achieve efficient white emission at the exciton scale [81,82]. Nonetheless, these device designs often introduce complexity and increase production costs. In light of this, a promising approach involves inserting an ultrathin phosphorescent emission layer into a blue fluorescent emissive layer, offering advantages such as simplified processing and flexible structural design. Zhao et al. [83] successfully implemented this approach by inserting a host-free ultrathin phosphorescent emission layer into a bipolar transport blue TADF emissive layer. The resulting TADF/phosphorescent hybrid WOLEDs achieved a 41.6 cd/A maximum CE, a 41.0 lm/W maximum PE, and a 19.1% maximum EQE.

3.2. Conventional Fluorescent and Phosphorescent Doped Hybrid WOLEDs

The main blue fluorescent materials used in WOLEDs are severely quenched due to the existence of too many singlet and triplet excitons [84–86]. In 2022, Zhang et al. [87] addressed this issue by incorporating the excellent bipolar TADF material, 1-(4-(4,6-Diphenyl-1,3,5-triazine-2-yl)phenyl)-1H-spiro[[pyridine-9,9′-fluorene (SpiroAC-TRZ), as the blue emitter. To enhance device efficiency and simplify the fabrication process while mitigating exciton quenching, an ultrathin orange phosphorescent emitting layer was inserted at a suitable position between the blue bipolar TADF emitters.

The resulting WOLEDs exhibited an optimized structure of ITO/hexaazatriphenylene hexacabonitrile (HATCN)/1,1-bis[(di-4tolylamino)phenyl]cyclohexane (TAPC)/4,4′,4′′-Tri(9-carbamoyl) triphenylamine (TCTA)/SpiroAC-TRZ/bis(4-phenyl-thieno[3,2-c]pyridinato-C2,N) (acetylacetonate)iridium(III)(PO-01)/SpiroAC-TRZ/PO-01/SpiroAC-TRZ/PO-01/SpiroAC-TRZ/4,6-bis(3,5-di(pyridine-3-yl)phenyl)-2-(pyridine-3yl) pyrimidine (B_3PYMPM)/Liq/Al, as depicted in Figure 10. At maximum luminance, the device achieved a 47.4 lm/W maximum PE, a 46.2 cd/A maximum CE, and a 15.0% maximum EQE. Notably, even at a brightness level of 1000 cd/m^2, the device maintained a respectable PE of 41.7 lm/W, a CE of 45.8 cd/A, and an EQE of 14.7%. By incorporating the bipolar blue TADF material Spiro AC-TRZ and strategically inserting an ultrathin orange-phosphorescent-emitting layer, Zhang et al. successfully achieved efficient WOLEDs with a simplified, host-free device structure. These advancements not only reduced the efficiency roll-off but also mitigated exciton quenching, leading to improved overall device performance.

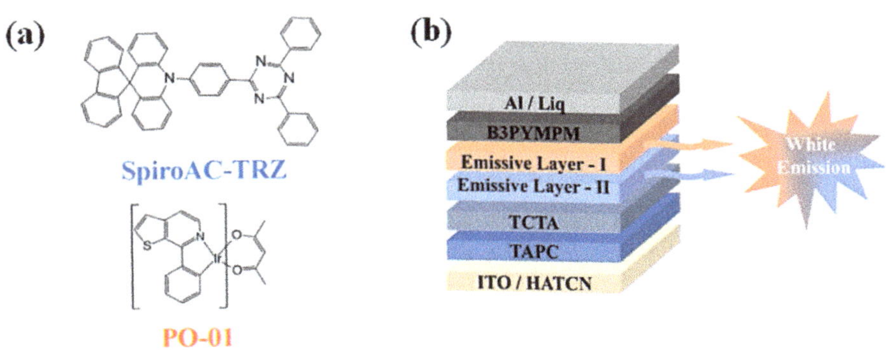

Figure 10. (a) Chemical structures of the emissive materials. (b) Structural representation of the hybrid WOLED.

The utilization of SpiroAC-TRZ as a bipolar transport material in WOLEDs not only benefits the device's electrical characteristics but also contributes to improved exciton diffusion and reduced exciton confinement, thereby simplifying efficiency roll-off. SpiroAC-TRZ's inherent bipolar transport properties enable efficient charge carrier migration in the device structure, resulting in a low turn-on voltage [83,88,89]. Furthermore, the incorporation of SpiroAC-TRZ as a blue TADF emissive layer facilitates balanced charge transport, expanded exciton distribution [90], reduced exciton quenching, and enhanced efficiency, increasing the device stability [91]. This is illustrated in Figure 10, which depicts the exciton energy transfer diagram between the blue TADF emissive layer and the orange phosphorescent emissive layer. Clearly, the SpiroAC-TRZ's bipolar transport characteristics not only streamline vehicle transportation within the device but also contribute to low turn-on voltages. The incorporation of SpiroAC-TRZ as a blue TADF emissive layer enhances charge balance, broadens the exciton distribution area, reduces the depth of exciton confinement, and simplifies efficiency roll-off. The exciton energy transfer diagram presented in Figure 11 provides a visual representation of these phenomena.

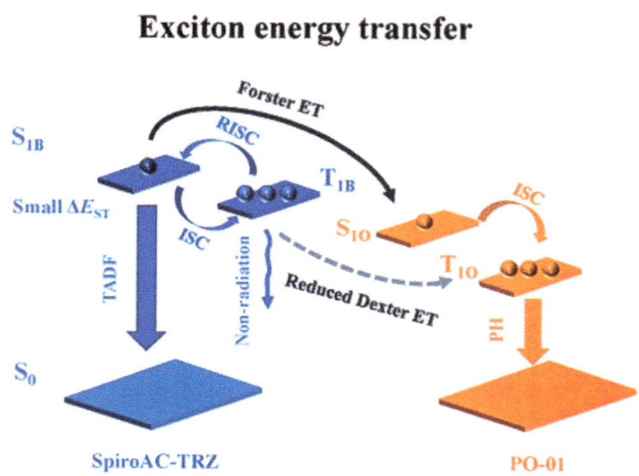

Figure 11. Exciton energy transfer schematic of SpiroAC-TRZ and PO-01.

4. WOLEDs Based on Host-Free Full-TADF Materials

Compared to WOLEDs based on conventional fluorescent materials or expensive phosphorescent materials containing heavy metals, all-TADF WOLEDs offer a promising alternative [92–94]. These all-TADF WOLEDs can achieve emission through RISC processes without the need for precious metals, making them cost-effective and environmentally friendly [95,96]. The combination of singlet and triplet excitons in TADF emitters yields high device efficiency at a lower cost. Both phosphorescence and TADF mechanisms have the potential to achieve 100% IQE in OLEDs [97,98]. However, the stability of phosphorescent materials due to their material characteristics remains a significant challenge [99,100]. Consequently, there is a growing interest in exploring all-TADF systems, although research in this area is still relatively limited.

Existing all-TADF WOLED studies are not in depth. In 2019, Tang et al. [101] designed and synthesized bis[3-(9,9-dimethyl-9,10-dihydropyridine)phenyl]sulfone (m-$ACSO_2$), which had a large triplet energy (T_1 = 2.94 eV), a tiny energy gap, and a photoluminescence quantum yield (PLQY) up to 76% in neat films. The synthesized 3,6,11-tris(9,9-dimethylacridin-10(9H)-yl)dibenzo[a,c]phenazine (3DMAC-BP) orange TADF material, with a 0.012 eV single-wire triple-state energy gap and an 89% PLQY and other excellent TADF characteristics, is shown in Figure 12a. The emission peaks of $ACSO_2$ and 3DMAC-BP were at 478 nm and 605 nm. Since m-$ACSO_2$ and m-$ACSO_2$ are in close contact at the ultrathin interface, the triplet energy of m-$ACSO_2$ (T_1 = 2.94 eV) is larger than 3DMACBP (T_1 = 1.98 ev), the triplet energy transfer (DET) between the two is easy to achieve. Figure 12b shows the exciton energy transfer process between m-$ACSO_2$ and 3DMAC-BP.

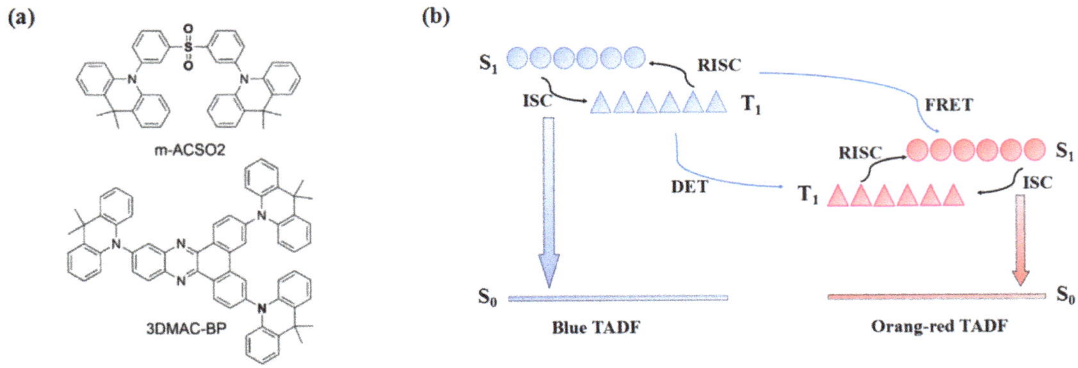

Figure 12. (**a**) Chemical structures of m-$ACSO_2$ and 3DMAC-BP. (**b**) Schematic diagram of energy transfer between m-$ACSO_2$ and 3DMAC-BP.

The host-free WOLEDs with full TADF were prepared as follows: ITO/PEDOT: PSS (60 nm)/m-$ACSO_2$ (40 nm)/3DMAC-BP (0.15 nm)/m-$ACSO_2$ (x = 0, 10, 20 and 30 nm)/TmPyPB (50 nm)/LiF (1 nm)/Al (120 nm). When X = 20, the device turn-on voltage is 2.6 eV, and bimodal white emission is achieved. In addition, at the 1000 cd/m^2 luminance, the device exhibits good warm white emission with International Lighting Committee CIE coordinates of (0.34, 0.43). When the voltage rises from 5 V to 8 V, the EL spectrum also shows a small change, and the device CRI value remains above 74. By controlling the position of the 3DMAC-BP ultrathin layer and comparing its different EL spectra, it can be proven that the orange–red ultrathin layer position controls the emission spectrum by controlling the distribution of excitons in the sky-blue-light-emitting layer. The results show that the thickness of 0.15 nm and x = 10's device has the best performance, has a 51.7 cd/a maximum CE, a 46.4 lm/W maximum PE, and a 24.2% maximum EQE. The device simplifies the device structure, reduces manufacturing complexity, and finally achieves high efficiency through the form of an ultrathin layer.

In a separate study, Zhong et al. [102] designed and synthesized 9,9-dimethyl-10-phenyl-acridan (BDMAc) as a deep blue luminescent layer in 2019. By incorporating 2-(10H-phenothiazin-10-yl)dibenzothiophene-S, S-dioxide (PVTZ-DBto2) as the yellow-emitting layer and host material in conjunction with BDMAc, they achieved a maximum EQE of 14.2%. Importantly, the construction of fully TADF-based single emission layer (SEL)-WOLEDs had led to significant advancements in achieving color-balanced white emission. These devices exhibited impressive performance characteristics, including high EQE, CE, and PE values. Notably, the incorporation of TADF material, such as 2,3,4,5,6-pentakis(3,6-dimethyl-9H-carbazol-9-yl)benzonitrile (5PCzCN), had demonstrated exceptional stability, with extended lifetimes at various brightness levels. Additionally, the design and synthesis of novel luminescent materials, such as 9,9-dimethyl-10-(3-vinylphenyl)-9,10-dihydropyridine (BDMAc), in combination with suitable hosts, had contributed to achieving high EQE values in specific color regions.

Figure 13 illustrates the polymers obtained through radical copolymerization of 2-(9,9-dimethy-acridine-10-yl)-8-vinyl dimethyl thioxanthene-S, S-dioxide (DMA-TXO_2), BDMAc, and 2-(10H-phenothiazine-10-yl)-8-vinyl dibenzothiophene-S, S-dioxide (PTZ-$DBTO_2$) monomers. The resulting polymers exhibit a mean RMS roughness within the narrow range, indicating a remarkably smooth and uniform film morphology. This suggests that these polymers possess desirable characteristics as EML materials for OLEDs. To optimize the device performance, a dual-layer device structure consisting of ITO/ PEDOT:

PSS (40 nm)/EML (30 nm)/TmPyPB (40 nm)/LiF (1 nm)/Al (100 nm) was employed, as depicted in Figure 14.

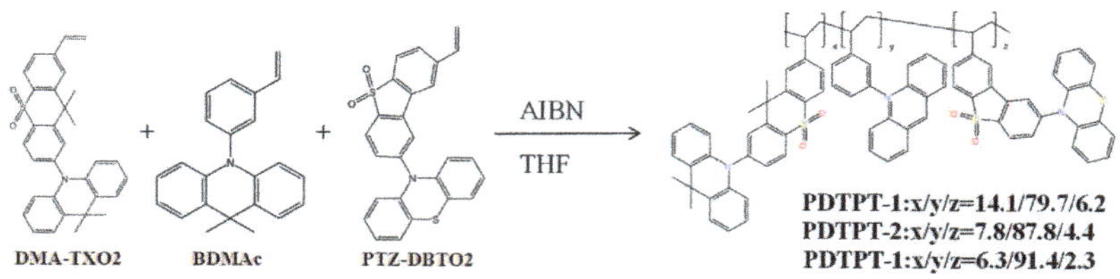

Figure 13. Molecular structures of polymers synthesized using DMA-TXO_2, BDMAc, and PVTZ-$DBTO_2$.

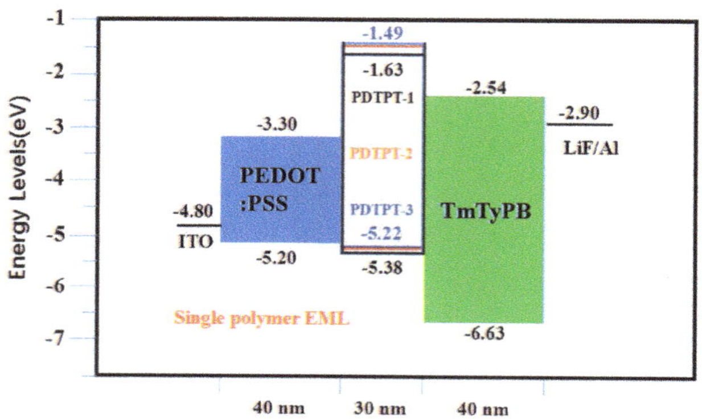

Figure 14. Schematic of the device structure diagram for a two-layer device configuration.

Notably, the radical copolymerization of DMA-TXO_2, BDMAc, and PTZ-$DBTO_2$ monomers yielded polymers with highly smooth and uniform film morphologies. These polymers hold great potential as EML materials in OLEDs. The optimized device structure, incorporating these polymers as the emitting layer, follows a dual-layer architecture and demonstrates promising prospects for achieving enhanced OLED performance.

The PDTPT-1-based device exhibited notable characteristics, including a 2900 cd/m^2 maximum brightness, a 38.8 cd/A maximum CE, a 20.3 lm/W maximum PE, and a (0.33, 0.42) CIE coordinates. The trap effect in the device arises from the host and adjacent layers having a significant energy gap [103], resulting in a large energy barrier for charge injection into the host layer. Additionally, the emitter possesses a deeper Lowest Unoccupied Molecular Orbital (LUMO) level and a shallower Highest Occupied Molecular Orbital (HUMO) level compared to the host, enabling direct charge injection and subsequent radiative decay in the emissive layer [104,105]. Notably, the HOMO level of BDMAc, an emitter in the optimized device, closely aligns with both the two emitters and the HOMO level of PEDOT: PSS, minimizing the energy barrier. In contrast, a substantial LUMO gap exists between BDMAc and the adjacent layer TmPyPB, while TMPYPB and DMA-TXO_2 or PGTZ-$DBTO_2$ exhibit only a small 0.1 eV barrier. Consequently, direct charge injection into the emissive layer is facilitated. The device offers several advantages:

1. The high T_1 and shallow HOMO level of BDMAC prevent triple energy back transfer (TEBT) from the blue and yellow emitters to the host. Moreover, it provides a minimal energy barrier for hole injection from the ITO/PEDOT:PSS anode into the polymer.

2. In comparison to DMA-TXO_2, DMA-TXO_2 retains 90% of the PLQY and the triple levels of 3.08 eV. However, the HOMO energy of DMA-TXO_2 is shifted from −6.10 eV to −5.50 eV, enabling better compatibility with the host.

3. The close physical contact of the PDTPT-1 intermolecular chains enhances the Förster resonance energy transfer (FRET) and DET processes, leading to efficient energy transfer. The trap-assisted recombination and Langevin recombination mechanisms ensure a balanced charge flux and a wide recombination zone, thereby reducing the formation of high-energy excitons on the host and minimizing exciton quenching. The utilization of dual-channel recombination contributes to enhanced device efficiency.

Clearly, the PDTPT-1 based device clearly demonstrates favorable characteristics, including high brightness, CE, and PE values. The trap effect, combined with optimized energy levels and intermolecular interactions, allows for efficient charge injection, reduced exciton quenching, and improved device performance.

5. Conclusions

In summary, this review discusses the performance of WOLEDs based on TADF in various configurations, including TADF-phosphorescence, TADF-fluorescence, and all-TADF devices. Among the different configurations, the efficiency of TADF-phosphorescent devices reaches 20.8%, while having good color stability and a high CRI [63]. The brightness of TADF-fluorescent devices is as high as 66,424 cd/m^2 while the CE of 67.6 cd/A is reached [76]. The host-free WOLED device composed of all TADF materials has the highest efficiency, reaching 24.2% EQE [101]. The TADF-phosphorescent host-free device reached 23.48%, with a low turn-on voltage of 2.5 V. By adjusting the thickness of the emitting layer, this device could emit white light with a tunable color temperature ranging from warm to cold. The all-TADF devices achieved superior efficiency, CE, and PE compared to solution-treated mixed WOLEDs. In some cases, they even approached the performance of all-phosphorescent devices. Notably, the host-free full-fluorescence optimized device achieved a record 130.7 lm/W PE and 31.1% EQE.

Although all phosphorescent WOLED structures can achieve high efficiency, but the high cost, low stability, and blue phosphorescent emitters remain significant challenges for large-scale applications. To solve this problem, researchers have explored the integration of complementary fluorescence or phosphorescence emitters with TADF in WOLEDs. Additionally, the solution treatment method has been investigated as an alternative to traditional thermal evaporation techniques, effectively addressing the dissolution issue between adjacent functional layers. However, further research is needed to optimize film design based on material and device requirements.

Over nearly three decades of research, WOLEDs have achieved remarkable advancements in operational longevity, luminous efficiency, device stability, and CRI. Fluorescent devices have demonstrated higher stability compared to all-TADF white devices. Nevertheless, all-TADF devices have rapidly advanced and now exhibit performance parameters comparable to phosphorescent WOLEDs, which have reached commercial viability. While phosphorescent materials remain costly, TADF-based white light devices offer promising alternatives. Despite current limitations in terms of cost, lifetime, and efficiency, considering the widespread adoption of OLED technology in displays and lighting, we can foresee continuous advancements in theoretical understanding, process development, and material research for WOLEDs. These advancements will help overcome technological barriers and ultimately benefit all of humanity.

Funding: This research was funded by the National Natural Science Foundation of China of funder (grant number 62374073), the program for the development of Science and Technology of Jilin province (grant number 20220101125JC, 20220101041JC, 20240101071JC, YDZ202301ZYTS278), National Key Research and Development Program (grant number 2022YFB3603802), Key Projects of Jilin Province Science and Technology Development Plan (grant number 20220201068GX) and the "Thirteenth five-year" science and technology research Project of Jilin province department of education (grant number JJKH20200420KJ).

Conflicts of Interest: The authors declare no conflict of interest.

References

1. Chen, B.; Liu, B.Q.; Zeng, J.J.; Nie, H.; Xiong, Y.; Zou, J.H.; Ning, H.L.; Wang, Z.M.; Zhao, Z.J.; Tang, B.Z. Efficient Bipc Blue AIEgens for High-Performance Nondoped Blue OLEDs and Hybrid White OLEDs. *Adv. Funct. Mater.* **2018**, *28*, 1803 [CrossRef]
2. Tang, X.; Liu, X.Y.; Jiang, Z.Q.; Liao, L.S. High-Quality White Organic Light-Emitting Diodes Composed of Binary Emitters w Color Rendering Index Exceeding 80 by Utilizing Color Remedy Strategy. *Adv. Funct. Mater.* **2019**, *29*, 1807541. [CrossRef]
3. Feng, C.; Zheng, X.J.; Xu, R.; Zhou, Y.G.; Hu, H.L.; Guo, T.L.; Ding, J.Q.; Ying, L.; Li, F.S. Highly efficient inkjet printed flexi organic light-emitting diodes with hybrid hole injection layer. *Org. Electron.* **2020**, *85*, 105822. [CrossRef]
4. Sun, Y.; Giebink, N.C.; Kanno, H.; Ma, B.; Thompson, M.E.; Forrest, S.R. Management of singlet and triplet excitons for effici white organic light-emitting devices. *Nature* **2006**, *440*, 908–912. [CrossRef] [PubMed]
5. Huang, Y.G.; Hsiang, E.L.; Deng, M.Y.; Wu, S.T. Mini-LED, Micro-LED and OLED displays: Present status and future perspecti *Light-Sci. Appl.* **2020**, *9*, 105. [CrossRef] [PubMed]
6. Sun, W.D.; Zhou, L.; Zhu, Q.; Wu, R.X.; Li, Z.Z.; Li, S.B.; Qin, D.S. Efficient multi-light-emitting layers warm and pure wh phosphorescent organic light-emitting diodes with excellent color stability. *J. Lumin.* **2020**, *228*, 117596. [CrossRef]
7. Hung, W.-Y.; Fang, G.-C.; Lin, S.-W.; Cheng, S.-H.; Wong, K.-T.; Kuo, Y.-Y.; Chou, P.-T. The First Tandem, All-exciplex-bas WOLED. *Sci. Rep.* **2014**, *4*, 5161. [CrossRef] [PubMed]
8. Yu, Y.; Lin, W.Y.; Peng, X.K.; Wu, Z.J.; Jin, Y.; Li, X.Y.; Zhang, X.N.; Yang, H.S.; Xie, G.H. Realizing an efficient warm white orga light-emitting device possessing excellent color-stability and color rendering index. *Org. Electron.* **2019**, *68*, 129–134. [CrossR
9. Li, X.K.; Liu, W.X.; Zhu, Q.; Wu, R.X.; Liu, G.J.; Zhou, L. High-performance full phosphorescent warm white organic light-emitt diodes with external quantum efficiency of 34.5%. *Opt. Mater.* **2022**, *124*, 112005. [CrossRef]
10. Zhang, C.; Lu, Y.; Liu, Z.Y.; Zhang, Y.W.; Wang, X.W.; Zhang, D.D.; Duan, L. A π-D and π-A Exciplex-Forming Host for Hi Efficiency and Long-Lifetime Single-Emissive-Layer Fluorescent White Organic Light-Emitting Diodes. *Adv. Mater.* **2020**, 2004040. [CrossRef]
11. Chen, K.; Wu, R.X.; Li, X.K.; Liu, W.Q.; Wei, Z.P.; Wang, J.H.; Zhou, L. High-quality all-fluorescent white organic light-emitti diodes obtained by balancing carriers with hole limit layer. *Opt. Mater.* **2022**, *123*, 1119117. [CrossRef]
12. Luo, W.; Wang, T.T.; Chen, X.; Tong, K.N.; He, W.; Sun, S.Q.; Zhang, Y.J.; Liao, L.S.; Fung, M.K. High-performance orga light-emitting diodes with natural white emission based on thermally activated delayed fluorescence emitters. *J. Mater. Chem* **2020**, *8*, 10431–10437. [CrossRef]
13. Zhang, M.; Zheng, C.J.; Wang, K.; Shi, Y.Z.; Yang, H.Y.; Lin, H.; Tao, S.L.; Zhang, X.H. Efficient and stable single-emitting-a white organic light-emitting diodes by employing all thermally activated delayed fluorescence emitters. *Org. Electron.* **2022**, *1* 106415. [CrossRef]
14. Wang, Q.; Zhang, Y.X.; Yuan, Y.; Hu, Y.; Tian, Q.S.; Jiang, Z.Q.; Liao, L.S. Alleviating Efficiency Roll-Off of Hybrid Single-Emitti Layer WOLED Utilizing Bipolar TADF Material as Host and Emitter. *Acs Appl. Mater. Interfaces* **2019**, *11*, 2197–2204. [CrossR
15. Du, X.Y.; Zhao, J.W.; Yuan, S.L.; Zheng, C.J.; Lin, H.; Tao, S.L.; Lee, C.S. High-performance fluorescent/phosphorescent (F hybrid white OLEDs consisting of a yellowish-green phosphorescent emitter. *J. Mater. Chem. C* **2016**, *4*, 5907–5913. [CrossRef
16. Wu, Z.B.; Yu, L.; Zhao, F.C.; Qiao, X.F.; Chen, J.S.; Ni, F.; Yang, C.L.; Ahamad, T.; Alshehri, S.M.; Ma, D.G. Precise Excit Allocation for Highly Efficient White Organic Light-Emitting Diodes with Low Efficiency Roll-Off Based on Blue Therma Activated Delayed Fluorescent Exciplex Emission. *Adv. Opt. Mater.* **2017**, *5*, 1100415. [CrossRef]
17. Xiao, P.; Dong, T.; Xie, J.N.; Luo, D.X.; Yuan, J.; Liu, B.Q. Emergence of White Organic Light-Emitting Diodes Based on Therma Activated Delayed Fluorescence. *Appl. Sci.* **2018**, *8*, 299. [CrossRef]
18. Baldo, M.A.; Lamansky, S.; Burrows, P.E.; Thompson, M.E.; Forrest, S.R. Very high-efficiency green organic light-emitting devi based on electrophosphorescence. *Appl. Phys. Lett.* **1999**, *75*, 4–6. [CrossRef]
19. Uoyama, H.; Goushi, K.; Shizu, K.; Nomura, H.; Adachi, C. Highly efficient organic light-emitting diodes from delay fluorescence. *Nature* **2012**, *492*, 234–248. [CrossRef]
20. Danyliv, I.; Ivaniuk, K.; Danyliv, Y.; Helzhynskyy, I.; Andruleviciene, V.; Volyniuk, D.; Stakhira, P.; Baryshnikov, G.V.; Grazule cius, J.V. Derivatives of 2-Pyridone Exhibiting Hot-Exciton TADF for Sky-Blue and White OLEDs. *Acs Appl. Electron. Mater.* **20** *5*, 4174–4186. [CrossRef]
21. Hu, B.; Lue, Z.Y.; Tang, Z.K.; Wu, Y.; Ji, W.Y.; Wang, J.L. Stabilized Huang-Rhys factor in non-doped exciplex-based Fir phosphorescent organic light emitting diodes via quantum well-like multiple emissive layer structure and investigation of carr behaviors by transient electroluminescence. *Org. Electron.* **2023**, *123*, 106919. [CrossRef]
22. Li, J.M.; Gong, P.; Han, D.M. Theoretical insight on the electronic structure and photophysical properties of three blue cyclome lated Ir(III) complexes based on the benchmark complex FIrpic. *Mol. Cryst. Liq. Cryst.* **2023**, *753*, 18–26. [CrossRef]
23. Tang, M.C.; Lee, C.H.; Lai, S.L.; Ng, M.; Chan, M.Y.; Yam, V.W.W. Versatile Design Strategy for Highly Luminescent Vacuu Evaporable and Solution-Processable Tridentate Gold(III) Complexes with Monoaryl Auxiliary Ligands and Their Applicatio for Phosphorescent Organic Light Emitting Devices. *J. Am. Chem. Soc.* **2017**, *139*, 9341–9349. [CrossRef] [PubMed]
24. Duan, Y.; Mazzeo, M.; Maiorano, V.; Mariano, F.; Qin, D.; Cingolani, R.; Gigli, G. Extremely low voltage and high bright p-i fluorescent white organic light-emitting diodes. *Appl. Phys. Lett.* **2008**, *92*, 113304. [CrossRef]
25. Khan, M.A.; Xu, W.; Cao, J.; Bai, Y.; Zhu, W.Q.; Jiang, X.Y.; Zhang, Z.L. Spectral studies of white organic light-emitting devic based on multi-emitting layers. *Displays* **2007**, *28*, 26–30. [CrossRef]

Rosenow, T.C.; Furno, M.; Reineke, S.; Olthof, S.; Lüssem, B.; Leo, K. Highly efficient white organic light-emitting diodes based on fluorescent blue emitters. *J. Appl. Phys.* **2010**, *108*, 113113. [CrossRef]
Xu, B.J.; Mu, Y.X.; Mao, Z.; Xie, Z.L.; Wu, H.Z.; Zhang, Y.; Jin, C.J.; Chi, Z.G.; Liu, S.W.; Xu, J.R.; et al. Achieving remarkable mechanochromism and white-light emission with thermally activated delayed fluorescence through the molecular heredity principle. *Chem. Sci.* **2016**, *7*, 2201–2206. [CrossRef] [PubMed]
Jeon, Y.P.; Kong, B.K.; Lee, E.J.; Yoo, K.-H.; Kim, T.W. Ultrahighly-efficient and pure deep-blue thermally activated delayed fluorescence organic light-emitting devices based on dimethylacridine/thioxanthene-S,S-dioxide. *Nano Energy* **2019**, *59*, 560–568. [CrossRef]
Yang, Z.Y.; Mao, Z.; Xie, Z.L.; Zhang, Y.; Liu, S.W.; Zhao, J.; Xu, J.R.; Chi, Z.G.; Aldred, M.P. Recent advances in organic thermally activated delayed fluorescence materials. *Chem. Soc. Rev.* **2017**, *46*, 915–1016. [CrossRef]
Xue, C.; Lin, H.; Zhang, G.; Hu, Y.; Jiang, W.L.; Lang, J.H.; Wang, D.D.; Xing, G.Z. Recent advances in thermally activated delayed fluorescence for white OLEDs applications. *J. Mater. Sci.-Mater. Electron.* **2020**, *31*, 4444–4462. [CrossRef]
Cai, X.Y.; Li, X.L.; Xie, G.Z.; He, Z.Z.; Gao, K.; Liu, K.K.; Chen, D.C.; Cao, Y.; Su, S.J. "Rate-limited effect" of reverse intersystem crossing process: The key for tuning thermally activated delayed fluorescence lifetime and efficiency roll-off of organic light emitting diodes. *Chem. Sci.* **2016**, *7*, 4264–4275. [CrossRef] [PubMed]
Luo, D.X.; Li, X.L.; Zhao, Y.; Gao, Y.; Liu, B.Q. High-Performance Blue Molecular Emitter-Free and Doping-Free Hybrid White Organic Light-Emitting Diodes: An Alternative Concept To Manipulate Charges and Excitons Based on Exciplex and Electroplex Emission. *Acs Photonics* **2017**, *4*, 1566–1575. [CrossRef]
Luo, D.X.; Chen, Q.Z.; Gao, Y.; Zhang, M.L.; Liu, B.Q. Extremely Simplified, High-Performance, and Doping-Free White Organic Light-Emitting Diodes Based on a Single Thermally Activated Delayed Fluorescent Emitter. *ACS Energy Lett.* **2018**, *3*, 1531–1538. [CrossRef]
Zhou, D.L.; Cheng, G.; Liu, W.Q.; Wu, S.P.; Che, C.M. Solution-processed single-emissive-layer WOLEDs with high efficiency and ultra-high color rendering index beyond 90. *J. Mater. Chem. C* **2023**, *11*, 3936–3943. [CrossRef]
Gao, T.; Wang, H.; Chen, Z.; Wang, Z.; Zhou, M.; Zhao, B.; Chen, F.; Zou, Q. Performance improved white full-organic light emitting diodes (WOLED) based on tricolor dual-layer of hyperfluorescence (HF). *Org. Electron.* **2023**, *120*, 106845. [CrossRef]
Pu, J.R.; Nie, X.W.; Li, D.H.; Peng, X.M.; Qiu, W.D.; Li, W.; Li, D.L.; Sun, G.W.; Shen, C.Y.; Ji, S.M.; et al. Multi-Sensitization Strategy for High Efficiency and Low Efficiency Roll-off Solution-Processed Single-Emission-Layer All-Fluorescence White Organic Light-Emitting Diodes. *Chem. Eng. J.* **2023**, *471*, 114508. [CrossRef]
Lee, J.H.; Chen, C.T. Hybrid white organic light-emitting diodes based on platinum complex. *J. Chin. Chem. Soc.* **2022**, *69*, 1276–1288. [CrossRef]
Han, H.; Hu, S.; Zhang, S.; Li, X.; Sun, H.; Chen, J.; Liu, B.; Liu, C.; Chen, W.; Zhang, Q. Achieving Solution-Processed Non-Doped Single-Emitting-Layer White Organic Light-Emitting Diodes through Adjusting Pyrene-Based Polyaromatic Hydrocarbon. *Chem.-A Eur. J.* **2022**, *28*, e202201741. [CrossRef]
Xue, C.; Zhang, G.; Jiang, W.; Lang, J.; Jiang, X. High performance non-doped blue-hazard-free hybrid white organic light-emitting diodes with stable high color rendering index and low efficiency roll-off. *Opt. Mater.* **2020**, *106*, 109991. [CrossRef]
Yang, F.; Kou, Z.; Yang, L.; Tang, Y. Influence of the hole transport layer on spectral stability in the white phosphorescent organic light emitting diode with non-doped structure. *Opt. Mater.* **2018**, *82*, 130–134. [CrossRef]
Huang, T.; Liu, D.; Li, D.; Jiang, W.; Jiang, J. Novel yellow thermally activated delayed fluorescence emitters for highly efficient full-TADF WOLEDs with low driving voltages and remarkable color stability. *New J. Chem.* **2019**, *43*, 13339–13348. [CrossRef]
Hua, J.; Zhan, Z.; Cheng, Z.; Cao, W.; Chai, Y.; Wang, X.; Wei, C.; Dong, H.; Wang, J. High-efficiency all-fluorescent white organic light-emitting diode based on TADF material as a sensitizer. *RSC Adv.* **2023**, *13*, 31632–31640. [CrossRef]
Wang, L.; Kou, Z.; Wang, B.; Zhou, J.; Lu, Z.; Li, L. Realizing high efficiency/CRI/color stability in the hybrid white organic light emitting diode by manipulating exciton energy transfer. *Opt. Mater.* **2021**, *115*, 111059. [CrossRef]
Yao, J.; Liu, W.; Lin, C.; Sun, Q.; Dai, Y.; Qiao, X.; Yang, D.; Chen, J.; Ma, D. High efficiency and long lifetime fluorescent white organic light-emitting diodes by phosphor sensitization to strategically manage singlet and triplet excitons. *J. Mater. Chem. C* **2021**, *9*, 3626–3634. [CrossRef]
Yao, J.; Xiao, S.; Zhang, S.; Sun, Q.; Dai, Y.; Qiao, X.; Yang, D.; Chen, J.; Ma, D. High efficiency, low efficiency roll-off and long lifetime fluorescent white organic light-emitting diodes based on strategic management of triplet excitons via triplet–triplet annihilation up-conversion and phosphor sensitization. *J. Mater. Chem. C* **2020**, *8*, 8077–8084. [CrossRef]
Lv, M.; Zhang, W.; Ning, Y.; Jing, W.; Song, D.; Zhao, S.; Xu, Z.; Fan, L.; Qiao, B. Interfacial Exciplex Host to Release Interfacial Accumulated Charges for Highly Efficient and Bright Solution-Processed White Organic Light-Emitting Diodes. *Adv. Mater. Interfaces* **2022**, *9*, 2200093. [CrossRef]
Chen, L.; Lv, J.; Wang, S.; Shao, S.; Wang, L. Dendritic Interfacial Exciplex Hosts for Solution-Processed TADF-OLEDs with Power Efficiency Approaching 100 lm W^{-1}. *Adv. Opt. Mater.* **2021**, *9*, 202100752. [CrossRef]
Li, Z.; Yu, J.; Liao, X.; Li, L. Flexible all fluorescence white organic light emitting device with over 22% EQE by stepped reverse intersystem crossing channels based on ternary exciplex. *Org. Electron.* **2021**, *91*, 106076. [CrossRef]
Chen, Y.; Zhang, D.D.; Zhang, Y.W.; Zeng, X.; Huang, T.Y.; Liu, Z.Y.; Li, G.M.; Duan, L. Approaching Nearly 40% External Quantum Efficiency in Organic Light Emitting Diodes Utilizing a Green Thermally Activated Delayed Fluorescence Emitter with an Extended Linear Donor-Acceptor-Donor Structure. *Adv. Mater.* **2021**, *33*, 2103293. [CrossRef]

50. Liu, H.; Fu, Y.; Tang, B.Z.; Zhao, Z.J. All-fluorescence white organic light-emitting diodes with record-beating power efficienc over 130 lm W^{-1} and small roll-offs. *Nat. Commun.* **2022**, *13*, 5154. [CrossRef]
51. Zhu, Z.L.; Chen, W.C.; Ni, S.F.; Yan, J.; Wang, S.F.; Fu, L.W.; Tsai, H.Y.; Chi, Y.; Lee, C.S. Constructing deep-blue bis-trident Ir(iii) phosphors with fluorene-based dianionic chelates. *J. Mater. Chem. C* **2021**, *9*, 1318–1325. [CrossRef]
52. Wu, Z.B.; Yu, L.; Zhou, X.K.; Guo, Q.X.; Luo, J.J.; Qiao, X.F.; Yang, D.Z.; Chen, J.S.; Yang, C.L.; Ma, D.G. Management of Sing and Triplet Excitons: A Universal Approach to High-Efficiency All Fluorescent WOLEDs with Reduced Efficiency Roll-Off Usi a Conventional Fluorescent Emitter. *Adv. Opt. Mater.* **2016**, *4*, 1067–1074. [CrossRef]
53. Miao, Y.Q.; Wei, X.Z.; Gao, L.; Wang, K.X.; Zhao, B.; Wang, Z.Q.; Zhao, B.; Wang, H.; Wu, Y.C.; Xu, B.S. Tandem white orga light-emitting diodes stacked with two symmetrical emitting units simultaneously achieving superior efficiency/CRI/co stability. *Nanophotonics* **2019**, *8*, 1783–1794. [CrossRef]
54. Chang, Y.Y.; Ren, Q.J.; Zhang, R.; Zhao, Y.; Wang, H.B. A low cost and simple structured WOLED device based on exciplex h and full fluorescent materials. *Solid-State Electron.* **2023**, *208*, 108753. [CrossRef]
55. Yin, M.N.; Wei, X.Z.; Miao, Y.Q.; Zhang, D.; Wang, G.L.; Xu, H.X.; Wang, H.; Chen, F. Combining complementary emissions hole- and electron-transport layers for ultra-simple white organic light-emitting diodes achieving high device performance *Lumin.* **2021**, *239*, 118343. [CrossRef]
56. Wang, Z.; Li, X.L.; Ma, Z.; Cai, X.; Cai, C.; Su, S.J. Exciton-Adjustable Interlayers for High Efficiency, Low Efficiency Roll-Off, a Lifetime Improved Warm White Organic Light-Emitting Diodes (WOLEDs) Based on a Delayed Fluorescence Assistant H *Adv. Funct. Mater.* **2018**, *28*, 1706922. [CrossRef]
57. Zhang, J.; Han, C.; Du, F.; Duan, C.; Wei, Y.; Xu, H. High-Power-Efficiency White Thermally Activated Delayed Fluorescer Diodes Based on Selectively Optimized Intermolecular Interactions. *Adv. Funct. Mater.* **2020**, *30*, 2005165. [CrossRef]
58. Li, Y.; Li, Z.; Zhang, J.; Han, C.; Duan, C.; Xu, H. Manipulating Complementarity of Binary White Thermally Activated Delay Fluorescence Systems for 100% Exciton Harvesting in OLEDs. *Adv. Funct. Mater.* **2021**, *31*, 2011169. [CrossRef]
59. Zhao, J.; Yang, Z.; Chen, X.; Xie, Z.; Liu, T.; Chi, Z.; Yang, Z.; Zhang, Y.; Aldred, M.P.; Chi, Z. Efficient triplet harvesting fluorescence–TADF hybrid warm-white organic light-emitting diodes with a fully non-doped device configuration. *J. Ma Chem. C* **2018**, *6*, 4257–4264. [CrossRef]
60. Hu, J.; Li, Q.; Wang, X.D.; Shao, S.Y.; Wang, L.X.; Jing, X.B.; Wang, F.S. Developing Through-Space Charge Transfer Polymers a General Approach to Realize Full-Color and White Emission with Thermally Activated Delayed Fluorescence. *Angew. Chem.- Ed.* **2019**, *58*, 8405–8409. [CrossRef]
61. Wu, Z.X.; Zhu, Y.J.; Wang, H.Z.; Wang, J.N.; He, Y. Methyl-modified Carbazole/Diphenyl Sulfone-basedAIE-TADF Blue Emit and Its OLEDs. *Chem. J. Chin. Univ.-Chin.* **2022**, *43*, 20220371. [CrossRef]
62. Ma, Z.W.; Wang, Y.Y.; Pu, J.R.; Li, G.Y.; Mao, Z.; Zhao, J.; Yang, Z.Y.; Su, S.J.; Chi, Z.G. Versatile Thermally Activated Delay Fluorescence Emitters via Non-conjugated Extension Strategy Enabling OLEDs with Efficiency Over 37%. *Adv. Opt. Mater.* **20** *2023*, 2302386. [CrossRef]
63. Khan, F.; Urbonas, E.; Volyniuk, D.; Grazulevicius, J.V.; Mobin, S.M.; Misra, R. White hyperelectrofluorescence from solutic processable OLEDs based on phenothiazine substituted tetraphenylethylene derivatives. *J. Mater. Chem. C* **2020**, *8*, 13375–133 [CrossRef]
64. Stanitska, M.; Mahmoudi, M.; Pokhodylo, N.; Lytvyn, R.; Volyniuk, D.; Tomkeviciene, A.; Keruckiene, R.; Obushak, M Grazulevicius, J.V. Exciplex-Forming Systems of Physically Mixed and Covalently Bonded Benzoyl-1*H*-1,2,3-Triazole a Carbazole Moieties for Solution-Processed White OLEDs. *J. Org. Chem.* **2022**, *87*, 4040–4050. [CrossRef] [PubMed]
65. Panoy, P.; Prakanpo, N.; Chasing, P.; Chatanop, N.; Sudyoadsuk, T.; Promarak, V. Stable Spiro-Linked Terfluorenes wi Triphenylamine Pendants as Deep Blue Emitters for Solution-Processed Organic Light-Emitting Diodes. *Chemistryselect* **2023**, e202302865. [CrossRef]
66. Sun, M.Z.; Ma, C.L.; Xie, M.L.; Chu, L.Z.; Wang, X.; Sun, Q.K.; Yang, W.J.; Xue, S.F. An efficient blue electro-fluorescence mater with high electron and balanced carrier mobilities based on effective p-stacking between acceptors. *Org. Chem. Front.* **2023**, 4878–4886. [CrossRef]
67. Zhan, X.J.; Wu, Z.B.; Lin, Y.X.; Xie, Y.J.; Peng, Q.; Li, Q.Q.; Ma, D.G.; Li, Z. Benzene-cored AIEgens for deep-blue OLEDs: Hig performance without hole-transporting layers, and unexpected excellent host for orange emission as a side-effect. *Chem. Sci.* **20** *7*, 4355–4363. [CrossRef] [PubMed]
68. Reineke, S.; Seidler, N.; Yost, S.R.; Prins, F.; Tisdale, W.A.; Baldo, M.A. Highly efficient, dual state emission from an orga semiconductor. *Appl. Phys. Lett.* **2013**, *103*, 093302. [CrossRef]
69. Baldo, M.A.; O'Brien, D.F.; You, Y.; Shoustikov, A.; Sibley, S.; Thompson, M.E.; Forrest, S.R. Highly efficient phosphoresce emission from organic electroluminescent devices. *Nature* **1998**, *395*, 151–154. [CrossRef]
70. Sato, K.; Shizu, K.; Yoshimura, K.; Kawada, A.; Miyazaki, H.; Adachi, C. Organic Luminescent Molecule with Energetica Equivalent Singlet and Triplet Excited States for Organic Light-Emitting Diodes. *Phys. Rev. Lett.* **2013**, *110*, 247401. [CrossRef]
71. Udagawa, K.; Sasabe, H.; Cai, C.; Kido, J. Low-Driving-Voltage Blue Phosphorescent Organic Light-Emitting Devices wi External Quantum Efficiency of 30%. *Adv. Mater.* **2014**, *26*, 5062–5066. [CrossRef] [PubMed]
72. Zhang, D.D.; Cai, M.H.; Zhang, Y.G.; Zhang, D.Q.; Duan, L. Highly Efficient Simplified Single-Emitting-Layer Hybrid WOLE with Low Roll-off and Good Color Stability through Enhanced Forster Energy Transfer. *Acs Appl. Mater. Interfaces* **2015**, 28693–28700. [CrossRef] [PubMed]

Liang, J.; Li, C.L.; Zhuang, X.M.; Ye, K.Q.; Liu, Y.; Wang, Y. Novel Blue Bipolar Thermally Activated Delayed Fluorescence Material as Host Emitter for High-Efficiency Hybrid Warm-White OLEDs with Stable High Color-Rendering Index. *Adv. Funct. Mater.* **2018**, *28*, 1707002. [CrossRef]
Zhao, J.; Wang, Z.J.; Wang, R.; Chi, Z.G.; Yu, J.S. Hybrid white organic light-emitting devices consisting of a non-doped thermally activated delayed fluorescent emitter and an ultrathin phosphorescent emitter. *J. Lumin.* **2017**, *184*, 287–292. [CrossRef]
Tan, J.-H.; Chen, W.-C.; Ni, S.-F.; Qiu, Z.; Zhan, Y.; Yang, Z.; Xiong, J.; Cao, C.; Huo, Y.; Lee, C.-S. Aggregation-state engineering and emission switching in D-A-D′ AIEgens featuring dual emission, MCL and white electroluminescence. *J. Mater. Chem. C* **2020**, *8*, 8061–8068. [CrossRef]
Wang, C.W.; Yang, N.; Fang, X.J.; Tian, Q.Y.; Zhang, J.C.; Fan, X.D.; Lan, B.F.; Wu, X.M.; Chu, W.Y.; Sun, Z.Z.; et al. Design and synthesis of D-A-D′ type compounds with AIE characteristics for hybrid WOLED with stable chromaticity. *Dyes Pigments* **2022**, *205*, 110511. [CrossRef]
Liao, X.Q.; An, K.G.; Cheng, J.; Li, Y.; Meng, X.; Yang, X.; Li, L. High efficiency ultra-thin doping-free WOLED based on blue thermally activated delayed fluorescence inter-layer switch. *Appl. Surf. Sci.* **2019**, *487*, 610–615. [CrossRef]
Miao, Y.Q.; Wang, G.L.; Yin, M.N.; Guo, Y.Y.; Zhao, B.; Wang, H. Phosphorescent ultrathin emitting layers sensitized by TADF interfacial exciplex enables simple and efficient monochrome/white organic light-emitting diodes. *Chem. Eng. J.* **2023**, *461*, 141921. [CrossRef]
Wang, Y.X.; Bai, X.C.; Zhang, Y.; Li, G.Q. Influence of high-frequency localized surface plasmonpolariton effect of Al nanoparticles on luminescenceefficiency of deep-blue BCzVBi OLED. *Acta Phys. Sin.* **2024**, *73*, 037802. [CrossRef]
Tankeleviciute, E.; Samuel, I.D.W.; Zysman-Colman, E. The Blue Problem: OLED Stability and Degradation Mechanisms. *J. Phys. Chem. Lett.* **2024**, *15*, 1034–1047. [CrossRef]
Ying, S.A.; Xiao, S.; Yao, J.W.; Sun, Q.O.; Doi, Y.F.; Yang, D.Z.; Qiao, X.F.; Chen, J.S.; Zhu, T.F.; Ma, D.G. High-Performance White Organic Light-Emitting Diodes with High Efficiency, Low Efficiency Roll-Off, and Superior Color Stability/Color Rendering Index by Strategic Design of Exciplex Hosts. *Adv. Opt. Mater.* **2019**, *7*, 1901291. [CrossRef]
Huang, C.C.; Zhang, Y.J.; Zhou, J.G.; Sun, S.Q.; Luo, W.; He, W.; Wang, J.N.; Shi, X.B.; Fung, M.K. Hybrid Tandem White OLED with Long Lifetime and 150 Lm W^{-1} in Luminous Efficacy Based on TADF Blue Emitter Stabilized with Phosphorescent Red Emitter. *Adv. Opt. Mater.* **2020**, *8*, 2000727. [CrossRef]
Zhao, C.Y.; Zhang, T.M.; Chen, J.S.; Yan, D.H.; Ma, D.G. High-performance hybrid white organic light-emitting diodes with simple emitting structures and low efficiency roll-off based on blue thermally activated delayed fluorescence emitters with bipolar transport characteristics. *J. Mater. Chem. C* **2018**, *6*, 9510–9516. [CrossRef]
Wu, Z.; Ma, D. Recent advances in white organic light-emitting diodes. *Mater. Sci. Eng. R Rep.* **2016**, *107*, 1–42. [CrossRef]
Banevicius, D.; Puidokas, G.; Kreiza, G.; Jursenas, S.; Orentas, E.; Kazlauskas, K. Prolonging blue TADF-OLED lifetime through ytterbium doping of electron transport layer. *J. Ind. Eng. Chem.* **2023**, *128*, 515–520. [CrossRef]
Xue, K.W.; Han, G.G.; Duan, Y.; Chen, P.; Yang, Y.Q.; Yang, D.; Duan, Y.H.; Wang, X.; Zhao, Y. Doping-free orange and white phosphorescent organic light-emitting diodes with ultra-simply structure and excellent color stability. *Org. Electron.* **2015**, *18*, 84–88. [CrossRef]
Zhao, L.A.; Zhang, W.X.; Zhang, Y.Y.; Wang, L.Y.; Zhang, G.; Jiang, W.L.; Lang, J.H. High performance doping-free hybrid WOLED utilizing bipolar TADF material as emitters with an ultra-low efficiency roll-off. *Opt. Mater.* **2022**, *134*, 113067. [CrossRef]
Wang, Z.H.; Su, S.J. Molecular and Device Design Strategies for Ideal Performance White Organic Light-Emitting Diodes. *Chem. Rec.* **2019**, *19*, 1518–1530. [CrossRef] [PubMed]
Zhou, N.L.; Wang, S.R.; Xiao, Y.; Li, X.G. A Novel *trans*-1-(9-Anthryl)-2-phenylethene Derivative Containing a Phenanthroimidazole Unit for Application in Organic Light-Emitting Diodes. *Chem.-Asian J.* **2018**, *13*, 81–88. [CrossRef]
Mahmoudi, M.; Keruckas, J.; Volyniuk, D.; Andruleviciene, V.; Keruckiene, R.; Narbutaitis, E.; Chao, Y.C.; Rutkis, M.; Grazulevicius, J.V. Bis(N-naphthyl-N-phenylamino)benzophenones as exciton-modulating materials for white TADF OLEDs with separated charge and exciton recombination zones. *Dyes Pigments* **2022**, *197*, 109868. [CrossRef]
Sych, G.; Guzauskas, M.; Volyniuk, D.; Simokaitiene, J.; Starykov, H.; Grazulevicius, J.V. Exciplex energy transfer through spacer: White electroluminescence with enhanced stability based on cyan intermolecular and orange intramolecular thermally activated delayed fluorescence. *J. Adv. Res.* **2020**, *24*, 379–389. [CrossRef] [PubMed]
Dong, R.Z.; Liu, D.; Li, J.Y.; Ma, M.Y.; Mei, Y.Q.; Li, D.L.; Jiang, J.Y. Acceptor modulation for blue and yellow TADF materials and fabrication of all-TADF white OLED. *Mater. Chem. Front.* **2021**, *6*, 40–51. [CrossRef]
Zhang, D.; Cao, X.D.; Wu, Q.J.; Zhang, M.C.; Sun, N.; Zhang, X.P.; Tao, Y.T. Purely organic materials for extremely simple all-TADF white OLEDs: A new carbazole/oxadiazole hybrid material as a dual-role non-doped light blue emitter and highly efficient orange host. *J. Mater. Chem. C* **2018**, *6*, 3675–3682. [CrossRef]
Ma, M.Y.; Li, J.Y.; Liu, D.; Mei, Y.Q.; Dong, R.Z. Rational Utilization of Intramolecular Hydrogen Bonds to Achieve Blue TADF with EQEs of Nearly 30% and Single Emissive Layer All-TADF WOLED. *Acs Appl. Mater. Interfaces* **2021**, *13*, 44615–44627. [CrossRef] [PubMed]
Chen, X.K.; Kim, D.; Brédas, J.L. Thermally Activated Delayed Fluorescence (TADF) Path toward Efficient Electroluminescence in Purely Organic Materials: Molecular Level Insight. *Acc. Chem. Res.* **2018**, *51*, 2215–2224. [CrossRef] [PubMed]
Teng, J.M.; Wang, Y.F.; Chen, C.F. Recent progress of narrowband TADF emitters and their applications in OLEDs. *J. Mater. Chem. C* **2020**, *8*, 11340–11353. [CrossRef]

97. Zhang, Q.; Li, B.; Huang, S.; Nomura, H.; Tanaka, H.; Adachi, C. Efficient blue organic light-emitting diodes employing therm activated delayed fluorescence. *Nat. Photonics* **2014**, *8*, 326–332. [CrossRef]
98. Im, Y.; Kim, M.; Cho, Y.J.; Seo, J.A.; Yook, K.S.; Lee, J.Y. Molecular Design Strategy of Organic Thermally Activated Dela Fluorescence Emitters. *Chem. Mater.* **2017**, *29*, 1946–1963. [CrossRef]
99. Liu, H.; Liu, F.T.; Lu, P. Multiple strategies towards high-efficiency white organic light-emitting diodes by the vacuum deposit method. *J. Mater. Chem. C* **2020**, *8*, 5636–5661. [CrossRef]
100. Xie, F.M.; Zou, S.J.; Li, Y.Q.; Lu, L.Y.; Yang, R.; Zeng, X.Y.; Zhang, G.H.; Chen, J.D.; Tang, J.X. Management of Delayed Fluorop Sensitized Exciton Harvesting for Stable and Efficient All-Fluorescent White Organic Light-Emitting Diodes. *Acs Appl. M Interfaces* **2020**, *12*, 16736–16742. [CrossRef]
101. Zhang, G.H.; Xie, F.M.; Wu, K.; Li, Y.Q.; Xie, G.; Zou, S.J.; Shen, Y.; Zhao, X.; Yang, C.; Tang, J.X. High-Efficiency White Orga Light-Emitting Diodes Based on All Nondoped Thermally Activated Delayed Fluorescence Emitters. *Adv. Mater. Interfaces* 2 *7*, 1901758. [CrossRef]
102. Li, C.S.; Xu, Y.W.; Liu, Y.C.; Ren, Z.J.; Ma, Y.G.; Yan, S.K. Highly efficient white-emitting thermally activated delayed fluoresce polymers: Synthesis, non-doped white OLEDs and electroluminescent mechanism. *Nano Energy* **2019**, *65*, 104057. [CrossRef]
103. Wen, L.L.; Zang, C.X.; Gao, Y.; Li, G.F.; Shan, G.G.; Wang, X.L.; Shao, K.Z.; Xie, W.F.; Su, Z.M. Rational Design of Ir(III) Phosph to Strategically Manage Charge Recombination for High-Performance White Organic Light-Emitting Diodes. *Inorg. Chem.* 2 *61*, 3736–3745. [CrossRef] [PubMed]
104. Kanno, H.; Holmes, R.J.; Sun, Y.; Kena-Cohen, S.; Forrest, S.R. White Stacked Electrophosphorescent Organic Light-Emitt Devices Employing MoO3 as a Charge-Generation Layer. *Adv. Mater.* **2006**, *18*, 339–342. [CrossRef]
105. Qu, B.N.; Zhao, H.N.; Forrest, S.R. Modeling the Charge and Exciton Distributions in Phosphorescent White Organic Lig Emitting Diodes. *Acs Photonics* **2023**, *10*, 3042–3051. [CrossRef]

micromachines

MDPI

view

trategies for Enhancing the Stability of Lithium Metal Anodes n Solid-State Electrolytes

nbyeol Lee [1], Taeho Yoon [2,*] and Oh B. Chae [1,*]

[1] School of Chemical, Biological and Battery Engineering, Gachon University, Seongnam-si 13120, Republic of Korea; byulbbang99@gachon.ac.kr
[2] Department of Chemical Engineering, Kyung Hee University, Yongin-si 17104, Republic of Korea
* Correspondence: tyoon@khu.ac.kr (T.Y.), obchae@gachon.ac.kr (O.B.C.)

Abstract: The current commercially used anode material, graphite, has a theoretical capacity of only 372 mAh/g, leading to a relatively low energy density. Lithium (Li) metal is a promising candidate as an anode for enhancing energy density; however, challenges related to safety and performance arise due to Li's dendritic growth, which needs to be addressed. Owing to these critical issues in Li metal batteries, all-solid-state lithium-ion batteries (ASSLIBs) have attracted considerable interest due to their superior energy density and enhanced safety features. Among the key components of ASSLIBs, solid-state electrolytes (SSEs) play a vital role in determining their overall performance. Various types of SSEs, including sulfides, oxides, and polymers, have been extensively investigated for Li metal anodes. Sulfide SSEs have demonstrated high ion conductivity; however, dendrite formation and a limited electrochemical window hinder the commercialization of ASSLIBs due to safety concerns. Conversely, oxide SSEs exhibit a wide electrochemical window, but compatibility issues with Li metal lead to interfacial resistance problems. Polymer SSEs have the advantage of flexibility; however their limited ion conductivity poses challenges for commercialization. This review aims to provide an overview of the distinctive characteristics and inherent challenges associated with each SSE type for Li metal anodes while also proposing potential pathways for future enhancements based on prior research findings.

Keywords: Li-ion batteries; Li metal; energy density; solid-state electrolytes

1. Introduction

Since the commercialization of lithium (Li)-ion batteries (LIBs) in the 1990s, they have been widely applied in various fields, such as small electronic devices, Electric Vehicles (EVs), and Energy Storage Systems (ESSs), owing to their high energy density and stable lifespan [1–4]. Graphite, which is mainly used as the anode material in current LIBs, demonstrates the reversible intercalation/de-intercalation of Li ions between graphene layers, providing stable charge and discharge cycles. However, the theoretical capacity of graphite is limited to approximately 372 mAh/g, posing challenges for high-capacity energy storage [5–8]. Transition metal oxides (e.g., MnO_2, ZnO, Fe_2O_3, and SnO_2) also have gained widespread use as electrode materials in the field of LIBs due to their more substantial theoretical capacities than graphite. Nevertheless, poor cycling performance owing to low electric conductivity and structural collapse as result of large volume expansion during lithiation/delithiation results in poor rate performance and a limited lifespan for LIBs [9]. To overcome this limitation, researchers worldwide are exploring next-generation anode materials for energy storage, including silicon and Li metal.

As Li metal is the lightest member of the alkali metal group and has the lowest reduction voltage (−3.04 V vs. standard hydrogen electrode), extensive research has been reported on highly promising next-generation anode materials [10–12]. However, the use of Li metal in liquid electrolytes faces many challenges owing to its high reactivity to nonuniform Li dendrite formation, which results in hazardous safety concerns (e.g., fire and explosion). Most of the graphite surface is covered by the reductive decomposition products of electrolyte components created during the initial few cycles. In the case of graphite, owing to the small volume change during the charge and discharge cycle, solid electrolyte interface (SEI) crack generation is mitigated, which minimizes the exposure of the fresh graphite surface during cycling, preventing continuous electrolyte decomposition. On the contrary, the initial SEI of Li metal is generated as soon as the Li metal is exposed to the electrolyte since the electrochemical potential of Li metal is lower than the potential window for the electrolyte components [13–15]. The uncontrolled Li deposition causes the inhomogeneous exposure of fresh Li metal, resulting in low cycling performance and Li dendrite formation.

To overcome the critical issues associated with Li metal, various attempts have been made to suppress Li dendrite growth by using electrolyte additives [16], forming a protective layer [17], and changing the Li salt [18]. However, these approaches present notable challenges in the practical implementation of Li metal batteries. Incorporating solid-state electrolytes (SSEs) into all-solid-state lithium-ion batteries (ASSLIBs) has promise as a method to replace traditional liquid electrolytes. This transition in physical composition suppresses dendrite growth, with SSEs serving dual roles as separators [19,20]. Therefore, when Li metal meets SSEs, the convergence of Li metal with SSEs offers the potential for high energy density and exceptional safety in ASSLIBs.

To realize highly promising ASSLIBs, SSEs should satisfy the following requirements: (1) high ion conductivity at room temperature, (2) a wide electrochemical window, and (3) high chemical compatibility with the Li metal and cathode [21]. They are typically classified into two main categories: inorganic and polymeric forms. Inorganic SSEs can be further categorized into sulfide and oxide types. Understanding the electrochemical characteristics and properties of these three materials is crucial owing to their differences. This review focuses on examining the characteristics and challenges associated with sulfide, oxide, and polymer SSEs when paired with lithium metal, in addition to potential solutions based on previous research.

2. Sulfide SSEs

Sulfide SSEs have demonstrated superior ionic conductivities compared with other solid electrolytes. The high ion conductivity of sulfide SSEs, which is similar to that of organic liquid electrolytes, makes them promising electrolytes for solid-state batteries ($Li_{10}GeP_2S_{12}$, 1.2×10^{-3} S cm^{-1} [22]; Li_2S-P_2S_5, 1.7×10^{-2} S cm^{-1} [23]). In recent years, Kanno et al. doped the $Li_{10}GeP_2S_{12}$-type structure LSiPSBrO and obtained $Li_{9.54}[Si_{0.6}Ge_{0.4}]_{1.74}P_{1.44}S_{11.1}Br_{0.3}O_{0.6}$, the bulk ion conductivity at room temperature of which was determined to be 3.2×10^{-2} S cm^{-1} [24]. Han et al. demonstrated a different type of sulfide SSE ($Li_7Si_2S_7I$) created using an ordering of sulfide and iodide [25]. It built a fast transport path of diverse lithium coordination geometries and anion neighbors (1.01×10^{-2} S cm^{-1}). However, the inherent limitations of sulfide SSEs, such as insufficient dendrite suppression, pose challenges, leading to potential short circuits [26]. Furthermore, their narrow electrochemical window can lead to various reactions with Li metal [27,28], emphasizing the importance of understanding dendrite growth mechanisms and interfacial reactions.

2.1. Challenges of Sulfide SSEs for Li Metal

In liquid electrolytes, Li-ion deposition occurs uniformly owing to consistent contact with Li metal, resulting in an even current density and ion deposition. In contrast, sulfide SSEs exhibit a localized high current density and Li-ion deposition owing to point contact with the Li metal electrode. Consequently, Li dendrite growth and crack formation along crystal boundaries have been reported (Figure 1) [29]. Porz et al. revealed that Li was preferentially deposited on existing cracks or defects, inducing crack-tip stresses and propagation beyond the critical current density [30]. Zhang et al. demonstrated that alloy materials such as lithium–indium (Li-In) alloys are widely used at the laboratory scale because of their (electro)chemical stability; however, Li-In dendrites grow along the pores and grain boundaries, with high compactness of the Li_6PS_5Cl SSE layer during the charge/discharge process [31].

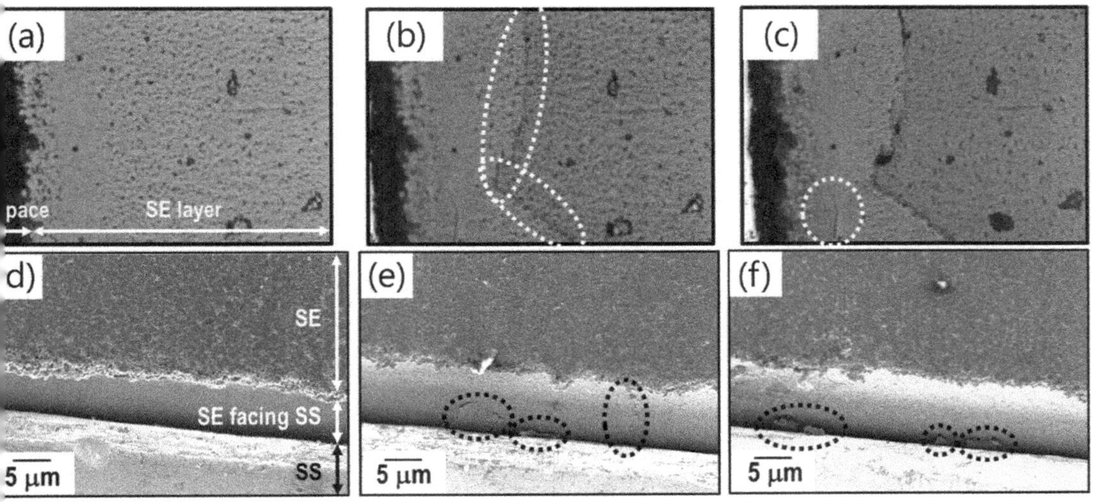

Figure 1. Optical microscope images of the solid electrolyte (SE) layer and the interface between SE and stainless steel (SS): (**a**) before the test; (**b**) Li deposition at the interface at 2 mA cm^{-2} for 730 s and (**c**) at 100 mA cm^{-2} for 2 h. SEM images of the interface between the SE layer and SS at different positions on the same cell (**d**) before and (**e**) after Li deposition for 600 s and (**f**) 1920 s [29]. White arrows showed cracks formed during the charge and discharge process, while black arrows showed inhomogeneous plating of Li. Reprinted with permission from Ref. [29]. Copyright 2013 Royal Society of Chemistry.

In addition, the strong reducing power of Li metal can trigger various reactions at the Li metal–sulfide SSE interface (Figure 2a,b) [27]. In Figure 2a,b, for example, the electrochemical windows of $Li_{10}GeP_2S_{12}$ and Li_3PS_4 are narrow compared to other SSEs. Thus, $Li_{10}GeP_2S_{12}$ and Li_3PS_4 begin to be lithiated and reduced at 1.71 V. Computational studies by Forero et al. using the density functional theory (DFT) and ab initio molecular dynamics (AIMD) demonstrated that sulfide SSE anions (such as PS_4^{3-}, $P_2S_6^{2-}$, $P_2S_7^{4-}$, and GeS_4^{4-}) react with Li to form Li-S, Li-P, and Li-Ge species (Figure 2c–e) [32]. These studies have demonstrated that sulfide SSEs generate byproducts such as Li_3P, Li_2S, and $Li_{15}Ge_4$ during discharge to 0 V, which causes the decomposition of SSEs. Wenzel et al. used XPS to identify compounds formed in sulfide-based SSEs, with $Li_7P_3S_{11}$ exhibiting the lowest resistance among the studied SSEs, resembling the values found in liquid electrolytes [33]. Argyrodite-type sulfide-based SSEs produce SEI layers comprising Li_3P, Li_2S, and decomposed LiX (X = Cl or Br) [34].

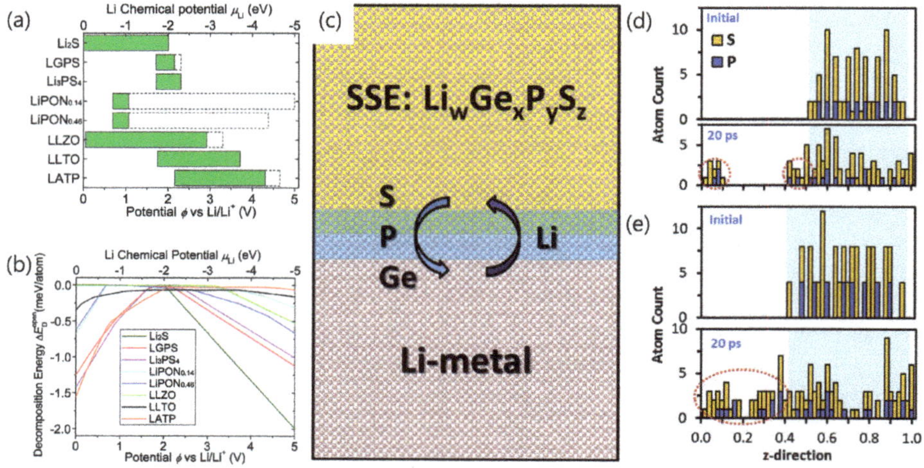

Figure 2. (**a**) Electrochemical window and (**b**) decomposition energy [27]. Reprinted with permission from Ref. [27]. Copyright 2013 Royal Society of Chemistry. (**c**) Schematic representation of interfacial phenomena between sulfide SSEs and Li metal anode. Sulfur and phosphorus atom profiles along z-direction for (**d**) $Li_7P_3S_{11}$-(100)/Li-slab and (**e**) $Li_2P_2S_6$-(100)/Li slab [32]. At 20 ps, some peaks moved outside the SSE space, which means that some S and P are present at SSE/Li metal interface. Reprinted/adapted with permission from Ref. [32]. Copyright 2018 ELSEVIER.

2.2. Strategies to Overcome Disadvantages of Sulfide SSEs for Li Metal

Wang et al. used a LiF (or LiI) layer at the interface between Li metal and a sulfide SSE and penetrated the sulfide SSE with methoxyperfluorobutane (HFE) (or an I solution) to suppress Li dendrite growth [21]. Li dendrite formation in sulfide SSEs is closely related to the interface stability of the electrolyte with Li metal. The LiF and LiI layers were electrochemically stable in both solid electrolytes and Li metal. The LiF (or LiI) interface layer ensured stability/compatibility at the Li metal/sulfide SSE interface and simultaneously inhibited Li's dendritic growth (Figure 3a,b). Consequently, the interfacial resistance of the cells using LiF- and LiI-coated Li metals was effectively reduced (Figure 3c).

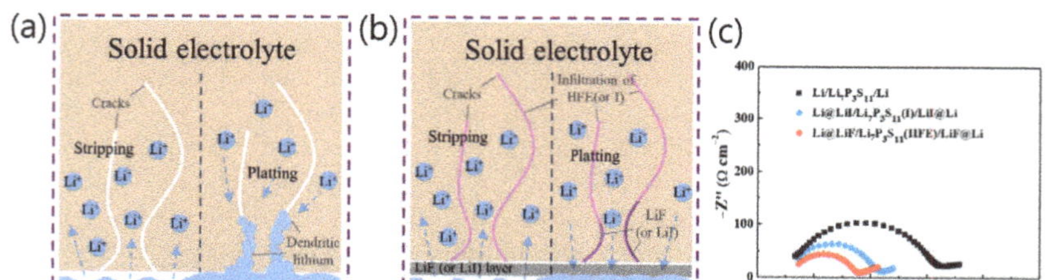

Figure 3. Schematic diagrams of the Li stripping and plating behavior of (**a**) bare Li metal with an untreated SSE and (**b**) LiF- or LiI-coated Li metal with HFE- or I-infiltrated electrolyte. (**c**) Nyquist plots of symmetric cells [21]. Reprinted with permission from Ref. [21]. Copyright 2018 ELSEVIER.

To mitigate dendrite formation and other undesirable reactions, it is crucial to minimize sub-reactions between Li metal and SSEs and SEI formation. The introduction of an artificial SEI has shown promise for inhibiting dendrite growth and ensuring uniform Li deposition. Connell et al. utilized Atomic Layer Deposition (ALD) to create a coating layer on Li_6PS_5Cl sulfide-based SSE powders, resulting in improved stability and a two-fold increase in ion conductivity (Figure 4) [35]. In addition, ALD is a method of depositing thin films which relies on sequential gas-phase chemical processes. It falls within the broader category of Chemical Vapor Deposition (CVD). The enhanced ion conductivity was attributed to a decrease in the Arrhenius activation energy owing to the coating. Additionally, optimizing the coating thickness is essential to prevent hindered Li^+ conductivity caused by excessively thick Al_2O_3 layers.

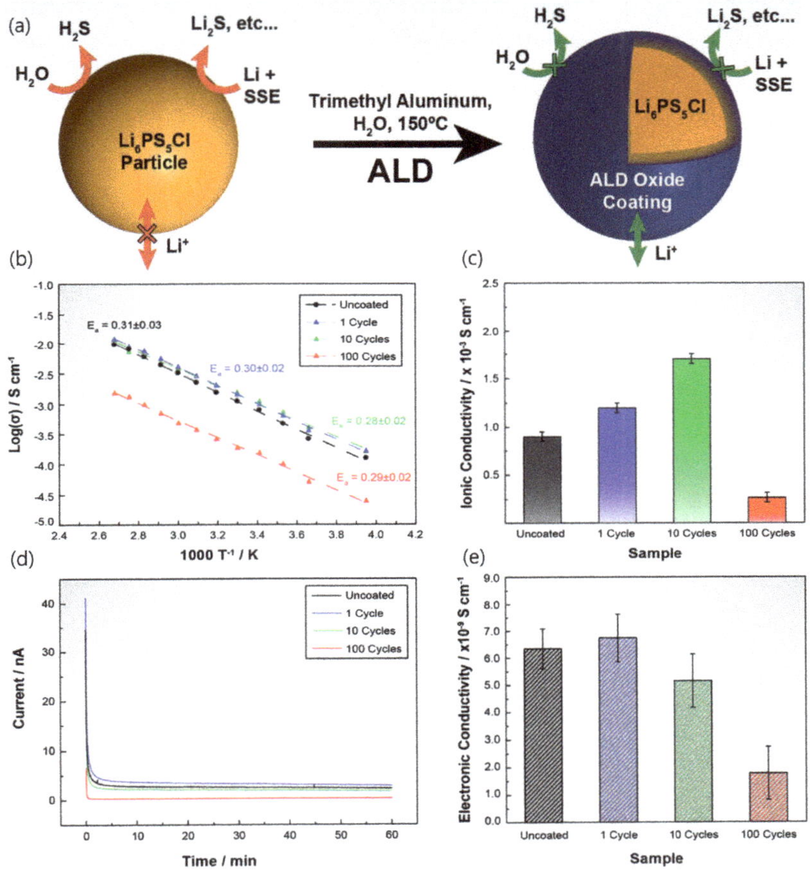

Figure 4. (**a**) Schematic of the coating strategy based on ALD to produce oxide−coated Li_6PS_5Cl powders. (**b**) Arrhenius plots, (**c**) ionic conductivity at 25 °C, (**d**) current versus time (DC polarization at 200 mV; 25 °C), and (**e**) electronic conductivity at 25 °C for Li_6PS_5Cl pellets pressed from powders coated by 1, 10, and 100 ALD alumina cycles in comparison to pellets pressed from uncoated powders [35]. Reprinted with permission from Ref. [35]. Copyright 2023 Wiley-VCH.

Furthermore, altering the composition of sulfide SSEs is an effective strategy. Composition tuning has proven to be effective in enhancing stability and controlling decomposition products, thereby preventing material propagation and reducing interfacial resistance. Sulfide SSE tuning is categorized into cation substitution and anion substitution, depending on the atoms comprising the structure. In a previous study, research was conducted to improve the performance of sulfide SSEs through cation substitution. Substituting a cation (e.g., Ge with Si, Sn, Al, P, Ba, Zn, or Y in LGPS) can adjust the stability of sulfide SSEs with minimal impact on material properties such as ion conductivity, as indicated by first-principles calculations [36]. However, this study demonstrates that Si, Sn, and Ge are also unstable versus Li and tend to form non-passivating degradation products on the Li metal. Research has also been conducted on anion substitution in addition to cation substitution. The partial substitution of an anion, especially sulfur by oxygen, has been shown to increase the thermodynamic stability of $Li_{10}GeP_2S_{12}$ by providing stronger chemical bonding, and oxygen substitution was recently used to stabilize the $Li_{10}GeP_2S_{12}$-phase of Li_3PS_4 [37,38].

3. Oxide SSEs

Oxide SSEs exhibit substantial energy gaps between their valence and conduction bands which provide enhanced stability at elevated voltages. In addition, the ionic mobility of oxide SSEs surpasses that of polymer electrolytes. Among the various oxide SSEs, garnet-type SSEs have the highest ion conductivity. The garnet-type $Li_7La_3Zr_2O_{12}$ (LLZO) oxide SSE synthesized by Weppner et al. is particularly attractive because of its good stability against Li metal, large electrochemical window, and easy handling in ambient environments [39–41]. The LLZO SSEs exhibit the best resistance to Li reduction and have the lowest reduction potential (approximately 0.05 V vs. Li/Li^+) and the most stable interface.

3.1. Challenges of Oxide SSEs for Li Metal

Several studies have shown that the absorbed CO_2 reacts directly with Li in the garnet structure to form Li_2CO_3 (Equation (1)) [42]. The Li_2CO_3 on the surface of the Li metal increases the interfacial resistance and decreases the ion conductivity (Figure 5) [43]. Cheng et al. reported that the large impedance at the LLZO/Li metal interface is not inherent to the material; however, it originates from the Li_2CO_3 due to ambient air exposure [44].

$$Li_7La_3Zr_2O_{12} + x \cdot CO_2 \rightarrow Li_{7-2x}La_3Zr_2O_{12} + x \cdot Li_2CO_3 \quad (1)$$

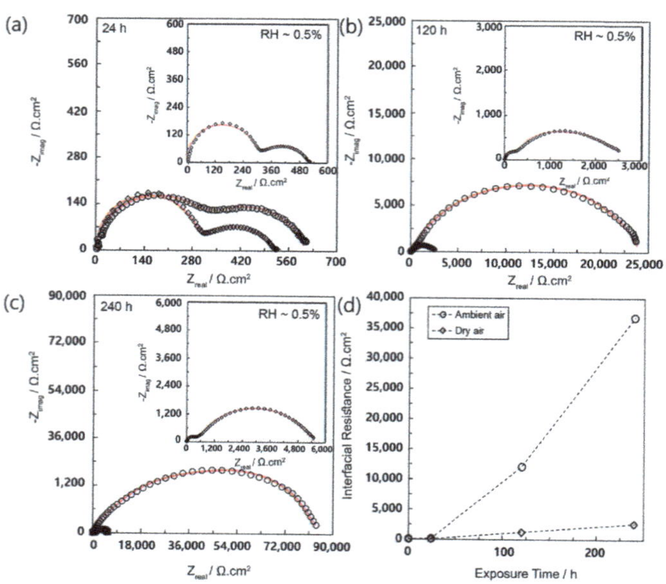

Figure 5. Impedance spectra measured at room temperature for Li metal symmetric cells after various exposure times to ambient (○) and dry air (◇). Schematic depicting the asymmetric Li metal symmetric cell before and after air exposure and equivalent circuit model used for fitting EIS data at (**a**) 24 h, (**b**) 120 h, and (**c**) 240 h. The insets display the impedance spectra for Li-LLZO-Li for samples exposed to dry air. (**d**) The LLZO/Li metal interfacial resistance versus time after exposure to ambient and dry air [43]. Reprinted with permission from Ref. [43]. Copyright 2017 Royal Society of Chemistry.

Another interface problem is the low wettability between garnet-type oxide SSEs and Li metal. When LLZO SSEs were manufactured through compression at room temperature, the particles did not agglomerate, and void spaces were present within the resulting structures (Figure 6) [45]. The rigid connection between the electrode and the SSEs and the existence of an inert surface layer of Li_2CO_3 resulted in poor interface compatibility, which led to an interfacial impedance of 10^2~10^3 Ω cm^2 and thus unstable Li plating and stripping [46,47].

Figure 6. Fracture cross-section SEM images of (**a**) LLZO and (**b**) sulfide electrolyte particles glass-pelletized at room temperature [45]. Reprinted with permission from Ref. [45]. Copyright 2013 Nature Publishing Group.

3.2. Strategies to Overcome Disadvantages of Oxide SSEs for Li Metal

Many studies have reported methods to eliminate Li_2CO_3, a factor that increases interfacial resistance (e.g., by carbothermal reactions, high-temperature calcination, or acid treatment) [48–50]. Furthermore, both dry and wet polishing have been reported to be effective approaches for removing Li_2CO_3 from a surface [13,51]. Sakamoto et al. noted that among the two approaches, dry and wet polishing, the latter was mentioned as the more effective method for removing Li_2CO_3 (Figure 7a–d) [51].

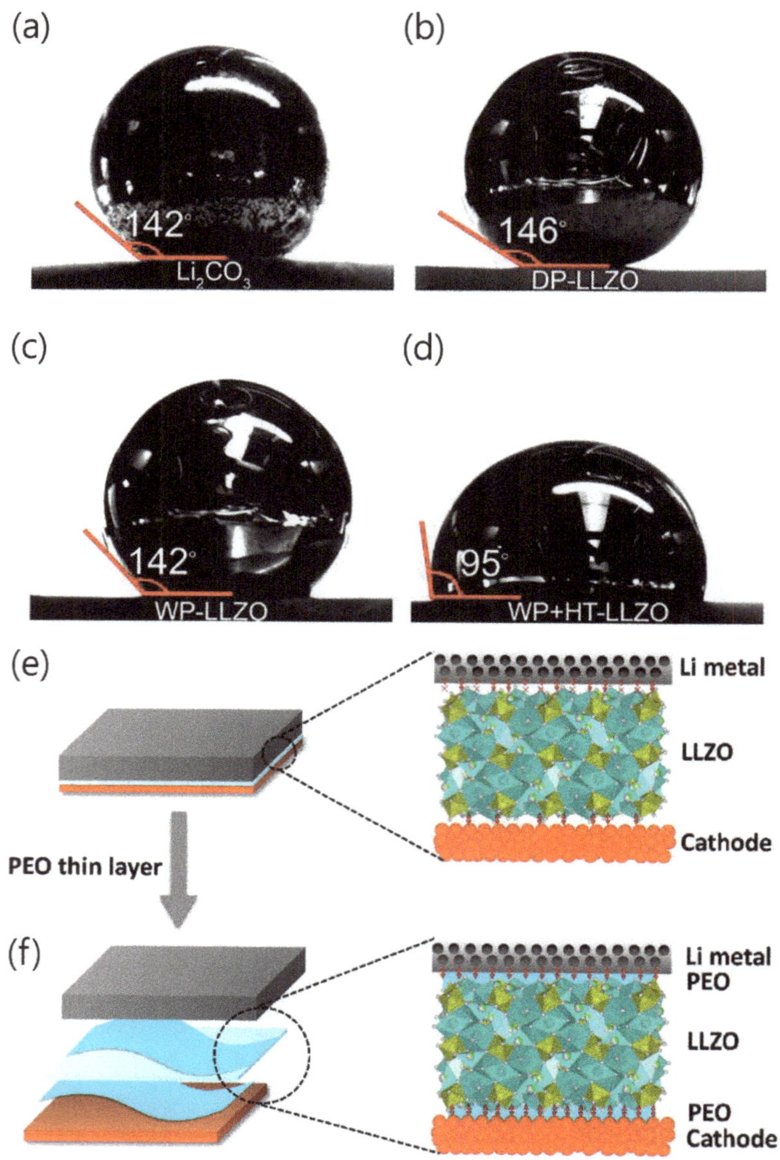

Figure 7. Contact angle measurements of molten metallic on (**a**) Li_2CO_3, (**b**) dry polishing of LLZO, (**c**) wet polishing of LLZO, and (**d**) wet polishing of LLZO after heat treatment at 500 °C [51]. Reprinted with permission from Ref. [51]. Copyright 2017 American Chemical Society. Schematic of electrode–LLZO interface for solid-state battery. (**e**) Hard-to-hard interfaces with bad contact between electrodes and LLZO. (**f**) Soft-to-hard interfaces with improved connection [52]. Reprinted with permission from Ref. [52]. Copyright 2020 ELSEVIER.

A common strategy for overcoming point contacts with Li metal is to use an artificial interface layer. The excellent flexibility and softness of polymers can enhance the low interface contact between LLZO and Li metal. Flexible polymers can serve as coatings for LLZO and electrodes, resembling the characteristics found in liquid electrolytes, and are advantageous for scalable processing. Yang et al. developed a thin polyethylene oxide (PEO) buffer layer to modify LLZO (Figure 7e,f) [52]. In their study, the interfacial resistance decreased from 1360 Ω cm^2 to 175 Ω cm^2 and exhibited good stability to Li metal at 0.2 mA cm^{-2}. PEO-Poly(acrylamide-2-methyl-1-propane-sulfonate) (PAS)-coated $Li_{6.5}La_3Zr_{1.5}Ta_{0.5}O_{12}$ (LLZTO) was demonstrated by Zhou et al. [53]. PAS is a Li^+-conducting polymer which carries out the transportation of high amounts of Li^+. The interfacial impedance of Li/LLZTO/Li was around 5000 Ω, while a Li/PEO-PAS/LLZTO/PEO-PAS/Li cell exhibited an interfacial impedance of less than 400 Ω. Owing to good Li ion transport and contact with Li metal, the capacity retention was around 137 mAh/g at 0.2 C after 160 cycles.

In addition, Goodenough et al. designed a cell utilizing a cross-linked polymer, poly(ethylene glycol) methyl ether acrylate (CPMEA), as a buffer layer between the cathode and Li metal (Figure 8a,b) [54]. The polymer layer not only improved contact at the interface because of its flexibility and softness but also increased the Li^+ flux, thereby improving the cycling performance and preventing dendrite formation in the cell.

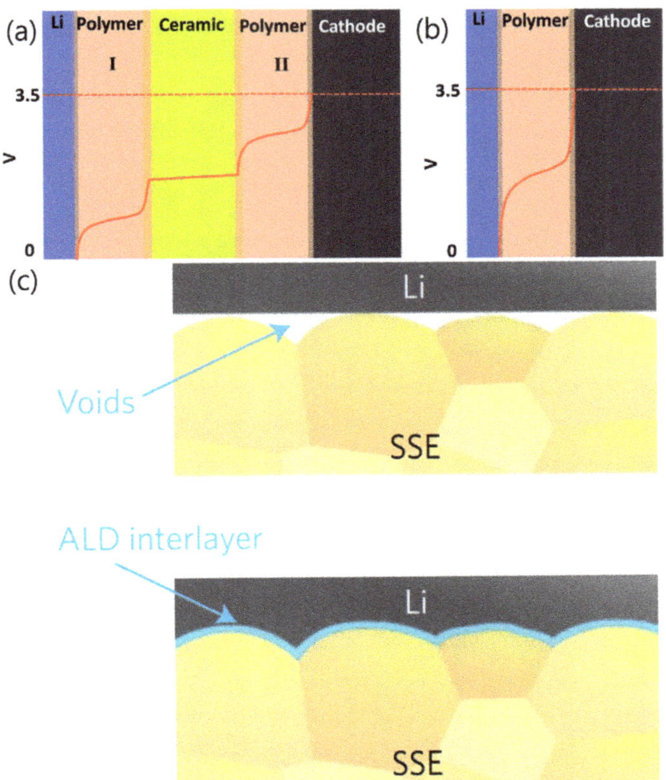

Figure 8. Schematic of the electric potential profile across the (**a**) sandwich electrolyte and (**b**) individual polymer electrolyte in the charge process of a full cell [54]. Both polymer I and II represent CPMEA. Reprinted with permission from Ref. [54]. Copyright 2016 American Chemical Society. (**c**) Schematic of the behavior of LLZO with molten Li metal [55]. Reprinted with permission from Ref. [55]. Copyright 2017 Nature Publishing Group.

Similar to the approach described for sulfide SSEs, there is a method employing ALD to form a thin film at the Li metal and LLZO interface, effectively improving wettability and reducing resistance. Hu et al. demonstrated that the wettability of SSEs against molten Li was significantly improved by applying an alumina ALD coating [55]. The LLZO/Li metal interfacial resistance decreased from 1710 Ω cm^2 to 34 Ω cm^2 owing to Li metal on a lithiated pure alumina layer with a varied Li stoichiometry, $Li_xAl_2O_{3+x/2}$; a strong chemical bond with the Li metal formed, and it exhibited good ion conductivity (Figure 8c). Various metals and metal oxides, including Si, ZnO, and Au, have been reported in studies in which ALD was employed for coating. These materials demonstrated enhanced wettability with molten Li and concurrently reduced interfacial resistance [56–58].

4. Polymer SSEs

Polymer SSEs are predominantly composed of thermally stable polymers, such as the previously mentioned polymers PEO, poly(vinylidene fluoride) (PVDF), and polyacrylonitrile (PAN). Among these, PEO, which can form various salts through interactions between ether oxygen atoms and cations, has been extensively studied as a dry polymer electrolyte host. Its ability to dissolve various salts makes it the most widely studied dry polymer electrolyte host. Consequently, they can be utilized in the design of all-solid-state batteries with self-supporting films without altering current battery assembly processes [59]. In addition, gel-formulated polymer electrolytes have been reported as various types of polymer solid-state electrolytes. In this configuration, the liquid components within the gel facilitate ion conduction. In this review, no detailed exposition is provided on this subject.

4.1. Challenges of Polymer SSEs for Li Metal

PEO-based electrolytes generally exhibit low ion conductivity at an order of 10^{-8}~10^{-7} S cm^{-1} at room temperature because the ion conduction in the PEO-based electrolyte mainly occurs in the amorphous part of the PEO matrix, while the crystalline part provides very limited ion motion (note that the PEO chain is mainly crystalline below 65 °C) [60,61]. The Li^+ ion conduction mechanisms of both the amorphous and crystalline polymers are shown in Figure 9 [62].

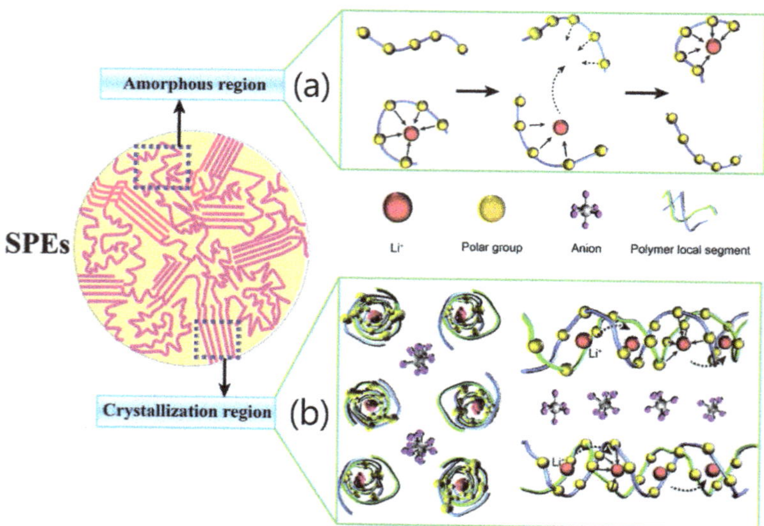

Figure 9. Schematic of conduction mechanism in two kinds of polymer electrolytes. (**a**) Amorphous polymer electrolytes and (**b**) crystalline polymer electrolytes [62]. Reprinted with permission from Ref. [62]. Copyright 2016 Royal Society of Chemistry.

Moreover, most polymer electrolytes are dual-ion conductors in which Li^+ and its counter anions are both mobile. Therefore, Li^+ is coupled with Lewis basic sites in the polymer matrix; thus, Li^+ is usually less mobile than anions. Therefore, the Li^+ transference number of ion conductors with dual ions is generally lower than 0.5. Li^+; in addition, counter anions move in opposite directions during discharge cycles, leading to the accumulation of anions at the anode side and causing concentration gradients and cell polarization (Figure 10) [63,64]. The transference number is defined as the ratio of the electric current derived from the cation to the total electric current. If the number is close to one, it implies that the ion-conducting performance of the polymer electrolyte is mainly accomplished by the cation [65].

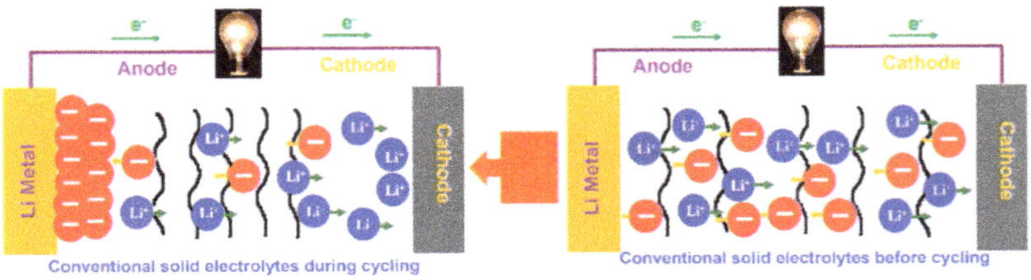

Figure 10. The operational phenomena of conventional SPEs [64]. Reprinted with permission from Ref. [64]. Copyright 2021 ELSEVIER.

4.2. Strategies to Overcome Disadvantages of Polymer SSEs for Li Metal

Among the various methods available, polymer blending is the most feasible. Polymer blending involves mixing at least two polymers, with or without chemical bonding [66]. Poly(methyl methacrylate) (PMMA)- and PVDF-blended polymer electrolytes were demonstrated by Nicotera et al. [67]. Oscillatory rheological tests showed better mechanical properties for the intermediate composition of the blend (Figure 11a,b). Subsequently, Ganessan et al. [49] studied the design of blended polymer electrolytes using machine learning [68]. Electrolyte performance was measured using a combination of ionic transport and electrolyte mechanical properties. In this study, through the utilization of a machine learning approach known as Bayesian optimization, a trade-off between ion transport and electrolyte mechanical properties was identified as a function of various design parameters, including host molecular weight and polarity.

Copolymer electrolytes may be considered a method for reducing crystalline PEO. A copolymer is a polymer composed of multiple monomer species. Copolymer electrolytes exhibit conductivities exceeding 10^{-4} S cm^{-1} at room temperature which result from the highly amorphous structure of the studied polymer hosts and their high flexibility, as evidenced by a T_g value below 210 K [69]. Recently, Li et al. reported the synthesis of copolymers through cationic ring-opening polymerization using 1,3-dioxolane (DOL) and trioxymethylene (TOM) as monomers [70]. The ion conductivity from the study by Li et al. reached 3.56×10^{-4} S cm^{-1} at room temperature (Figure 11c).

The use of single–ion conducting polymer electrolytes (SICPEs), ideally defined as polymer electrolytes with Li^+ transference numbers similar to unity (approximately 1), can be an effective method for reducing concentration polarization. With a higher transference number than traditional polymer electrolytes, SICPEs exhibit remarkably decreased polarization, which should theoretically allow them to suppress Li dendrites [71–74]. Conventional SICPEs are fabricated by covalently bonding anions to the polymer backbone, allowing only single-ion species to be mobile. Gohy et al. reported a copolymer electrolyte combining a single-ion conducting anionic group (poly(lithium methacrylate-*co*-oligoethylene glycol methacrylate)) [75]. Its ion conductivity reached up to 2×10^{-5} S cm^{-1}. Mecerreyes et al. developed SICPEs based on UV-cross-linkable polyurethanes [76]. A coulombic efficiency near 100% and capacity retention exceeding 72.8% were measured after 80 cycles at 0.1 C. The ion conductivities and properties of the other SICPEs are displayed in Figure 12 and Table 1 [77].

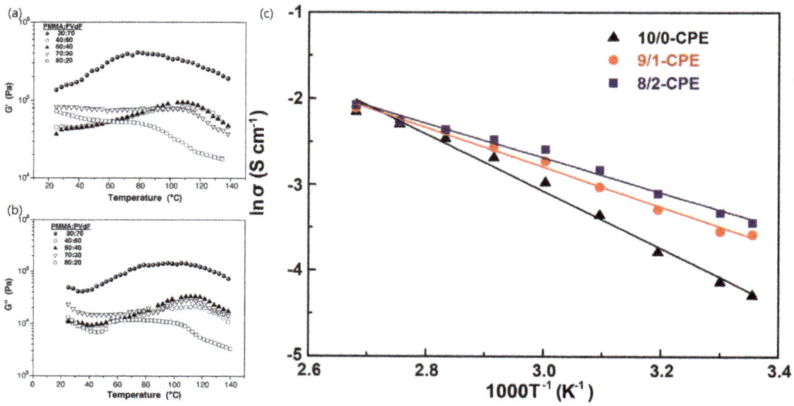

Figure 11. Temperature sweep test from 25 to 140 °C of (**a**) G′ vs. temperature and (**b**) G″ vs. temperature [67]. Reprinted with permission from Ref. [67]. Copyright 2006 ELSEVIER. (**c**) The ion conductivity of DOL and TOM copolymer electrolytes in a temperature range of 25–110 °C (black, red, and blue lines represent the DOL/TOM molar ratio as 10/0, 9/1, and 8/2, respectively) [70]. Reprinted with permission from Ref. [70]. Copyright 2020 Wiley-VCH.

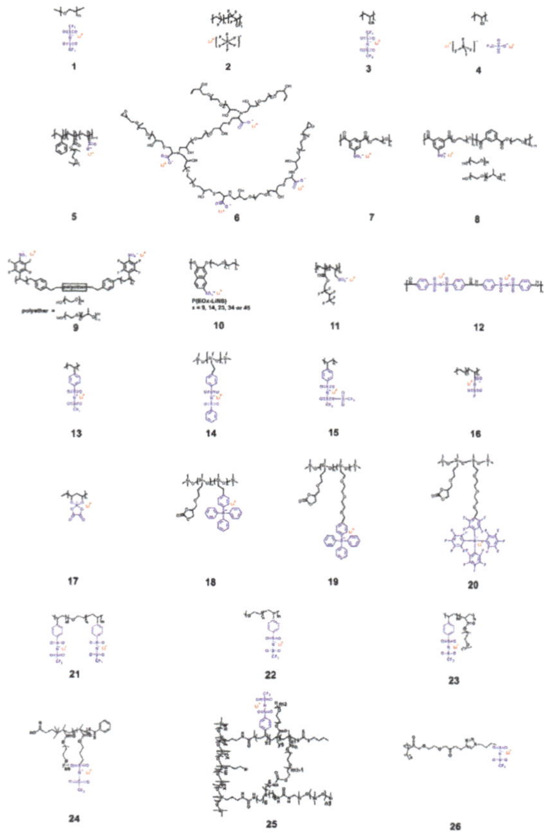

Figure 12. Representative structures of the reported SICPEs [77]. Reprinted with permission from Ref. [77]. Copyright 2021 Wiley-VCH.

Table 1. Summary of various SICPEs with corresponding property parameters [77]. Reused with permission from Ref. [77]. Copyright 2021 Wiley-VCH.

Materials	T_g [°C]	t_{Li}^+	Ionic Conductivity [S cm^{-1}]	Testing Temperature [°C]	Mechanical Strength [MPa]
1	−57	0.29	N/A	N/A	N/A
2	N/A	0.46	8.9×10^{-4}	25	N/A
3	−105.2	N/A	2.5×10^{-3}	23	N/A
4	N/A	N/A	2.6×10^{-3}	25	N/A
5	N/A	Close to unity	$\approx 10^{-8}$	20	N/A
6	<0	0.86	1.2×10^{-4}	85	N/A
7	−37	N/A	10^{-6}	25	N/A
8	−20	N/A	$\approx 10^{-5}$	25	N/A
9	<0	N/A	1.4×10^{-5}	60	N/A
10	−38	N/A	5.5×10^{-6}	120	N/A
11	17	0.92	10^{-4}	80	7.1
12	N/A	0.92	1.4×10^{-4}	25	N/A
13	152	N/A	$\approx 10^{-5}$	41	N/A
14	−110	0.89	7.2×10^{-4}	25	5.8
15	−47	>0.9	7.6×10^{-6}	25	N/A
16	−30	0.91	5.84×10^{-4}	25	N/A
17	N/A	N/A	6.11×10^{-6}	25	N/A
18	10	N/A	$10^{-8.2}$	25	N/A
19	−17	N/A	$10^{-7.0}$	25	N/A
20	−16	N/A	$10^{-6.9}$	25	N/A
21	N/A	0.85	1.35×10^{-5}	60	10
22	N/A	0.95	3.8×10^{-4}	90	N/A
23	<0	>0.9	10^{-4}	60	N/A
24	−61	0.83	2.3×10^{-6}	25	N/A
25	−17.5	0.79	4.5×10^{-7}	30	0.37
26	−27 to −14	0.79–0.99	10^{-5}–10^{-4}	90	N/A

5. Conclusions

In this review, the properties, challenges, and previous studies of sulfide, oxide, and polymer SSEs for Li metal anodes were examined. Despite the high theoretical capacity and energy density of lithium metal, its commercialization is challenging when using liquid electrolytes. This challenge has led to extensive research on SSEs. Due to the varying physical and chemical properties of each SSE, the challenges and improvement strategies for Li metal anodes also differ. At the interfaces between sulfide SSEs and lithium metal, challenges such as lithium dendrite formation and side reactions are prevalent. Various methods can be employed to tackle these issues, including applying interlayers, utilizing ALD coatings, and optimizing the electrolyte composition. In the case of oxide SSEs, significant hurdles arise from substantial interfacial resistance due to point contact with lithium metal and impurity generation from reactions with CO_2. These challenges can be alleviated through techniques such as polishing, employing buffer layers, and implementing ALD coatings similar to those used for sulfide SSEs. Polymer SSEs encounter challenges associated with low ionic conductivity and concentration polarization, which can potentially lead to lithium dendrite formation. Addressing these issues involves strategies like polymer blending, copolymerization, and the utilization of solid ion-conducting polymer electrolytes. Despite these efforts, numerous technical and commercial challenges remain in the commercialization of solid-state electrolytes (SSEs). However, with a thorough understanding of SSEs and ongoing research on Li metal anodes and cathodes, the commercialization of ASSLIBs is expected to be achieved.

Funding: This work was supported by the Gachon University Research Fund of 2022 (GCU-202301000001).

Conflicts of Interest: The authors declare no conflicts of interest.

References

1. Yoon, D.H.; Park, Y.J. Electrochemical Properties of Cathode According to the Type of Sulfide Electrolyte and the Application Surface Coating. *J. Electrochem. Sci. Technol.* **2021**, *12*, 126–136. [CrossRef]
2. Zhao, W.; Choi, W.; Yoon, W.-S. Nanostructured electrode materials for rechargeable lithium-ion batteries. *J. Electrochem. S Technol.* **2020**, *11*, 195–219. [CrossRef]
3. Adiraju, V.A.K.; Chae, O.B.; Robinson, J.R.; Lucht, B.L. Highly Soluble Lithium Nitrate-Containing Additive for Carbonate-Bas Electrolyte in Lithium Metal Batteries. *ACS Energy Lett.* **2023**, *8*, 2440–2446. [CrossRef]
4. Kwon, K.; Kim, J.; Han, S.; Lee, J.; Lee, H.; Kwon, J.; Lee, J.; Seo, J.; Kim, P.J.; Song, T. Low-Resistance $LiFePO_4$ Thick Fi Electrode Processed with Dry Electrode Technology for High-Energy-Density Lithium-Ion Batteries. *Small Sci.* **2024**, 23003 [CrossRef]
5. Lee, G.; Kim, I.T.; Hur, J. Highly conductive and robust telluride-carbon hybrid matrix for enhanced copper diphosphide ano in Li-ion batteries. *J. Alloys Compd.* **2023**, *950*, 169914. [CrossRef]
6. Thieu, Q.Q.V.; Kidanu, W.G.; Nguyen, H.D.; Nguyen, T.L.T.; Le, M.L.P.; Nguyen, D.Q.; Tran, N.T.; Nguyen, X.V.; Kim, I. Nguyen, T.L. Spinel Ni-ferrite advanced high-capacity anode for Li-ion batteries prepared via coprecipitation route. *Ceram. I* **2022**, *48*, 31470–31477. [CrossRef]
7. Nguyen, Q.H.; Kim, I.T.; Hur, J. Core-shell Si@ c-PAN particles deposited on graphite as promising anode for lithium-ion batteri *Electrochim. Acta* **2019**, *297*, 355–364. [CrossRef]
8. Lim, Y.E.; Choi, W.S.; Kim, J.H.; Ahn, Y.N.; Kim, I.T. The Sn–red P–Fe–based alloy materials for efficient Li–ion battery anodes *Ind. Eng. Chem.* **2023**, *121*, 299–311. [CrossRef]
9. Tran, Q.N.; Park, C.H.; Le, T.H. Nanocrystalline Cellulose-Supported Iron Oxide Composite Materials for High-Performan Lithium-Ion Batteries. *Polymers* **2024**, *16*, 691. [CrossRef]
10. Cheng, X.B.; Zhang, R.; Zhao, C.Z.; Zhang, Q. Toward Safe Lithium Metal Anode in Rechargeable Batteries: A Review. *Chem. R* **2017**, *117*, 10403–10473. [CrossRef]
11. Yang, S.; Kim, J.; Lee, S.; Seo, J.; Choi, J.; Kim, P.J. Uniform Li Deposition through the Graphene-Based Ion-Flux Regulator f High-Rate Li Metal Batteries. *ACS Appl. Mater. Interfaces* **2024**, *16*, 3416–3426. [CrossRef]
12. Paul, S. Materials and electrochemistry: Present and future battery. *J. Electrochem. Sci. Technol.* **2016**, *7*, 115–131. [CrossRef]
13. Chae, O.B.; Lucht, B.L. Interfacial Issues and Modification of Solid Electrolyte Interphase for Li Metal Anode in Liquid and Sol Electrolytes. *Adv. Energy Mater.* **2023**, *13*, 2203791. [CrossRef]
14. Chae, O.B.; Kim, J.; Lucht, B.L. Modification of lithium electrodeposition behavior by variation of electrode distance. *J. Pow Sources* **2022**, *532*, 231338. [CrossRef]
15. Kim, M.; Park, H.G.; Park, K. A strategy of enhancing the ionic conductivity of $Li_7La_3Zr_2O_{12}$ under accurate sintering conditior *Phys. Chem. Chem. Phys.* **2022**, *24*, 29159–29164. [CrossRef] [PubMed]
16. Zhang, X.-Q.; Cheng, X.-B.; Chen, X.; Yan, C.; Zhang, Q. Fluoroethylene Carbonate Additives to Render Uniform Li Deposits Lithium Metal Batteries. *Adv. Funct. Mater.* **2017**, *27*, 1605989. [CrossRef]
17. Fan, L.; Zhuang, H.L.; Gao, L.; Lu, Y.; Archer, L.A. Regulating Li deposition at artificial solid electrolyte interphases. *J. Mat Chem. A* **2017**, *5*, 3483–3492. [CrossRef]
18. Jurng, S.; Brown, Z.L.; Kim, J.; Lucht, B.L. Effect of electrolyte on the nanostructure of the solid electrolyte interphase (SEI) ar performance of lithium metal anodes. *Energy Environ. Sci.* **2018**, *11*, 2600–2608. [CrossRef]
19. Sun, C.; Liu, J.; Gong, Y.; Wilkinson, D.P.; Zhang, J. Recent advances in all-solid-state rechargeable lithium batteries. *Nano Energ* **2017**, *33*, 363–386. [CrossRef]
20. Gao, Z.; Sun, H.; Fu, L.; Ye, F.; Zhang, Y.; Luo, W.; Huang, Y. Promises, challenges, and recent progress of inorganic solid-sta electrolytes for all-solid-state lithium batteries. *Adv. Mater.* **2018**, *30*, 1705702. [CrossRef]
21. Xu, R.; Han, F.; Ji, X.; Fan, X.; Tu, J.; Wang, C. Interface engineering of sulfide electrolytes for all-solid-state lithium batteries. *Nar Energy* **2018**, *53*, 958–966. [CrossRef]
22. Kamaya, N.; Homma, K.; Yamakawa, Y.; Hirayama, M.; Kanno, R.; Yonemura, M.; Kamiyama, T.; Kato, Y.; Hama, S.; Kawamot K. A lithium superionic conductor. *Nat. Mater.* **2011**, *10*, 682–686. [CrossRef] [PubMed]
23. Seino, Y.; Ota, T.; Takada, K.; Hayashi, A.; Tatsumisago, M. A sulphide lithium super ion conductor is superior to liquid io conductors for use in rechargeable batteries. *Energy Environ. Sci.* **2014**, *7*, 627–631. [CrossRef]
24. Li, Y.; Song, S.; Kim, H.; Nomoto, K.; Kim, H.; Sun, X.; Hori, S.; Suzuki, K.; Matsui, N.; Hirayama, M. A lithium superioni conductor for millimeter-thick battery electrode. *Science* **2023**, *381*, 50–53. [CrossRef] [PubMed]
25. Han, G.; Vasylenko, A.; Daniels, L.M.; Collins, C.M.; Corti, L.; Chen, R.; Niu, H.; Manning, T.D.; Antypov, D.; Dyer, M. Superionic lithium transport via multiple coordination environments defined by two-anion packing. *Science* **2024**, *383*, 739–74 [CrossRef] [PubMed]

Chen, S.; Xie, D.; Liu, G.; Mwizerwa, J.P.; Zhang, Q.; Zhao, Y.; Xu, X.; Yao, X. Sulfide solid electrolytes for all-solid-state lithium batteries: Structure, conductivity, stability and application. *Energy Storage Mater.* **2018**, *14*, 58–74. [CrossRef]

Zhu, Y.; He, X.; Mo, Y. First principles study on electrochemical and chemical stability of solid electrolyte–electrode interfaces in all-solid-state Li-ion batteries. *J. Mater. Chem. A* **2016**, *4*, 3253–3266. [CrossRef]

Lee, J.Y.; Park, Y.J. Li_3PO_4 Coated $Li[Ni_{0.75}Co_{0.1}Mn_{0.15}]O_2$ Cathode for All-Solid-State Batteries Based on Sulfide Electrolyte. *J. Electrochem. Sci. Technol.* **2022**, *13*, 407–415. [CrossRef]

Nagao, M.; Hayashi, A.; Tatsumisago, M.; Kanetsuku, T.; Tsuda, T.; Kuwabata, S. In situ SEM study of a lithium deposition and dissolution mechanism in a bulk-type solid-state cell with a Li_2S–P_2S_5 solid electrolyte. *Phys. Chem. Chem. Phys.* **2013**, *15*, 18600–18606. [CrossRef] [PubMed]

Porz, L.; Swamy, T.; Sheldon, B.W.; Rettenwander, D.; Frömling, T.; Thaman, H.L.; Berendts, S.; Uecker, R.; Carter, W.C.; Chiang, Y.M. Mechanism of Lithium Metal Penetration through Inorganic Solid Electrolytes. *Adv. Energy Mater.* **2017**, *7*, 1701003. [CrossRef]

Luo, S.; Wang, Z.; Li, X.; Liu, X.; Wang, H.; Ma, W.; Zhang, L.; Zhu, L.; Zhang, X. Growth of lithium-indium dendrites in all-solid-state lithium-based batteries with sulfide electrolytes. *Nat. Commun.* **2021**, *12*, 6968. [CrossRef]

Camacho-Forero, L.E.; Balbuena, P.B. Exploring interfacial stability of solid-state electrolytes at the lithium-metal anode surface. *J. Power Sources* **2018**, *396*, 782–790. [CrossRef]

Wenzel, S.; Weber, D.A.; Leichtweiss, T.; Busche, M.R.; Sann, J.; Janek, J. Interphase formation and degradation of charge transfer kinetics between a lithium metal anode and highly crystalline $Li_7P_3S_{11}$ solid electrolyte. *Solid State Ion.* **2016**, *286*, 24–33. [CrossRef]

Wenzel, S.; Sedlmaier, S.J.; Dietrich, C.; Zeier, W.G.; Janek, J. Interfacial reactivity and interphase growth of argyrodite solid electrolytes at lithium metal electrodes. *Solid State Ion.* **2018**, *318*, 102–112. [CrossRef]

Hood, Z.D.; Mane, A.U.; Sundar, A.; Tepavcevic, S.; Zapol, P.; Eze, U.D.; Adhikari, S.P.; Lee, E.J.; Sterbinsky, G.E.; Elam, J.W.; et al. Multifunctional Coatings on Sulfide-Based Solid Electrolyte Powders with Enhanced Processability, Stability, and Performance for Solid-State Batteries. *Adv. Mater.* **2023**, *35*, 2300673. [CrossRef] [PubMed]

Ong, S.P.; Mo, Y.F.; Richards, W.D.; Miara, L.; Lee, H.S.; Ceder, G. Phase stability, electrochemical stability and ionic conductivity of the $Li_{10\pm1}MP_2X_{12}$ (M = Ge, Si, Sn, Al or P, and X = O, S or Se) family of superionic conductors. *Energy Environ. Sci.* **2013**, *6*, 148–156. [CrossRef]

Sun, Y.; Suzuki, K.; Hara, K.; Hori, S.; Yano, T.-A.; Hara, M.; Hirayama, M.; Kanno, R. Oxygen substitution effects in $Li_{10}GeP_2S_{12}$ solid electrolyte. *J. Power Sources* **2016**, *324*, 798–803. [CrossRef]

Suzuki, K.; Sakuma, M.; Hori, S.; Nakazawa, T.; Nagao, M.; Yonemura, M.; Hirayama, M.; Kanno, R. Synthesis, structure, and electrochemical properties of crystalline Li–P–S–O solid electrolytes: Novel lithium-conducting oxysulfides of $Li_{10}GeP_2S_{12}$ family. *Solid State Ion.* **2016**, *288*, 229–234. [CrossRef]

Murugan, R.; Thangadurai, V.; Weppner, W. Fast Lithium Ion Conduction in Garnet-Type $Li_7La_3Zr_2O_{12}$. *Angew. Chem. Int. Ed.* **2007**, *46*, 7778–7781. [CrossRef]

Thangadurai, V.; Narayanan, S.; Pinzaru, D. Garnet-type solid-state fast Li ion conductors for Li batteries: Critical review. *Chem. Soc. Rev.* **2014**, *43*, 4714. [CrossRef]

Choi, H.M.; Jun, S.J.; Lee, J.; Ryu, M.-H.; Shin, H.; Jung, K.-N. UV-cured Polymer Solid Electrolyte Reinforced using a Ceramic-Polymer Composite Layer for Stable Solid-State Li Metal Batteries. *J. Electrochem. Sci. Technol.* **2023**, *14*, 85–95. [CrossRef]

Xia, W.H.; Xu, B.Y.; Duan, H.A.; Tang, X.Y.; Guo, Y.P.; Kang, H.M.; Li, H.; Liu, H.Z. Reaction mechanisms of lithium garnet pellets in ambient air: The effect of humidity and CO_2. *J. Am. Chem. Soc.* **2017**, *100*, 2832–2839. [CrossRef]

Sharafi, A.; Yu, S.H.; Naguib, M.; Lee, M.; Ma, C.; Meyer, H.M.; Nanda, J.; Chi, M.F.; Siegel, D.J.; Sakamoto, J. Impact of air exposure and surface chemistry on Li-$Li_7La_3Zr_2O_{12}$ interfacial resistance. *J. Mater. Chem. A* **2017**, *5*, 13475–13487. [CrossRef]

Cheng, L.; Crumlin, E.J.; Chen, W.; Qiao, R.M.; Hou, H.M.; Lux, S.F.; Zorba, V.; Russo, R.; Kostecki, R.; Liu, Z.; et al. The origin of high electrolyte-electrode interfacial resistances in lithium cells containing garnet type solid electrolytes. *Phys. Chem. Chem. Phys.* **2014**, *16*, 18294–18300. [CrossRef]

Sakuda, A.; Hayashi, A.; Tatsumisago, M. Sulfide Solid Electrolyte with Favorable Mechanical Property for All-Solid-State Lithium Battery. *Sci. Rep.* **2013**, *3*, 2261. [CrossRef] [PubMed]

Chen, X.Z.; He, W.J.; Ding, L.X.; Wang, S.Q.; Wang, H.H. Enhancing interfacial contact in all solid state batteries with a cathode-supported solid electrolyte membrane framework. *Energy Environ. Sci.* **2019**, *12*, 938–944. [CrossRef]

Shen, X.; Zhang, Q.; Ning, T.; Liu, T.; Luo, Y.; He, X.; Luo, Z.; Lu, A. Critical challenges and progress of solid garnet electrolytes for all-solid-state batteries. *Mater. Today. Chem.* **2020**, *18*, 100368. [CrossRef]

Li, Y.T.; Chen, X.; Dolocan, A.; Cui, Z.M.; Xin, S.; Xue, L.G.; Xu, H.H.; Park, K.; Goodenough, J.B. Garnet Electrolyte with an Ultralow Interfacial Resistance for Li-Metal Batteries. *J. Am. Chem. Soc.* **2018**, *140*, 6448–6455. [CrossRef]

Wu, J.F.; Pu, B.W.; Wang, D.; Shi, S.Q.; Zhao, N.; Guo, X.X.; Guo, X. In Situ Formed Shields Enabling Li_2CO_3-Free Solid Electrolytes: A New Route to Uncover the Intrinsic Lithiophilicity of Garnet Electrolytes for Dendrite-Free Li-Metal Batteries. *ACS Appl. Mater. Interfaces* **2019**, *11*, 898–905. [CrossRef] [PubMed]

Ruan, Y.; Lu, Y.; Huang, X.; Su, J.; Sun, C.; Jin, J.; Wen, Z. Acid induced conversion towards a robust and lithiophilic interface for Li–$Li_7La_3Zr_2O_{12}$ solid-state batteries. *J. Mater. Chem. A* **2019**, *7*, 14565–14574. [CrossRef]

51. Sharafi, A.; Kazyak, E.; Davis, A.L.; Yu, S.; Thompson, T.; Siegel, D.J.; Dasgupta, N.P.; Sakamoto, J. Surface Chemistry Mechani of Ultra-Low Interfacial Resistance in the Solid-State Electrolyte $Li_7La_3Zr_2O_{12}$. *Chem. Mater.* **2017**, *29*, 7961–7968. [CrossRef]
52. Ma, X.; Xu, Y.; Zhang, B.; Xue, X.; Wang, C.; He, S.; Lin, J.; Yang, L. Garnet Si–$Li_7La_3Zr_2O_{12}$ electrolyte with a durable, lo resistance interface layer for all-solid-state lithium metal batteries. *J. Power Sources* **2020**, *453*, 227881. [CrossRef]
53. Zhou, W.; Zhu, Y.; Grundish, N.; Xin, S.; Wang, S.; You, Y.; Wu, N.; Gao, J.; Cui, Z.; Li, Y. Polymer lithium-garnet interphase for all-solid-state rechargeable battery. *Nano Energy* **2018**, *53*, 926–931. [CrossRef]
54. Zhou, W.D.; Wang, S.F.; Li, Y.T.; Xin, S.; Manthiram, A.; Goodenough, J.B. Plating a Dendrite-Free Lithium Anode with Polymer/Ceramic/Polymer Sandwich Electrolyte. *J. Am. Chem. Soc.* **2016**, *138*, 9385–9388. [CrossRef] [PubMed]
55. Han, X.; Gong, Y.; Fu, K.; He, X.; Hitz, G.T.; Dai, J.; Pearse, A.; Liu, B.; Wang, H.; Rubloff, G.; et al. Negating interfacial impedar in garnet-based solid-state Li metal batteries. *Nat. Mater.* **2017**, *16*, 572–579. [CrossRef] [PubMed]
56. Wang, C.; Gong, Y.; Liu, B.; Fu, K.; Yao, Y.; Hitz, E.; Li, Y.; Dai, J.; Xu, S.; Luo, W.; et al. Conformal, Nanoscale ZnO Surfa Modification of Garnet-Based Solid-State Electrolyte for Lithium Metal Anodes. *Nano Lett.* **2017**, *17*, 565–571. [CrossRef] [PubMe
57. Luo, W.; Gong, Y.; Zhu, Y.; Fu, K.K.; Dai, J.; Lacey, S.D.; Wang, C.; Liu, B.; Han, X.; Mo, Y.; et al. Transition from Superlithiophobic to Superlithiophilicity of Garnet Solid-State Electrolyte. *J. Am. Chem. Soc.* **2016**, *138*, 12258–12262. [CrossRef] [PubMed]
58. Wakasugi, J.; Munakata, H.; Kanamura, K. Effect of Gold Layer on Interface Resistance between Lithium Metal Anode a $Li_{6.25}Al_{0.25}La_3Zr_2O_{12}$ Solid Electrolyte. *J. Electrochem. Soc.* **2017**, *164*, A1022–A1025. [CrossRef]
59. Bocharova, V.; Sokolov, A.P. Perspectives for polymer electrolytes: A view from fundamentals of ionic conductivity. *Macromolecu* **2020**, *53*, 4141–4157. [CrossRef]
60. Meyer, W.H. Polymer electrolytes for lithium-ion batteries. *Adv. Mater.* **1998**, *10*, 439–448. [CrossRef]
61. Jiang, Y.; Yan, X.; Ma, Z.; Mei, P.; Xiao, W.; You, Q.; Zhang, Y. Development of the PEO based solid polymer electrolytes f all-solid state lithium ion batteries. *Polymers* **2018**, *10*, 1237. [CrossRef]
62. Chen, R.; Qu, W.; Guo, X.; Li, L.; Wu, F. The pursuit of solid-state electrolytes for lithium batteries: From comprehensive insight emerging horizons. *Mat. Horiz.* **2016**, *3*, 487–516. [CrossRef]
63. Sun, X.-G.; Kerr, J.B. Synthesis and characterization of network single ion conductors based on comb-branched polyepoxi ethers and lithium bis (allylmalonato) borate. *Macromolecules* **2006**, *39*, 362–372. [CrossRef]
64. Irfan, M.; Atif, M.; Yang, Z.; Zhang, W. Recent advances in high performance conducting solid polymer electrolytes for lithium-i batteries. *J. Power Sources* **2021**, *486*, 229378. [CrossRef]
65. Ye, L.; Feng, Z. Polymer electrolytes as solid solvents and their applications. In *Polymer electrolytes*; Elsevier: Amsterdam, T Netherlands, 2010; pp. 550–582.
66. Ramesh, S.; Liew, C.-W.; Morris, E.; Durairaj, R. Effect of PVC on ionic conductivity, crystallographic structural, morphologic and thermal characterizations in PMMA–PVC blend-based polymer electrolytes. *Thermochim. Acta* **2010**, *511*, 140–146. [CrossRe
67. Nicotera, I.; Coppola, L.; Oliviero, C.; Castriota, M.; Cazzanelli, E. Investigation of ionic conduction and mechanical properties PMMA–PVdF blend-based polymer electrolytes. *Solid State Ion.* **2006**, *177*, 581–588. [CrossRef]
68. Wheatle, B.K.; Fuentes, E.F.; Lynd, N.A.; Ganesan, V. Design of polymer blend electrolytes through a machine learning approa *Macromolecules* **2020**, *53*, 9449–9459. [CrossRef]
69. Przyłuski, J.; Wieczorek, W. Copolymer electrolytes. *Solid State Ion.* **1992**, *53*, 1071–1076. [CrossRef]
70. Liu, F.; Li, T.; Yang, Y.; Yan, J.; Li, N.; Xue, J.; Huo, H.; Zhou, J.; Li, L. Investigation on the Copolymer Electrolyte of Poly 3-dioxolane-co-formaldehyde). *Macromol. Rapid Commun.* **2020**, *41*, 2000047. [CrossRef]
71. Chen, Z.; Steinle, D.; Nguyen, H.-D.; Kim, J.-K.; Mayer, A.; Shi, J.; Paillard, E.; Iojoiu, C.; Passerini, S.; Bresser, D. High-energ lithium batteries based on single-ion conducting polymer electrolytes and $Li[Ni_{0.8}Co_{0.1}Mn_{0.1}]O_2$ cathodes. *Nano Energy* **2020**, 105129. [CrossRef]
72. Gao, J.; Wang, C.; Han, D.-W.; Shin, D.-M. Single-ion conducting polymer electrolytes as a key jigsaw piece for next-generatic battery applications. *Chem. Sci.* **2021**, *12*, 13248–13272. [CrossRef] [PubMed]
73. Porcarelli, L.; Shaplov, A.S.; Bella, F.; Nair, J.R.; Mecerreyes, D.; Gerbaldi, C. Single-ion conducting polymer electrolytes f lithium metal polymer batteries that operate at ambient temperature. *ACS Energy Lett.* **2016**, *1*, 678–682. [CrossRef]
74. Zhang, H.; Li, C.; Piszcz, M.; Coya, E.; Rojo, T.; Rodriguez-Martinez, L.M.; Armand, M.; Zhou, Z. Single lithium-ion conductir solid polymer electrolytes: Advances and perspectives. *Chem. Soc. Rev.* **2017**, *46*, 797–815. [CrossRef] [PubMed]
75. Rolland, J.; Poggi, E.; Vlad, A.; Gohy, J.-F. Single-ion diblock copolymers for solid-state polymer electrolytes. *Polymer* **201** *68*, 344–352. [CrossRef]
76. Porcarelli, L.; Manojkumar, K.; Sardon, H.; Llorente, O.; Shaplov, A.S.; Vijayakrishna, K.; Gerbaldi, C.; Mecerreyes, D. Single ic conducting polymer electrolytes based on versatile polyurethanes. *Electrochim. Acta* **2017**, *241*, 526–534. [CrossRef]
77. Zhu, J.; Zhang, Z.; Zhao, S.; Westover, A.S.; Belharouak, I.; Cao, P.F. Single-ion conducting polymer electrolytes for solid-sta lithium–metal batteries: Design, performance, and challenges. *Adv. Energy Mater.* **2021**, *11*, 2003836. [CrossRef]

MDPI AG
Grosspeteranlage 5
4052 Basel
Switzerland
Tel.: +41 61 683 77 34

Micromachines Editorial Office
E-mail: micromachines@mdpi.com
www.mdpi.com/journal/micromachines

www.ingramcontent.com/pod-product-compliance
Lightning Source LLC
Chambersburg PA
CBHW061601210926
51850CB00009B/2175
* 9 7 8 3 7 2 5 8 6 3 5 6 3 *